Sustainable Development of STEAM and Mathematics Education with Active and Innovative Methodology

Sustainable Development of STEAM and Mathematics Education with Active and Innovative Methodology

Editors

Jin Su Jeong
David González-Gómez

MDPI

Basel • Beijing • Wuhan • Barcelona • Belgrade • Novi Sad • Cluj • Manchester

Editors

Jin Su Jeong
Departamento de Didáctica
de las Ciencias
Experimentales y
Matemáticas
Universidad de Extremadura
Cáceres
Spain

David González-Gómez
Departamento de Didáctica
de las Ciencias
Experimentales y
Matemáticas
Universidad de Extremadura
Cáceres
Spain

Editorial Office
MDPI
St. Alban-Anlage 66
4052 Basel, Switzerland

This is a reprint of articles from the Special Issue published online in the open access journal *Sustainability* (ISSN 2071-1050) (available at: www.mdpi.com/journal/sustainability/special_issues/steam_mathematics_education_active_innovative_methodology).

For citation purposes, cite each article independently as indicated on the article page online and as indicated below:

Lastname, A.A.; Lastname, B.B. Article Title. *Journal Name* **Year**, *Volume Number*, Page Range.

ISBN 978-3-7258-0024-7 (Hbk)
ISBN 978-3-7258-0023-0 (PDF)
doi.org/10.3390/books978-3-7258-0023-0

Contents

About the Editors

Jin Su Jeong

Jin Su Jeong is currently a Professor and Academic Secretary at the Department of Science and Mathematics Education, the Universidad de Extremadura (Spain), and has worked with the Juan de la Cierva Program at the Universidad Politécnica de Madrid (Spain). He received his MS in Architecture from the Texas A&M University, College Station (United States); his first PhD with International Mention and Suma Cum Laude in Graphic Engineering, Geomatics, and Projects; and his second PhD with Suma Cum Laude and Outstanding Thesis Award in Teaching Education from the Universidad de Extremadura and Universidad de Huelva (Spain) with a PhD qua;lification from University of Washington, Seattle (United States). For his doctoral dissertation, he has conducted different doctoral research stays and collaborations. He has published more than 280 papers in journals, books, and conference proceedings, including 64 papers in journals indexed in the Web of Science (JCR) and/or SJR. He has participated in the plenary as a keynote speaker, as well as scientific and organizing committees, and has been the chairperson of several international conferences and associations. He has also served more than 555 reviews in a wide range of international journals. He also serves as an associate editor (*Heliyon, Education Section and Frontiers in Psychology/Education,* and *Educational Psychology*), an invited editor for special volumes, and an Editorial Board Member for a couple of prestigious international journals.

David González-Gómez

David González-Gómez received his PhD in analytical chemistry at the University of Extremadura (2005) after different doctoral research stays at the National University of Rosario, Argentina, and at the University of Central Florida, the United States. After graduating, Dr. González obtained a two-year postdoctoral fellowship at the University of Washington, working with Professor Norman J. Dovichi in proteomic analysis. After returning to Spain, he has worked at the Technological Institute of Food and Agriculture (INTAEX) as a research group director. During this time, he also served as an associate professor at the University of Extremadura. Currently, he is a professor at the Department of Science and Mathematics Education at the University of Extremadura and the Dean of the Teaching Training School of the University of Extremadura (Cáceres). He is the author of more than 100 international journal papers, has assisted in a great number of international conferences with peer-reviewed proceedings, and was the director of five doctoral theses. Dr. González has been involved in several national and European grants, as well as collaborations with local industries. Due to his outstanding research, he has been awarded the Outstanding Thesis Award (2006) and the Juan Jesus Morales Young Researcher Award (2011), both from the University of Extremadura (Spain).

sustainability

Editorial

Sustainable Development of STEAM and Mathematics Education with Active and Innovative Methodology

Jin Su Jeong * and David González-Gómez

Departamento de Didáctica de las Ciencias Experimentales y Matemáticas, Universidad de Extremadura, 10004 Cáceres, Spain; dggomez@unex.es
* Correspondence: jin@unex.es

1. Introduction

There is a broad consensus in recognizing the importance of having a citizenry competent in mathematics which, in this way, contributes to ensuring both their socioeconomic progress and their employability, as well as their personal fulfillment, social inclusion, and active participation as citizens [1–3]. However, there are numerous indicators, both nationally and internationally, that reveal the existence of a deterioration not only in the acquisition of skills but also in the motivation and attitude that students have toward learning mathematical content in particular and mathematicians in a broader sense [4,5]. In science, technology, engineering, arts, and mathematics (STEAM) studies, along with mathematics education, in recent years, active and innovative methods have recognized increasing consecration of specifying the apposite arrangements of students' capacities [6,7]. In daily circumstances, a STEAM education offers various combinations of mathematics disciplines as concrete objects, the teaching/learning of which are merged and ordered so they may be exploited for problem solving [1,8,9].

In varied educational divisions, educational sustainability development (ESD) has pursued lifetime consciousness and features for each individual [10,11]. Thus, in the context of higher education, it should be a fragment of a universal organization which champions sustainability education [12,13]. Precisely, it was elevated to a better understanding of the sustainability concept and reoriented to instructive syllabuses [14,15]. It was also designated with respect to the acquisition of skills, information, worth, and knowledge [14–16]. Sterling [17] stated that sustainable education exists for directing transformative learning, which is an amendment of educational philosophy for the potential fulfilment and economic, social, and ecological interdependence of each individual. Conclusively, through an instructional culture, teaching resolutions are deliberated with the goal of affirming pre-service teachers and capabilities, communications, criteria, and thinking procedures, which perform as transition representatives for sustainable development [18–20].

This Special Issue, "Sustainable Development of STEAM and Mathematics Education with Active and Innovative Methodology", is intended to provide a solid/concrete research corpus. It presents the challenges/skills required for distributing a proper active, innovative, and sustainable STEAM and mathematics education, which is for scholars and/or professionals of dissimilar educational backgrounds and practical fields. In particular, a STEAM and mathematics education can contour the fragments, or is beginning to contour the fragments, of numerous educational organizations' curricula with feasible presentations for numerous categorizations/disciplines. Nevertheless, in higher education institutions (HEIs), efforts should be made toward the Sustainable Development Goals (SDGs), in addition to making ongoing efforts in the working environment to validate a suitable arrangement and advance sustainability-oriented problems.

Citation: Jeong, J.S.; González-Gómez, D. Sustainable Development of STEAM and Mathematics Education with Active and Innovative Methodology. *Sustainability* **2024**, *16*, 591. https://doi.org/10.3390/su16020591

Received: 1 January 2024
Accepted: 4 January 2024
Published: 10 January 2024

2. Scientific Contributions

2.1. Assessment of Students' Mathematical Skills with Respect to the European Qualifications Framework

Stańdo et al. focus on the outcomes of association amongst factors related to students' mathematical proficiencies through socio-economic, demographic, and ontogenetic settings. This article attempts to answer the query of how the mathematical proficiencies of students are developed with regard to their strengths/weaknesses. It also addresses mathematics utilization in daily life. Here, this article scrutinizes the association of results, which comprise a simulated final mathematics exam for eighth-grade elementary students and/or final-year students in high school. In further mathematics classes, mathematics is applied to daily life and the greatest problems with particular extents of material thought.

2.2. Computer Tools for the Sustainability of Knowledge and Developing Active and Innovative Methods in STEAM and Mathematics Education

Körtesi et al. concentrate on learning/teaching mathematics and analyze the functions of information, communication, and technology (ICT) tools, computer algebra systems (CASs) and dynamic geometry systems (DGSs), in education, which allow for the implementation of active and innovative instruction methodologies correlated with sustainable science, technology, engineering, arts, and mathematics (STEAM) education. Similarly, this article emphasizes the requirement of learners to embrace a knowledge of mathematical theory, an indispensable qualification for guaranteeing the reliable/effective use of current version of mathematical software provided. Throughout practical research, this work underlines a mixed teaching method which can meaningfully develop mathematical knowledge sustainability.

2.3. Sustainable Learning, Cognitive Gains, and Improved Attitudes in College Algebra Flipped Classrooms

Karjanto and Acelajado deliberate on the implementation of curricula which nurture the cross-cutting/transversal key competencies for sustainability. This article aims to organize students to become sustainability residents. In numerous conditions and across dissimilar regulations, they can maintain learning throughout their lives. Also, through their tendencies, approaches, and proficiencies, they participate productively and conscientiously toward any challenges for the future world. The flipped classroom, which is an active learning, flexible, student-oriented, and multidimensional instruction, is acknowledged as a sustainable teaching methodology Finally, toward mathematics learned by means of quantitative/qualitative methods, the objective of this article is to examine the effect of pedagogy on the academic accomplishments of learners and their attitudes.

2.4. Technology-Based Pedagogy for Mathematics Education in the Sustainable Development of Mathematics Education

Naidoo contemplates this qualitative study of 38 postgraduate students, which investigated the experiences/perceptions of technology-based instruction during the COVID-19 pandemic. Particularly, the participants are mathematics education students who are practicing mathematics teachers in South Africa. The virtual communities theory was implemented in this practical study. Here, in two interactive virtual workshops, participants were asked for using numerous technology-based instructions. Consequently, these participants were requested to partake in online interviews which concentrated on their experiences/perceptions of technology-based instruction for acquiring a mathematics education. Therefore, this study discloses the challenges/strengths of using technology-based instruction for studying mathematics education during COVID-19.

2.5. Implementation of STEM Education Using Partial Least Squares Approach

Wijaya et al. focus on the determination of the elements prompting pre-service teachers' intents and the outcomes of gender/age on the science, technology, engineering, and mathematics (STEM) implementation of education as well. Among pre-service secondary school teachers, the theory of planned behavior (TPB) was embraced to envisage the affiliation between the social influence, knowledge, attitude, apparent effectiveness, control, and behavioral intention (BI) of STEM education. Here, the results displayed that observed effectiveness had an affirmative importance and a connection with the pre-service teachers' attitudes toward STEM education. Habit had an affirmative importance in motivating interactive intentions and the implementation of teachers. Subjective standards did not have a substantial correlation with BI and employment.

2.6. *Design Framework for Mathematical Activities in Hong Kong Mathematics Education*

Lo et al. concentrate on the establishment of a design framework in Hong Kong for mathematical modeling activities which are appropriate for teachers and students. Here, the study explores the scheme and content of certified mathematical modeling paradigms announced by the Hong Kong Education Bureau, exploiting a document analysis approach. For an emerging framework, the outcomes stipulate the foundation to be expended in the future design of activities in mathematical modeling. Therefore, four examples were identified and examined in terms of their fundamental constituents, level of learning involvement in mathematical modeling, and design features.

2.7. *Students' Perceptions of Online Learning during the COVID-19 Pandemic*

Curelaru et al. deliberate on a thematic examination of the views/perspectives of university students regarding online learning. Specifically, this article describes their clarifications and understandings of the alteration from face-to-face courses in a traditional manner to online teaching during the COVID-19 pandemic. Some of the main themes obtained from the information denote the negative facets of online learning described by participants. These relate to its drawbacks, such as health/psychosocial difficulties and learning procedure glitches. Examples include a lack of feedback and challenges, misinterpretations, extra academic necessities, and disengagement. Other recurring premises refer to the positive features of online learning and its advantages, such as contentment, accessibility, and psychological/medical safety.

2.8. *Educational Innovations for STEAM Education in Digital Technology Environments*

Lavicza et al. focus on three instances of projects on both pedagogical and technological advances to elucidate the influence of fast technological modifications on research. The research team members established and expended technological applications in the research projects, exploiting design-based research (DBR). Here, it was not only necessary to redesign the methods proposed on the basis of the research results; in addition, the technological modifications were so fast that the materials/pedagogies also needed to be adjusted while considering how to redesign the project for the next sequence on the basis of the analysis of the obtained data. Therefore, this study reveals an extra facet to be measured in DBR in addition to examining technology integration or innovative technologies.

2.9. *Service Learning as an ESD in a STEM University Course*

Martín-Sánchez, González-Gómez, and Jeong deliberate on a service learning (SL) methodology for the design, implementation, and evaluation of as a STEM education for sustainable development (ESD) approach in a university course. A pre- and post-test methodology was expended to evaluate the SL suitability of such an education for a sustainable development (ESD) approach. It resulted in a substantial escalation in the students' knowledge for the strategies of innovative teaching, along with appropriate contents, as well as in their comprehension of SDGs. Furthermore, in the SL project, the students' contribution afforded them an awareness of the community repercussions of sustaining the atmosphere and producing an advantage for the whole community. Thus, this research displays how the SL teaching methodology is a significant instrument for the accomplishment of both curricular competences and environmental awareness. Here, theoretical experience pertains to concrete work to accomplish an actual community service and is therefore a very appropriate instruction strategy pertaining to EDS.

2.10. SEM Statistical Software Packages for a Sustainable Mathematics Education

Sakaria et al. focus on the ideal selection of patented statistical software packages for structural equation modeling (SEM) methods, which are insufficient despite its immense significance in sustainable education. Here, this study provides a systematic review which is obliged to examine the experimental studies to fill this existing gap. Web of Science (WoS), Scopus, and Education Resources Information Center (ERIC) databases were searched to identify publications from 2018 to 2022. Current version of Lisrel, Amos, Mplus, SmartPLS, R package (plspm), and WarpPLS statistical software provided was utilized to determine an optimal choice. Despite the extensive practice of a diversity of statistical programs, SmartPLS and AMOS were thoroughly applied in CB-SEM and VB-SEM/PLS-SEM, respectively.

3. Conclusions

This Special Issue encompasses articles, reviews, short notes, etc., on active and innovative approaches toward methodologies and advanced research related to sustainable STEAM and mathematics education [21,22]. Here, it encourages theoretical, methodological, and empirical research on teaching and learning, competencies and assessment, policy, program development and implementation, instructor preparation, community- and project-based learning, institutional collaborations and partnerships, and other relevant topics. Particular focus was directed toward active and innovative teaching and learning approaches and methodologies that have been substantiated to be relevant to sustainable STEAM and mathematics education, such as flipped classrooms, blended learning, escape rooms, gamification, technology-based classrooms, future classrooms, virtual reality, e-learning and online learning, project-based learning, service-learning, and inclusive learning.

Funding: The authors want to thank the Consejería de Economía y Agenda Digital de la Junta de Extremadura (Spain), FEDER funds (Projects IB18004 and GR21047), and the Ministry of Science and Innovation research project (PID2020-115214RB-I00/AEI/10.13039/501100011033).

Institutional Review Board Statement: Not applicable.

Informed Consent Statement: Not applicable.

Conflicts of Interest: The authors declare no conflicts of interest.

List of Contributions:

1. Stańdo, J.; Kołodziejski, K.; Fechner, Ż. Assessment of students' mathematical skills in relation to their strengths and weaknesses, at different levels of the european qualifications framework. *Sustainability* **2023**, *15*, 8661.
2. Körtesi, P.; Simonka, Z.; Szabo, Z.K.; Guncaga, J.; Neag, R. Challenging examples of the wise use of computer tools for the sustainability of knowledge and developing active and innovative methods in STEAM and mathematics education. *Sustainability* **2022**, *14*, 12991.
3. Karjanto, N.; Acelajado, M.J. Sustainable learning, cognitive gains, and improved attitudes in college algebra flipped classrooms. *Sustainability* **2022**, *14*, 12500.
4. Naidoo, J. Technology-based pedagogy for mathematics education in South Africa: Sustainable development of mathematics education post COVID-19. *Sustainability* **2022**, *14*, 10735.
5. Wijaya, T.T.; Jiang, P.; Mailizar, M.; Habibi, A. Predicting factors influencing preservice teachers' behavior intention in the implementation of STEM education using partial least squares approach. *Sustainability* **2022**, *14*, 9925.
6. Lo, C.-K.; Huang, X.; Cheung, K.-L. Toward a design framework for mathematical modeling activities: An analysis of official exemplars in Hong Kong mathematics education. *Sustainability* **2022**, *14*, 9757.

7. Curelaru, M.; Curelaru, V.; Cristea, M. Students' perceptions of online learning during COVID-19 pandemic: A qualitative approach. *Sustainability* **2022**, *14*, 8138.
8. Lavicza, Z.; Weinhandl, R.; Prodromou, T.; Anđić, B.; Lieban, D.; Hohenwarter, M.; Fenyvesi, K.; Brownell, C.; Diego-Mantecón, J.M. Developing and evaluating educational innovations for STEAM education in rapidly changing digital technology environments. *Sustainability* **2022**, *14*, 7237.
9. Martín-Sánchez, A.; González-Gómez, D.; Jeong, J.S. Service learning as an education for sustainable development (ESD) teaching strategy: Design, implementation, and evaluation in a STEM university course. *Sustainability* **2022**, *14*, 6965.
10. Sakaria, D.; Maat, S.M.; Mohd Matore, M.E.E. Examining the optimal choice of SEM statistical software packages for sustainable mathematics education: A systematic review. *Sustainability* **2023**, *15*, 3209.

References

1. Jeong, J.S.; González-Gómez, D. Mathematics self-belief comparison and examination of pre-service teacher (PST) through a flipped-open calculation based on numbers (ABN) learning method. *Heliyon* **2022**, *8*, e09806. [CrossRef]
2. Jeong, J.S.; González-Gómez, D. Flipped-OCN method in mathematics learning to analyze the attitudes of pre-service teachers. *Mathematics* **2021**, *9*, 607. [CrossRef]
3. Jeong, J.S.; González-Gómez, D. Adapting to PSTs' pedagogical changes in sustainable mathematics education through flipped e-learning: Ranking its criteria with MCDA/F-DEMATEL. *Mathematics* **2020**, *8*, 858. [CrossRef]
4. Akınoğlu, O.; Tandoğan, R.Ö. The effects of problem-based active learning in science education on students' academic achievement, attitude and concept learning. *Eurasia J. Math. Sci. Technol. Educ.* **2007**, *3*, 71–81. [CrossRef] [PubMed]
5. Bego, C.R.; Ralston, P.A.S.; Thompson, A.K. Improving performance in a large flipped barrier mathematics course: A longitudinal case study. *Int. J. Math. Educ. Sci. Technol.* **2020**, *53*, 1916–1933. [CrossRef]
6. Jeong, J.S.; González-Gómez, D. A web-based tool framing a collective method for optimizing the location of a renewable energy facility and its possible application to sustainable STEM education. *J. Clean. Prod.* **2020**, *251*, 119747. [CrossRef]
7. Miller, K.; Sonnert, G.; Sadler, P. The influence of students' participation in STEM competitions on their interest in STEM careers. *Int. J. Sci. Educ.* **2018**, *8*, 95–114. [CrossRef]
8. Wiswall, M.; Stiefel, L.; Schwartz, A.E.; Boccardo, J. Does attending a STEM high school improve student performance? Evidence from New York City. *Econ. Educ. Rev.* **2014**, *40*, 93–105. [CrossRef]
9. González-Gómez, D.; Jeong, J.S. The flipped learning model in general science: Effects on students' learning outcomes and affective dimensions. In *Active Learning in College Science*; Mintzes, J.J., Walter, E.M., Eds.; Springer: Cham, Switzerland, 2020; pp. 541–549.
10. Eneroth, C. E-Learning for Environment. Improving e-Learning as a Tool for Cleaner Production Education. Licentiate. Ph.D. Thesis, Lund University, Lund, Sweden, 2000.
11. Mora-Granados, M.; González-Gómez, D.; Jeong, J.S.; Gallego-Picó, A. A molecularly imprinted polymer for selective extraction of phenolic acids from human urine. *Appl. Sci.* **2021**, *11*, 1577. [CrossRef]
12. Wiek, A.; Withycombe, L.; Redman, C.L. Key competencies in sustainability: A reference framework for academic program development. *Sustain. Sci.* **2011**, *6*, 203–218. [CrossRef]
13. Valcke, M.M. Teacher education in Logo-based environments: A handbook for teacher trainers. *Educ. Comput.* **1991**, *7*, 293–304. [CrossRef]
14. Jeong, J.S.; González-Gómez, D. A STEM course analysis during COVID-19: A comparison study in performance and affective domain of PSTs between F2F and F2S flipped classroom. *Front. Psychol.* **2021**, *23*, 669855. [CrossRef] [PubMed]
15. Lozano, R. Incorporation and institutionalization of SD into universities: Breaking through barriers to change. *J. Clean. Prod.* **2006**, *14*, 787–796. [CrossRef]
16. Jeong, J.S.; González-Gómez, D. Multi-criteria decision analysis methods for sustainability assessment of renewable energy systems and its potential application to sustainable STEM education. In *Energy Systems Evaluation (Volume 2). Green Energy and Technology*; Ren, J., Ed.; Springer: Cham, Switzerland, 2021; pp. 39–62.
17. Sterling, S. *Sustainable Education: Re-Visioning Learning and Change*; Schumacher Briefings No. 6; Green Books: Devon, UK, 2001.
18. Thomas, I. Critical thinking, transformative learning, sustainable education, and problem-based learning in universities. *J. Transform. Educ.* **2009**, *7*, 245–264. [CrossRef]
19. Fadeeva, Z.; Mochizuki, Y. Higher education for today and tomorrow: University appraisal for diversity, innovation and change towards sustainable development. *Sustain. Sci.* **2010**, *5*, 249–256. [CrossRef]
20. González-Gómez, D.; Gallego Picó, A.; Muñoz de la Peña, A.; Jeong, J.S. Implementation of a computer-aided learning toolbox for establishing an instrumental analysis calibration and the quality parameters of an analytical chemistry method. *Chem. Educ.* **2013**, *18*, 136–143.

21. Naidoo, J.; Reddy, S. Embedding sustainable mathematics higher education in the fourth industrial revolution era post-COVID-19: Exploring technology-based teaching methods. *Sustainability* **2023**, *15*, 9692. [CrossRef]
22. Hsiao, P.-W.; Su, C.-H. A Study on the impact of STEAM Education for sustainable development courses and its effects on student motivation and learning. *Sustainability* **2021**, *13*, 3772. [CrossRef]

sustainability

MDPI

Article

Assessment of Students' Mathematical Skills in Relation to Their Strengths and Weaknesses, at Different Levels of the European Qualifications Framework

Jacek Stańdo [1], Kamil Kołodziejski [2] and Żywilla Fechner [2,*]

[1] Centre of Mathematics and Physics, Lodz University of Technology, 90-924 Łódź, Poland; jacek.stando@p.lodz.pl
[2] Institute of Mathematics, Lodz University of Technology, al. Politechniki 8, 93-590 Łódź, Poland
* Correspondence: zywilla.fechner@p.lodz.pl

Abstract: Many scientific studies focus on finding the relationship between students' mathematical skills and socio-economic, demographic, and ontogenetic factors. In this publication, we answer the question of how students' mathematical skills are achieved in relation to their strengths and weaknesses, also with regard to the use of mathematics in everyday life. In this article, we examine the relationship between the results of the mock final math exam for eighth grade primary school students/final year high school students and additional math classes, the application of math in everyday life and the greatest difficulties with specific areas of taught material. The study was conducted in Poland on almost ten thousand eighth graders and high school leavers who took part in mock exams online, respectively: eighth-grader exam, and school-leaving maturity exam. The participants of these online exams were asked to respond to a survey that pertained to their math grades, attending additional math classes, their perceived most useful mathematical topics in everyday life and future professional work, and identification of their strengths and weaknesses. In the following paper, the relationships between the answers to the survey questions and the results of the mock online exam are analyzed. The results indicate that there are differences in the area of results of the mock exam and answers about strengths and weakness in mathematical literacy. The analysis of answers about use the mathematical knowledge are different for eight-graders and high-school students. Eight-graders indicate the importance of arithmetic operations while high -school students point out more abstract topics like probability, statistics and geometry. The results of the study are compared to the existing results.

Keywords: eighth-grade mock exam; school-leaving mock maturity exam; difficulties in math; everyday-life mathematics

Citation: Stańdo, J.; Kołodziejski, K.; Fechner, Ż. Assessment of Students' Mathematical Skills in Relation to Their Strengths and Weaknesses, at Different Levels of the European Qualifications Framework. *Sustainability* **2023**, *15*, 8661. https://doi.org/10.3390/su15118661

Academic Editors: David González-Gómez and Jin Su Jeong

Received: 30 March 2023
Revised: 23 May 2023
Accepted: 23 May 2023
Published: 26 May 2023

1. Introduction

The problem of students' difficulties with mathematics is discussed in many contexts, both from the perspective of the student and the teacher as discussed by Hamukwaya [1], Klymchuk et al. [2], Ramli et al. [3] among others. Researchers are also interested in the relationship of difficulties in acquiring mathematical knowledge in the context of the prospect of potential use of mathematics in everyday life, including professional work for example discussed by Ojose in [4]. Different levels of education are also taken into account: from primary education to higher education, which have been studied by Saeed [5], Udousoro [6] or Hamukwaya [1]. The aim of our research is to examine the relationship between the results of the electronic mock exam in mathematics at the eighth grade level and the final exam with answers to survey questions regarding the use of mathematical skills in the future, and the strengths and weaknesses in mathematics declared by the participants [4,7–24]. The problem of sustainability in mathematics education has been discussed for two decades. Li and Tsay [25] draw attention to the complexity of the problem of sustainability in teaching mathematics. They show little progress in answering fundamental questions about teaching and learning in the light of sustainable development. They also address the issue of sustainable development in the teacher education process. The problem of sustainability in mathematics in teacher education is addressed by Joutsenlahti and Perkkilä [26]. They discuss the problem of sustainable development both in the general education process and in teaching mathematics. The study concerned the understanding of the a/b symbol at different stages of education. A questionnaire was used during a first-year mathematics teaching course at two Finnish universities. The study indicated pedagogical limitations in teaching the fraction concept. The authors suggested ways to improve the teaching of the fraction concept for sustainable development in math education. Moreno-Pino et al. [27] They discussed the problem of teaching mathematics from the perspective of an academic teacher. The authors emphasize the role of sustainable development in teaching students who, as future teachers, will be responsible for social changes and transformations serving sustainable development. The results presented in our article indicate the need to emphasize the greater use of mathematics in everyday life and to point out the practical aspects of mathematics, which leads to conscious living for sustainable development.

This paper aims to find a connection between the results of mock exams for eighth grade students and a school living maturity exam and answers to the questions from the questionnaire attached to the mock exam. We will analyze the relationship between the total number of points obtained from the mock exam and the following issues: the use of additional mathematics classes in last two years, the final grade in mathematics, strengths and weaknesses in mathematical knowledge declared by the participants, and use of mathematical skills in future work and mathematical issues used in everyday life.

2. Literature Review

Phonapichat et al. [7] list five reasons why students have difficulty solving math problems. These are as follows: "(1) Students have difficulties in understanding the keywords appearing in problems, thus cannot interpret them in mathematical sentences. (2) Students are unable to figure out what to assume and what information from the problem is necessary to solving it, (3) Whenever students do not understand the problem, they tend to guess the answer without any thinking process, (4) Students are impatient and do not like to read mathematical problems, and (5) Students do not like to read long problems" (see [7]). The literature of the subject also discusses issues related to students' problems with specific concepts or topics covered in mathematics lessons. In Eisenberg's [8] review article the difficulties associated with the notion of function from both a historical and psychological point of view are described.

Sholeha et al. [9] discuss the results of a qualitative descriptive research conducted on eighth grade 1X students at SMPN (Sekolah Menengah Pertama Negeri) 2 Batang Tuaka. Authors point out several issues in this study, such as small number of research subjects and lack of other data sources e.g., learning test results. Zulfah et al. [10] describe ways to measure solving mathematical problems related to the Pythagorean theorem and solving a system of linear equations on the basis of research conducted on eighth-grade students of junior high school. Puspitarani and Retnawati [11] present their findings concerning problems with tasks related to the Pythagorean theorem based on a study of 8th grade students in SMP (Sekolah Menengah Pertama) 1 Todanan and SMP (Sekolah Menengah Pertama) Muhammadiyah 9 Todanan. Study shows similar results to [1], i.e., students had difficulty understanding and analyzing problems, were not careful when solving problems and too hasty in their rush to solve problems. Students' difficulties with geometry tasks have been discussed by Kuzniak and Rauscher [12], Retnawati et al. [13] and Smith [14]; and those connected with calculus of probability and statistics have been presented by Puspitasari et al. [15] or Garfield and Ahlgren [16]. Puspitasari [15] claims that improving ability to think logically is the key to better handling with statistical and probability problems. However, we cannot compare this result with students point of view, because research did not contain questionary. One of the ways of dealing with difficulties in mathematics is attending additional classes, outside of school. This topic was already taken up in the nineties by Levine and Zimmerman [17]. They discussed the impact of taking additional math classes on future earnings. The likelihood of choosing a specific type of profession traditionally assigned to one gender was also discussed. The results indicated a greater likelihood of better earnings and learning a non-traditional profession for women who were taking math classes. No significant effect of extra math classes was observed in the case of blue-collar workers. In [18] there are published the results of his study on the increase in the participation rate in additional, extracurricular math classes around the world. Differences were detected both between different countries as well as within countries. A dependency was observed between a higher demand for participation in additional mathematics classes and the weakness of the national education system. The research compared the case of Korea and the United States: in the former, private lessons are seen as a threat to the education system and should be subject to legal regulations, while in the USA as support for the education system. A similar issue was raised in the article by Zhang et al. [19]. The study was conducted on Chinese high school students. As in the case of Korea, research showed that parents should choose their children's extracurricular activities appropriately, and the government should issue proper regulations regarding their organization. Discussions about the effectiveness of private lessons at school in terms of future university success were examined by Guill and Boss [20]. The study was conducted in Germany. Most of the respondents asserted that additional classes have an impact on mathematical achievements. However, there was no significant difference between math grades or test scores depending on participation in extracurricular activities.

Regardless of the level of mathematical knowledge, there is a need to apply mathematics in everyday life to a greater or lesser extent. The issue of adequate preparation of students for the everyday use of mathematics has been widely discussed in the literature. Ojose [4] touches upon problems related to the knowledge of mathematics and its use in everyday life. He raises the issue of the essence of mathematics and indicates a list of necessary competences that comprise the knowledge of mathematics. He claims that the school does not provide proper knowledge of mathematics and seeks the reasons for this state of affairs. Putnam [21] describes two lessons taught by Valerie Taft in California. The first one is based on the official textbook and deals with the concept of average. The other one consists in hands-on finding the average based on data prepared by Valerie. Focusing on particular steps and lack of reflection on the calculations cause erroneous determination of averages. Kalchman [22] raises the problem of preparing students for final exams in isolation from the use of mathematics in everyday life. Jansen et al. [23] present conclusions from a study on a population of over 500 adult Dutch citizens on the relationship between fear of mathematics and mathematical skills and their use in everyday life, taking into account the gender of the respondents. Kang et al. [24] address the issue of using Augmented Reality to transmit knowledge in the form of everyday life problems.

3. Methodology

In this research we compare two studies carried out in Poland in 2022. They consisted in conducting mock exams for eighth graders and high school leavers using a platform. On the platform, we can monitor: the number of points, the number of entries, the time of solving the task.

The project was organized under the patronage of the Rector of Lodz University of Technology and Stowarzyszenie Nauczycieli Matematyki (The Association of Teachers of Mathematics). Dr. Jacek Stańdo created all the exam tasks, both for the eight-grader exam and the high school leaving exam.

The tasks in both exams were reviewed by specialists from the Regional Examination Board in Lodz.

In Poland, as a result of the reform of the education system in 1999, a system of external examinations was introduced, unified throughout the country. The exams verify the requirements written in the so-called A curriculum that covers the entire country. Work on exams is supervised by the Central Examination Commission. CKE is responsible for creating examination papers. Examinations are conducted in schools, which are supervised by eight Regional Examination Boards. Pupils take a compulsory exam in mathematics at the end of primary and secondary education. The lists of learning outcomes for primary and secondary schools presented below are consistent with the learning outcomes required by the core curriculum.

The first study was conducted among eighth-grade students between 10th and 20th May 2022. The mock eighth-grader exam included 19 tasks which had been reviewed by specialists from the Regional Examination Board in Lodz. Table 1. presents the assumed learning outcomes which were validated with the use of auto-generated math problems.

Table 1. The learning outcomes verified for eighth-grade students.

Task Number	The Learning Outcome
1	Analyzes operations with numbers
2	Finds the value of an angle
3	Constructs a perpendicular line
4	Performs operations with numbers
5	Raises numbers to a power
6	Analyzes the average
7	Applies percentages in practical situations
8	Applies counting methods
9	Calculates the probability of an event
10	Constructs figures with axial symmetry and central symmetry
11	Calculates the surface area of a figure in practical situations
12	Applies the Pythagorean theorem
13	Transforms algebraic expressions
14	Calculates the radius of a circle
15	Solves a linear inequality
16	Constructs a parallelogram in the coordinate plane
17	Validates the described situation
18	Analyzes the problem situation
19	Determines the volume of the pyramid

Definition 1. *Let X be a set of objects. A one-dimensional generator task (problem) Z(x) is called a linguistic expression which becomes a task (problem) if an element from the set X is substituted for x.*

Definition 2. *Let* $X_1 \times X_2 \times \ldots \times X_n$, *where* $X_1,\ X_2, ..X_n$ *are sets of objects. An n-dimensional generator problem Z(x) is called a linguistic expression which becomes a problem if an element from the set X is substituted for x.*

Example 1. *Eve has* x_1 *kg of strawberries and her brother Adam has* x_2 *kg* x_3. *How many kilograms of strawberries do Eve and Adam have together?*

Answer 1. *Eve and Adam have together {if* x_3 *="="less" " then* $2x_1 - x_2$ *if* x_3 *=="more" " then* $2x_1 + x_2$*} kg of strawberries.*

The list of learning outcomes are summarized in Table 1.

3.1. Description of the Study Group of Eighth Graders

An invitation to participate in a real-time online mock exam was sent to all primary schools in Poland. School data came from the database of the Education Information System (SIO). 261 primary schools from all voivodeships declared their participation, which constituted about 1.7% of all primary schools in Poland, Figure 1.

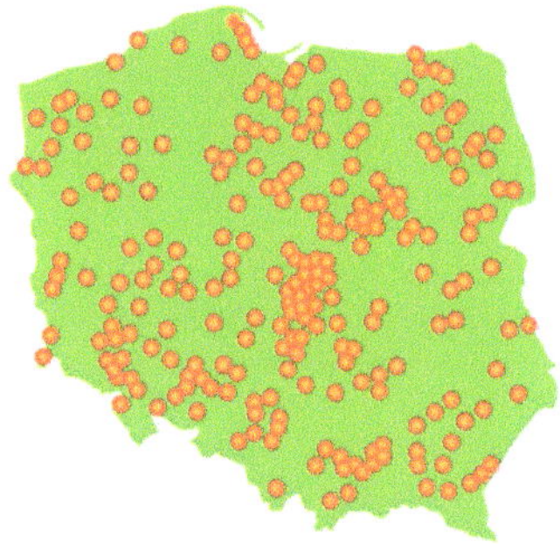

Figure 1. Map of schools participating in the project (eight–grade exam).

The participants of the study were 6827 students of the eighth grade of primary schools, who constituted 1.4% of the total population in Poland, Table 2.

Table 2. Population of the eighth grader mock exam.

City or Town of Students' Residence	Data from the Central Examination Board (2022) [28]		Mock Exam		% of Population
	Number	Percentage	Number	Percentage	
Town up to 20,000	262,576	54.23%	3248	47.57%	1.2%
City of 20–100,000	97,149	23.32%	1477	21.60%	1.5%
City over 100,000	118,263	24.75%	2102	30.70%	1.7%
Total	484,174		6827		

The analysis of specific tasks is presented in the Section 4.

3.2. The Study Group of High-School Leavers (Graduates)

Between 1–15 April 2022 a nationwide study was conducted that consisted in running a mock high school-leaving math exam online (basic level). The mock exam included 35 tasks. Table 3 presents the assumed learning outcomes.

Table 3. The learning outcomes verified for high school leavers.

Task Number	The Learning Outcome
1	Determines the equation of the quadratic function
2	Produces a graph of an exponential function
3	Applies operations on percentages in practical situations
4	Determines the equation of a straight line
5	Applies the abbreviated multiplication formula
6	Analyzes the arithmetic mean
7	Defines the domain of the function
8	Analyzes operations on numbers
9	Calculates a weighted average
10	Applies counting methods
11	Uses progression
12	Draws a quadratic function
13	Creates the canonical form of a quadratic function
14	Finds the value of an angle
15	Constructs the equation of a line through two points
16	Calculates the probability of an event
17	Determines the value of a function based on the graph
18	Uses properties of logarithms
19	Constructs a line perpendicular to a given straight line
20	Analyzes the properties of a prism

Table 3. *Cont.*

Task Number	The Learning Outcome
21	Solves a polynomial equation
22	Evaluates the trigonometric function for a triangle
23	Solves a rational equation
24	Applies theorems about the circumcircle of a triangle
25	Analyzes problematic situations
26	Performs operations with numbers
27	Analyzes the problem using similar figures
28	Creates figures with central and axial symmetry
29	Interprets a system of linear equations
30	Creates an inequality based on the data
31	Determines the largest and smallest value in an interval
32	Analyzes the problem situation
33	Uses arithmetic progression
34	Constructs a parallelogram with given properties
35	Determines the surface area and volume of a pyramid

3.3. Study Group

An invitation to participate in a real-time online mock exam was sent to all high schools in Poland. School details were obtained from the Education Information System (SIO) database. 188 high schools (general high schools and technical schools) from all voivodeships declared their participation, Figure 2.

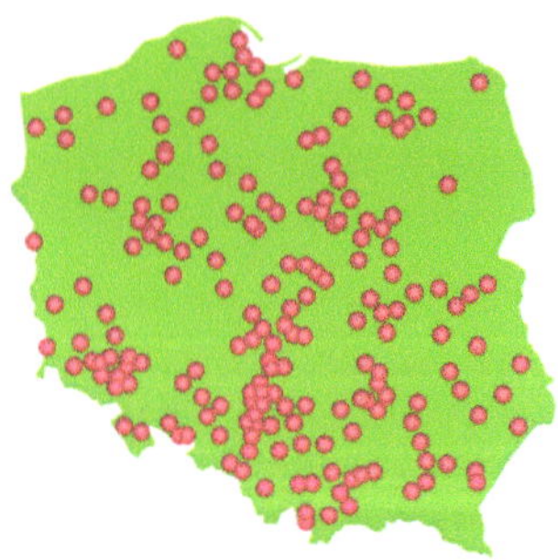

Figure 2. Map of schools participating in the project (high-school).

The participants of the study were 3388 students of the final year of high school, who constituted over 1.0% of the total population in Poland, Table 4.

Table 4. Population of the high-school leaving mock exam.

City or Town of Students' Residence	Data from the Central Examination Board (2021) [28]		Mock Exam		% of Population
	Number	Percentage	Number	Percentage	
Town up to 20,000	70,714	20.30%	821	24.23%	1.1%
City of 20–100,000	126,780	36.41%	1413	41.71%	1.1%
City over 100,000	150,742	43.29%	1154	36.06%	0.8%
Total	3,480,236		3388		

The aim of this research is to find a connection between the results of mock exams and answers from the questionnaire. Analysis of each answer separately was impossible due to the large sample size. Because of that, we decided to use statistical analysis. Survey responses were grouped in order to check dependencies between survey responses and exam results. In open questions, grouping based on specific keywords was applied. This method allowed to analyze a large number of possible answers and detect what were the main issues that students struggle with. In order to receive a good representation of the average score and to ignore potential outliers, a median of total exam points was calculated for each group. To check if the differences in medians between groups are statistically significant, Mann–Whitney Wilcoxon tests were performed, which is one of the most popular non-parametric tests for checking differences between non-normally distributed populations. The detailed discussion about the use of non-parametric Mann–Whitney test and using statistical tests in psychology and education can be found in Ferguson and Takane [29].

3.3.1. Questions 1–2

Questions 1 and 2 were closed and had 5 possible answers. After removing blank responses (no answer provided to the questionnaire), data was grouped. Each question was considered separately. Then, from the total number of points obtained in the test, medians, averages, and sample sizes were calculated for each of the 5 possible answers and for the entire sample. To show the statistical difference between the means in the groups, the Wilcoxon test was performed for each two pairs of the groups. As a result of the test, the p value was returned along with the information whether it is higher than the selected significance level of 0.05.

3.3.2. Questions 3–9

The same statistical analyzes were performed for open questions 3 to 9, the only difference being how the responses were grouped. At the beginning, blank responses were excluded, and the remaining responses were prepared by removing Polish characters and changing all capital letters to lowercase. Next, a division into groups: "A", "B", "C" and „D" was carried out according to keywords. A response that contained a word from the list corresponding to a given group was assigned to that group. In the case of conflicts and belonging to several groups at the same time, the most extreme group was selected; that is, for groups ABC, C would be selected, and for BD, D would be selected. Responses that were not classified into any group were placed in the "other" category.

Overall:

- Group A: operations with numbers, practical calculations, including percentages
- Group B: geometry of figures and their properties, calculating circuit length and surface areas (without the Pythagorean theorem)
- Group C: probability and statistics
- Group D: the Pythagorean theorem
- Group E: trigonometry
- Group F: function and differential calculus
- "Other" group: all others

In this paper we investigate six selected questions closely connected with the topic.

4. Data Collection and Analysis

The following section offers a detailed analysis of the responses to selected questionnaire items along with a comparison of the eighth graders' and high school leavers' responses. The more detailed description of the questionnaire is in Appendix A.

Question 1. Have you taken extra math classes in the last two years?

- I haven't (group 1)
- I haven't, but I wanted to (group 2)
- I have no opinion (group 3)
- I have, occasionally (group 4)
- I have regularly attended additional math lessons (group 5)

The group names introduced as above are used in Tables 5 and 6 below.

Table 5. Summary of the Mann–Whitney Wilcoxon test results—eighth grade students, question 1.

First Group	Second Group	*p*-Value	Conclusion	Median First Group	Median Second Group
1	all	0.0203	reject	9	9
1	3	0.0043	reject	9	8
1	2	0.0037	reject	9	8
1	4	0.0000	reject	9	8
2	all	0.0396	reject	8	9
3	2	0.9488	accept	8	8
3	4	0.6107	accept	8	8
3	all	0.0412	reject	8	9
4	2	0.6353	accept	8	8
4	all	0.0025	reject	8	9
5	1	0.5679	accept	9	9
5	all	0.1396	accept	9	9
5	3	0.0089	reject	9	8
5	2	0.0083	reject	9	8
5	4	0.0003	reject	9	8

Table 6. Summary of the Mann–Whitney Wilcoxon test results—high school graduates, question 1.

First Group	Second Group	*p*-Value	Conclusion	Median First Group	Median Second Group
1	3	0.0023	reject	25	21
1	2	0.0042	reject	25	21
1	4	0.1019	accept	25	24
1	all	0.1257	accept	25	24
2	all	0.0180	reject	21	24
3	all	0.0094	reject	21	24
3	2	0.7876	accept	21	21
4	3	0.0252	reject	24	21
4	2	0.0472	reject	24	21
4	all	0.5844	accept	24	24
5	3	0.0073	reject	24	21
5	2	0.0136	reject	24	21
5	4	0.4177	accept	24	24
5	1	0.5132	accept	24	25
5	all	0.6108	accept	24	24

The results for the first question are summarized in Figure 3. The results of Mann–Whitney tests are in Tables 5 and 6.

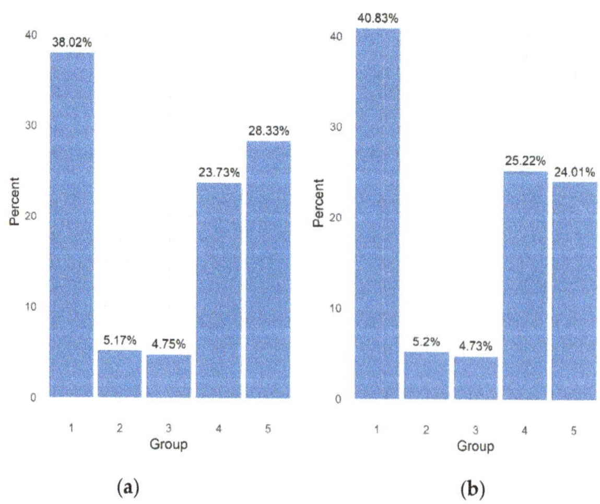

Figure 3. Summary of responses to question 1 (**a**) eighth graders (**b**) high school graduates.

Summary of question 1

Every fourth student regularly attends additional mathematics classes. On the other hand, about 45% of the students participating in the study do not take any extra classes.

Those eighth graders who did not attend any additional classes and those who attended regularly achieved higher scores than students in other groups. Those students who did not attend additional classes but wanted to and those who attended occasionally achieved lower scores than other students. Those eighth graders who attended extra classes regularly had better scores than those who did not attend but wanted to and better than those who attended occasionally. No statistically significant difference was observed between the median scores of students who regularly attended extra classes, either in relation to the group of students who did not attend additional classes at all, or in relation to the group of all students.

In the group of final year high school students, there were no significant differences between the scores of those who did not attend extra math classes and those who attended regularly (group 5), or those who attended occasionally (group 4) or in relation to all participants (groups 1–5 combined). The study showed that there are no significant differences among high school leavers who regularly attended additional classes in relation to combined groups 1–5.

Question 2. State your math grade on the school-leaving report card (Polish grades are expressed by numbers and their corresponding names: dopuszczający (2)—barely passing, dostateczny (3)—satisfactory, dobry (4)—good, bardzo dobry (5)—very good, celujący (6)—excellent)

- dopuszczający—barely passing (D) (group 1)
- dostateczny—satisfactory (C) (group 2)
- dobry—good (B) (group 3)
- bardzo dobry—very good (A) (group 4)
- celujący—excellent (A plus) (group 5)

The results for the second question are summarized in Figure 4. The results of Mann–Whitney tests are in Tables 7 and 8.

Summary of question 2

Table 7. Summary of the Mann–Whitney Wilcoxon test results—eighth grade students, question 2.

First Group	Second Group	*p*-Value	Conclusion	Median First Group	Median Second Group
1	2	0.0000	reject	3	6
1	all	0.0000	reject	3	9
2	all	0.0000	reject	6	9
3	all	0.0000	reject	10	9
3	2	0.0000	reject	10	6
3	1	0.0000	reject	10	3
4	3	0.0000	reject	14	10
4	all	0.0000	reject	14	9
4	2	0.0000	reject	14	6
4	1	0.0000	reject	14	3
5	4	0.0000	reject	15	14
5	3	0.0000	reject	15	10
5	all	0.0000	reject	15	9
5	2	0.0000	reject	15	6
5	1	0.0000	reject	15	3

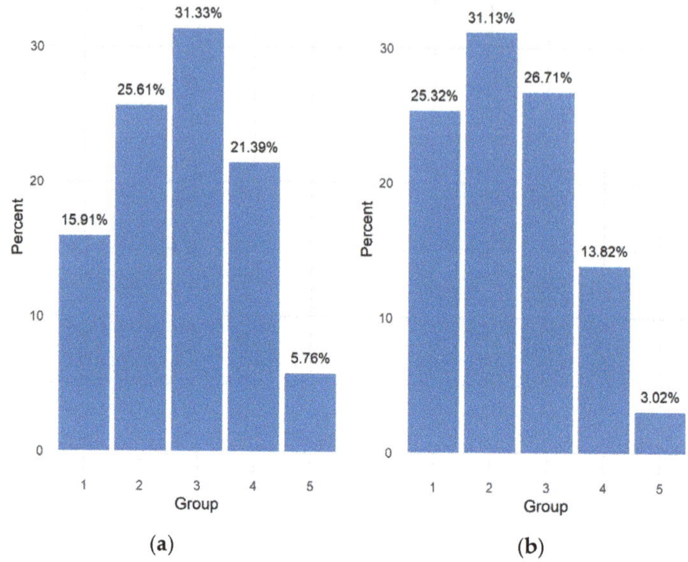

Figure 4. Summary of responses to question 2 (**a**) eighth graders (**b**) high school graduates.

Table 8. Summary of the Mann–Whitney Wilcoxon test results—high school graduates, question 2.

First Group	Second Group	*p*-Value	Conclusion	Median First Group	Median Second Group
1	all	0.0000	reject	14	24
2	4	0.0000	reject	21	33
2	1	0.0000	reject	21	14
2	5	0.0000	reject	21	34.5
2	all	0.0000	reject	21	24
3	1	0.0000	reject	29	14
3	2	0.0000	reject	29	21
3	all	0.0000	reject	29	24
3	4	0.0000	reject	29	33
3	5	0.0000	reject	29	34.5
4	1	0.0000	reject	33	14
4	all	0.0000	reject	33	24
4	5	0.1772	accept	33	34.5
5	1	0.0000	reject	34.5	14
5	all	0.0000	reject	34.5	24

Eight graders and high school leavers show the same relationship: the better their math grade at the end of school, the higher their test score. Groups of students considered in terms of each of the grades differ significantly (with the exception of "very good" and "excellent" in the case of high school leavers). It should be noted that, in contrast to question 1, homogeneous results are observed here.

Question 3. Which areas of mathematics that you learned in the math class do you consider the most useful in daily life?

The results for the third question are summarized in Figure 5. The results of Mann–Whitney tests are in Tables 9 and 10.

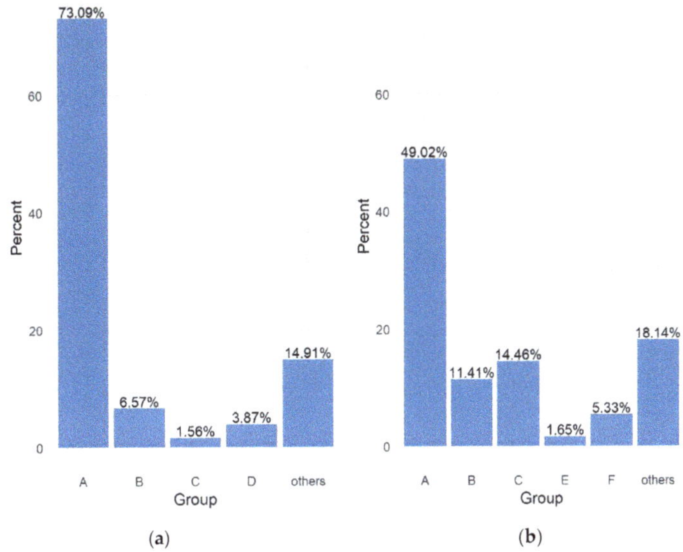

Figure 5. Summary of responses to question 3 (**a**) eighth graders (**b**) high school graduates.

Table 9. Summary of the Mann–Whitney Wilcoxon test results—eighth grade students, question 3.

First Group	Second Group	p-Value	Conclusion	Median First Group	Median Second Group
A	C	0.0000	reject	9	13
A	others	0.0000	reject	9	7
A	all	0.8482	accept	9	9
B	others	0.0000	reject	10	7
B	C	0.0003	reject	10	13
B	all	0.0086	reject	10	9
B	A	0.0124	reject	10	9
C	others	0.0000	reject	13	7
C	all	0.0000	reject	13	9
D	others	0.0005	reject	11	7
D	C	0.0009	reject	11	13
D	all	0.1306	accept	11	9
D	A	0.1502	accept	11	9
D	B	0.7938	accept	11	10
others	all	0.0000	reject	7	9

Table 10. Summary of the Mann–Whitney Wilcoxon test results—high school graduates, question 3.

First Group	Second Group	p-Value	Conclusion	Median First Group	Median Second Group
A	C	0.0000	reject	24	30
others	C	0.0000	reject	23	30
B	others	0.0001	reject	29	23
others	F	0.0002	reject	23	32
A	B	0.0003	reject	24	29
C	all	0.0003	reject	30	25
A	F	0.0003	reject	24	32
F	all	0.0046	reject	32	25
others	all	0.0098	reject	23	25
B	all	0.0131	reject	29	25

Table 10. *Cont.*

First Group	Second Group	*p*-Value	Conclusion	Median First Group	Median Second Group
A	all	0.0184	reject	24	25
others	E	0.0334	reject	23	27.5
A	E	0.0602	accept	24	27.5
E	all	0.1751	accept	27.5	25
B	F	0.2625	accept	29	32
A	others	0.3466	accept	24	23
B	C	0.5101	accept	29	30
C	F	0.5654	accept	30	32
B	E	0.7393	accept	29	27.5
F	E	0.8437	accept	32	27.5
C	E	0.9899	accept	30	27.5

Summary of question 3

Three in four students in the eighth grade say that operations with numbers and practical calculations, including percentages, are needed in everyday life. However, only one in two high school graduates says so.

The highest scores among eighth graders were achieved by those who considered topics from group C (Probability and Statistics) to be the most necessary in everyday life. The study showed a significant difference in the number of points compared to the group that indicated calculations, including percentages, as most needed in everyday life. Similarly, students indicating probabilistic and statistical issues obtained higher test scores in relation to the other groups, as well as in relation to the group indicating other concepts. Eighth grade students who indicated the topics from group A as the most important in everyday life, obtained lower results than students who indicated group C as the most useful. However, there is no statistically significant difference in the results obtained by eighth graders indicating the most important topics from group A in relation to the group "others".

Among high school leavers, those students who indicated areas from groups B, C and F obtained better results in comparison to students indicating topics other than the ones from groups B, C and F, respectively. However, there are no significant differences between the results achieved by high school graduates indicating topics from groups B, C and F as the most important. As in the case of eighth-graders, the high school graduates who indicated the concepts from group A as the most useful in everyday life achieved significantly lower results in the mock exam than those who indicated the terms from group C as the most useful.

Question 4. What areas of mathematics do you think will be most useful in your future work?

The results for the fourth question are summarized in Figure 6. The results of Mann–Whitney tests are in Tables 11 and 12.

Table 11. Summary of the Mann–Whitney Wilcoxon test results—eighth grade students, question 4.

First Group	Second Group	*p*-Value	Conclusion	Median First Group	Median Second Group
A	C	0.0005	reject	8	13
A	all	0.1827	accept	8	8
A	D	0.3867	accept	8	9
B	A	0.0001	reject	10	8
B	others	0.0005	reject	10	8
B	all	0.0007	reject	10	8
B	C	0.0188	reject	10	13
B	D	0.3825	accept	10	9

Table 11. *Cont.*

First Group	Second Group	*p*-Value	Conclusion	Median First Group	Median Second Group
C	all	0.0010	reject	13	8
C	D	0.0196	reject	13	9
D	all	0.5757	accept	9	8
others	C	0.0009	reject	8	13
others	D	0.4348	accept	8	9
others	all	0.4477	accept	8	8
others	A	0.8140	accept	8	8

Table 12. Summary of the Mann–Whitney Wilcoxon test results—high school graduates, question 4.

First Group	Second Group	*p*-Value	Conclusion	Median First Group	Median Second Group
A	B	0.0004	reject	23	27.5
A	C	0.0000	reject	23	27.5
A	F	0.0000	reject	23	33
A	all	0.0014	reject	23	25
B	C	0.6605	accept	27.5	27.5
B	F	0.0004	reject	27.5	33
B	all	0.0517	accept	27.5	25
C	F	0.0049	reject	27.5	33
C	all	0.0085	reject	27.5	25
E	A	0.0011	reject	31.5	23
E	B	0.1803	accept	31.5	27.5
E	C	0.3253	accept	31.5	27.5
E	F	0.2785	accept	31.5	33
E	all	0.0169	reject	31.5	25
F	all	0.0000	reject	33	25
others	E	0.0079	reject	25	31.5
others	A	0.1415	accept	25	23
others	B	0.0135	reject	25	27.5
others	C	0.0018	reject	25	27.5
others	F	0.0000	reject	25	33
others	all	0.2135	accept	25	25

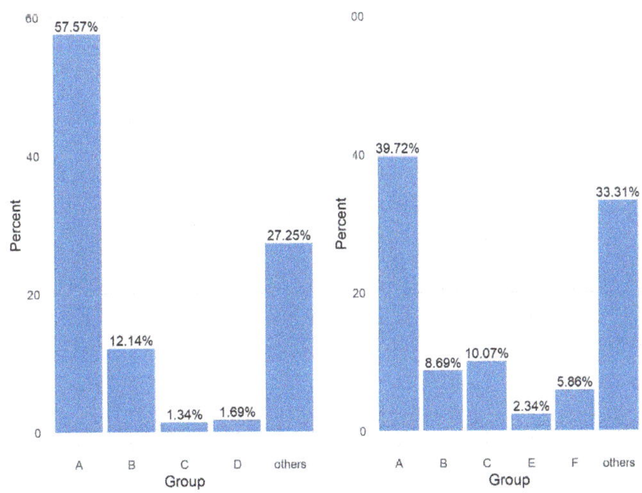

(**a**) (**b**)

Figure 6. Summary of responses to question 4 (**a**) eighth graders (**b**) high school graduates.

Summary of question 4

More than fifty per cent of the eighth-grade students say that operations with numbers and practical calculations, including percentages, will be necessary in their future work. However, this conviction is less firm among high school leavers, who demonstrated increasing awareness of the likely use of probability and statistics in their future work.

Those eighth-grade students who indicated concepts from group C as the most useful, achieved significantly better results in the mock exam compared to other separate groups and the combined group. The eighth graders who indicated the concepts from group B as the most needed ones scored better on the test than the eighth graders who indicated the concepts from group A. Similarly, the results of these study participants were better than those of other respondents who indicated concepts from a group other than B. However, the same students scored lower than eighth graders who indicated concepts from group C.

High school leavers indicating concepts from group C as the most useful in future work obtained better results than those indicating concepts from group A and group "other", while their results were lower than those of high school leavers indicating group F. High school leavers indicating group F as the most useful in future work obtained better results than those indicating groups: A, B, C, "other". High school leavers indicating the concepts from group A as the most needed in future work achieved significantly lower results in relation to participants indicating the issues from groups B, C and F.

Question 5. Which areas of mathematics do you think you would like to understand better, but were difficult for you?

The results for the fifth question are summarized in Figure 7. The results of Mann–Whitney tests are in Tables 13 and 14.

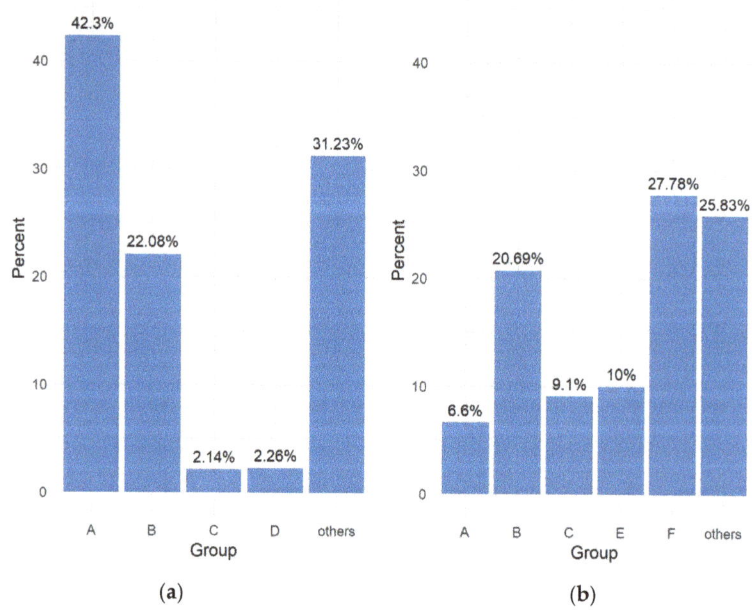

Figure 7. Summary of responses to question 5 (**a**) eighth graders (**b**) high school graduates.

Table 13. Summary of the Mann–Whitney Wilcoxon test results—eighth grade students, question 5.

Number of First Group	Number of Second Group	*p*-Value	Conclusion	Median First Group	Median Second Group
A	others	0.0841	accept	8	9
A	B	0.1357	accept	8	9
A	all	0.2385	accept	8	8.5
B	all	0.4784	accept	9	8.5
C	A	0.0003	reject	12	8
C	all	0.0008	reject	12	8.5
C	B	0.0020	reject	12	9
C	others	0.0026	reject	12	9
D	C	0.0000	reject	5	12
D	B	0.0001	reject	5	9
D	others	0.0002	reject	5	9
D	all	0.0003	reject	5	8.5
D	A	0.0007	reject	5	8
others	all	0.3457	accept	9	8.5
others	B	0.8911	accept	9	9

Table 14. Summary of the Mann–Whitney Wilcoxon test results—high school graduates, question 5.

First Group	Second Group	*p*-Value	Conclusion	Median First Group	Median Second Group
A	E	0.0000	reject	22	30
A	all	0.0011	reject	22	25
B	C	0.0001	reject	29	34
B	others	0.0000	reject	29	23
B	F	0.0000	reject	29	23
B	A	0.0000	reject	29	22
B	E	0.2979	accept	29	30
B	all	0.0033	reject	29	25
C	others	0.0000	reject	34	23
C	F	0.0000	reject	34	23
C	A	0.0000	reject	34	22
C	E	0.0019	reject	34	30
C	all	0.0000	reject	34	25
E	all	0.0004	reject	30	25
F	A	0.1830	accept	23	22
F	E	0.0000	reject	23	30
F	all	0.0004	reject	23	25
others	F	0.8759	accept	23	23
others	A	0.1951	accept	23	22
others	E	0.0000	reject	23	30
others	all	0.0029	reject	23	25

Summary of question 5

Eighth grade students most often want to better understand the concepts from group A, and least often the concepts from group C. In the case of high school graduates, the largest number of the respondents want to understand the concepts from group E, and the smallest number the concepts from group A.

Eighth grade students who would like to better understand the Pythagorean Theorem (Group D) obtain low results relative to the other groups, both when considering each group separately and when combined. Very high scores are achieved by eighth graders indicating Group C, who want to better understand probability and statistics. A higher result is observed here both in the case of individual groups and combined groups.

High school graduates who want to understand concepts from groups B and C achieve higher results than those indicating the other groups, while high school graduates indicating group C achieve significantly higher results than those indicating group B. Poor results are observed in the case of groups A and F.

Question 6. Which math topics do you consider to be your strongest point?

The results for the sixth question are summarized in Figure 8. The results of Mann–Whitney tests are in Tables 15 and 16.

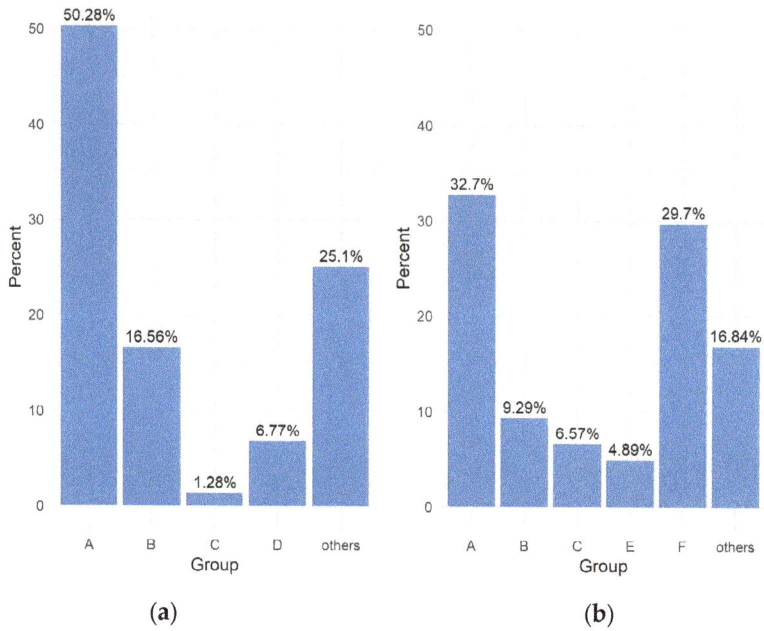

Figure 8. Summary of responses to question 6 (**a**) eighth graders (**b**) high school graduates.

Table 15. Summary of the Mann–Whitney Wilcoxon test results—eighth grade students, question 6.

First Group	Second Group	*p*-Value	Conclusion	Median First Group	Median Second Group
A	D	0.1186	accept	8	8
A	C	0.1571	accept	8	10
A	all	0.6822	accept	8	8
B	others	0.0000	reject	10	8
B	D	0.0000	reject	10	8
B	all	0.0000	reject	10	8
B	A	0.0001	reject	10	8
B	C	0.9021	accept	10	10
C	D	0.0371	reject	10	8
C	all	0.1757	accept	10	8
D	all	0.0675	accept	8	8
others	all	0.0470	reject	8	8
others	C	0.0697	accept	8	10
others	A	0.1318	accept	8	8
others	D	0.5558	accept	8	8

Table 16. Summary of the Mann–Whitney Wilcoxon test results—high school graduates, question 6.

First Group	Second Group	*p*-Value	Conclusion	Median First Group	Median Second Group
A	C	0.8356	accept	21	21
A	all	0.0000	reject	21	25
B	E	0.0013	reject	29	34
B	others	0.0019	reject	29	24
B	A	0.0000	reject	29	21
B	C	0.0003	reject	29	21
B	all	0.0295	reject	29	25
C	all	0.0102	reject	21	25
E	others	0.0000	reject	34	24
E	A	0.0000	reject	34	21
E	C	0.0000	reject	34	21
E	all	0.0000	reject	34	25
F	B	0.2229	accept	30	29
F	E	0.0059	reject	30	34
F	others	0.0000	reject	30	24
F	A	0.0000	reject	30	21
F	C	0.0000	reject	30	21
F	all	0.0000	reject	30	25
others	A	0.0389	reject	24	21
others	C	0.2481	accept	24	21
others	all	0.0600	accept	24	25

Summary of question 6

Half of the eighth graders state that their strongest point is operations with numbers and practical calculations, including percentages (group A). In the final year of high school, the number of students in this category significantly decreases.

Eighth grade students who declare geometry (group B) as their strongest point achieve higher results than groups: A, D, "others". The same students achieve results that are not statistically significant in relation to students who indicate group C as their strength.

Among high school graduates, higher results are observed in group B in relation to groups A, C and "others". High school graduates indicating group E score better than any other group. High school graduates indicating group A as their strongest point achieve worse results than those indicating a group other than A. In turn, high school graduates indicating group C as their strongest point achieve significantly worse results than those indicating group B, E or F.

5. Discussion

About half of the survey participants declare that they regularly or occasionally attend additional mathematics classes. This indicates that additional mathematics classes are also popular among students in Poland. In line with the assumptions presented by Guill and Boss [20], our study shows that participation in additional classes does not affect the result of the mock exam in mathematics. We do not observe any significant differences in the results of the exam in the case of students attending extra classes when it comes to eighth graders. In light of the research discussed in the introduction, it is worth rethinking whether a central policy for the organization of additional mathematics lessons is necessary. In view of this study, it is not clear whether participation in additional mathematics classes has a positive impact on the results of final exams. What is also worth considering are the reasons for the significant interest in additional lessons among students. The research cited in the introduction shows that in countries with a weaker education system there is more interest in additional mathematics classes than in countries with a better education system. This suggests looking for one of the possible causes in the system that calls for the improvement of mathematical education. Taking a closer look at these issues will make it possible to try to reduce the scale of additional private classes in Poland for a more balanced development.

The study showed a strong relationship between mock exam scores and math grades. Both in the group of eighth graders and high school graduates, those with higher grades achieved better exam results than those with lower grades. Similar considerations have been applied by Ha et al. [30] for pharmacy students during the pharmaceutical calculations course. About 93% of the students took advantage of the opportunity to take the mock exam. The results of the final exam were significantly higher than the results of the mock exam. This means that the test results were significantly lower than the final grade. In our research, students who performed better on the mock exam received higher math final grades.

The analysis of responses to questions about the use of mathematics in everyday life and in future work shows that at the primary school level, the largest percentage of respondents indicate that topics related to arithmetic calculations, including percentages, are the most useful in the future. However, this percentage decreases in the case of high school graduates who see a greater need, compared to primary school students, to apply topics from the field of probability, statistics, and geometry. Similar observations are presented by Ojose [4], pointing to the need to teach mathematics in a manner consistent with the guidelines of the Program for International Students Assessment (PISA).

The study shows that students of both primary and secondary schools indicate concepts related to arithmetic, practical calculation of percentages, or probability and statistics as the most useful topics applied in everyday life or work. However, the participants did not indicate responses related to logical thinking skills, creating mathematical models to describe the surrounding reality or other, more abstract concepts. For example, the responses did not include topics related to algorithms, widely used for example in programming.

The conducted study demonstrated a difference in the results of the mock exam depending on the areas that students defined as their strengths and weaknesses. It was noticed that students who indicate simple issues as their weaknesses which they would like to overcome, achieve lower results. Eighth graders who pointed to the Pythagorean theorem as the topic they wanted to understand, scored worse in the mock exam. In the case of high school leavers, those students who wanted to better understand issues related to arithmetic calculations, including percentages or functions, achieved low results in the mock exam. One possible explanation is that the difficulty in mastering simpler topics makes it impossible to understand more complex ones.

Analyzing the responses regarding the students' declared strengths in terms of mathematical knowledge, it can be seen that in the case of eighth graders, those declaring geometry or probability and statistics achieve better results than students declaring other categories of concepts. On the other hand, students who declare arithmetic as their strongest point obtain significantly lower results than students who consider other groups of concepts to be their strong points. This might hint at the need to explore in the future the relationship between the way of thinking that is required to analyze geometric or probabilistic problems and raising the general level of mathematical skills. As a continuation of the research, an experiment might be conducted in which two groups of students would be compared: the first one taught in accordance with the curricula set by the educational authorities without additional geometric tasks, and the second one taught fewer typically computational tasks (equivalent to topics from group A) but given extended time devoted to issues in geometry. The results of this study suggest that discussion of issues in geometry can be expected to contribute to the overall improvement of mathematical skills.

6. Conclusions

The Recommendation of the Council of the European Union indicates the need to build a European Qualifications Framework (https://eur-lex.europa.eu/legal-content/EN/TXT/PDF/?uri=CELEX:32017H0615(01), accessed on 17 March 2023) For example, an analysis of the answer to the question 4: "What areas of mathematics do you think will be most useful in your future work"? points out that percentage calculations should be attributed to the second level of the European Qualifications Framework, while the calculus of probabilities, statistics and differential calculus are topics included in at least the sixth level of the European Qualifications Framework for engineers, mathematicians, and economists.

The study has its strengths and weaknesses. One of the strengths of the study is its size. In total, in both groups, almost ten thousand respondents took part. Another asset of the study is the fact that the respondents represented about 1 percent of the group of eighth graders and high school graduates with regard to the size of the town where the student comes from. The electronic form of the test ensured its accessibility regardless of the place of residence. However, this also had a downside: students had unlimited time to solve tasks and complete the survey. The analysis of the time allocated to solve the tasks shows that in some cases it extended to several hours, which suggests that the exam was solved intermittently. As in all surveys, there is no way to verify the accuracy of the answers. In particular, it is impossible to check whether the declared final grade was in line with the facts.

The study was conducted on the population of Polish students and adapted to the Polish education system in the field of mathematics. Taking into account the strengths and weaknesses of the study, it may be the basis for extending the issues to the European Union, where the European Framework of Education applies. Possible future designs of the study should try to limit the weaknesses of the study, such as unlimited response time. It may be considered to extend the IT tools (platform) in a way that enables better verification of the person taking the test.

Author Contributions: Conceptualization, J.S. and Ż.F.; methodology, J.S. software, K.K.; validation, J.S., K.K. and Ż.F.; formal analysis, J.S. and Ż.F.; investigation, J.S.; resources, J.S.; data curation, J.S. and K.K.; writing—original draft preparation, J.S. and Ż.F.; writing—review and editing, J.S., K.K. and Ż.F.; visualization K.K.; supervision, J.S.; project administration, J.S. All authors have read and agreed to the published version of the manuscript.

Funding: This research received no external funding.

Institutional Review Board Statement: Not applicable.

Informed Consent Statement: Not applicable.

Data Availability Statement: The data was collected during the mock exams described in the paper and is not publicly available.

Conflicts of Interest: The authors declare no conflict of interest.

Appendix A

Questions in the questionnaire can be divided into two main categories. The first concerns the measurement of the student's knowledge level and the possibility of supplementing knowledge in extracurricular activities (Question one and two), and the second (other questions) concerns the student's subjective views on the strengths and weaknesses of his mathematical skills, the potential use of mathematical knowledge in everyday life, including professional work.

We combine the category concerning the assessment of the student's mathematical knowledge with the results of the mock exam. The final grade in mathematics is an average of the results obtained by the student throughout the year, issued by a given teacher on the basis of partial tests he has checked. The mock exam is centrally structured and the test comparing student skills is the same for all study participants. The results of the mock exam and the final assessment show convergence, so we decided to use the result of the final exam as a criterion for assessing the level of mathematical knowledge of the student. The argument for choosing the results of the trial exam as the criterion for assessing the level of knowledge is its greater objectivity than the final grades. Final scores are determined on the basis of mid-term tests, the level of difficulty of which may vary from school to school.

The second category of questions concerns the student's subjective assessment of the strengths and weaknesses of his mathematical knowledge and the possibility of using mathematics in everyday life and at work. The aim of the study is to find a relationship between the objectively assessed level of knowledge and the subjective assessment of the usefulness of mathematics and difficulties with the acquisition of mathematical knowledge. In general, the higher the level of mathematical knowledge, the more difficult topics are more complicated, and the mathematical tools potentially used in everyday life are more advanced.

Below we present the content of questions for high school and eighth grade.

Table A1. Eight grader questionnaire.

Question	Number of Valid Answers
Have you taken extra math classes in the last two years?	4627
State your math grade on the school-leaving report card	4581
How do you use the Internet to learn math?	4554
Which areas of mathematics that you learned in the math class do you consider the most useful in daily life?	3594
Which areas of mathematics that you learned in the math class do you consider the most useless in daily life?	3516
What areas of mathematics do you think will be most useful in your future work?	3427
What areas of mathematics do you think will be most useless in your future work?	3353
Which areas of mathematics do you think you would like to understand better, but were difficult for you?	3324
Which math topics do you consider to be your strongest point?	3351

Table A2. Secondary school questionnaire.

Question	Number of Valid Answers
Have you taken extra math classes in the last two years?	2593
State your math grade on the school-leaving report card	2621
Which areas of mathematics that you learned in the math class do you consider the most useful in daily life?	1588
Which areas of mathematics that you learned in the math class do you consider the most useless in daily life?	1562
What areas of mathematics do you think will be most useful in your future work?	1460
What areas of mathematics do you think will be most useless in your future work?	1420
Which areas of mathematics do you think you would like to understand better, but were difficult for you?	1449
Which math topics do you consider to be your strongest point?	1440
In what life situation did you need math?	1388
What is your authority?	1346

References

1. Hamukwaya, S.T. A Comparison of the Perceived Difficulties in Learning High School Mathematics from the Perspective of Teachers and Students. *Int. J. Math. Teach. Learn.* **2022**, *23*, 46–63. [CrossRef]
2. Klymchuk, S.; Zverkova, T.; Gruenwald, N.; Sauerbier, G. University Students' Difficulties in Solving Application Problems in Calculus: Student Perspectives. *Math. Educ. Res. J.* **2010**, *22*, 81–91. [CrossRef]
3. Ramli, F.; Shafie, N.; Tarmizi, R.A. Exploring Student's in-Depth Learning Difficulties in Mathematics through Teachers' Perspective. *Procedia Soc. Behav. Sci.* **2013**, *97*, 339–345. [CrossRef]
4. Ojose, B. Mathematics Literacy: Are We Able to Put the Mathematics We Learn into Everyday Use? *J. Math. Educ.* **2011**, *4*, 89–100.
5. Saeed, Z.M.S. Difficulty in Learning Mathematics for Primary School Students. *Coll. Basic Educ. Res. J.* **2020**, *16*, 77–96. [CrossRef]
6. Udousoro, U. Perceived and Actual Learning Difficulties of Students in Secondary School Mathematics. *Afr. Res. Rev.* **2011**, *5*, 357–366. [CrossRef]

7. Phonapichat, P.; Wongwanich, S.; Sujiva, S. An Analysis of Elementary School Students' Difficulties in Mathematical Problem Solving. *Procedia Soc. Behav. Sci.* **2014**, *116*, 3169–3174. [CrossRef]
8. Eisenberg, T. Functions and Associated Learning Difficulties. In *Advanced Mathematical Thinking*; Springer: Dordrecht, The Netherlands, 2002; pp. 140–152.
9. Sholeha, V.A.; Risnawati, R.; Habibullah, H. An Analysis of Student Difficulties in Mathematics Learning in Terms of Student Mathematical Connection Ability on Pythagoras Theorem. *Prism. Sains J. Pengkaj. Ilmu Pembelajaran Mat. IPA IKIP Mataram* **2021**, *9*, 12. [CrossRef]
10. Zulfah; Astuti; Surya, Y.F.; Marta, R.; Wijaya, T.T. Measurement of Mathematics Problems Solving Ability Using Problem Based Mathematics Question. *J. Phys. Conf. Ser.* **2020**, *1613*, 012026. [CrossRef]
11. Puspitarani; Retnawati, H. Diagnosis of Learning Difficulties in Mathematics for Students Resolving Problems Related to Material in the Pythagorean Theorem for 8th Grade Students in SMP 1 Todanan and SMP Muhammadiyah 9 Todanan, Academic Year 2018/2019. *J. Phys. Conf. Ser.* **2020**, *1581*, 012026. [CrossRef]
12. Kuzniak, A.; Rauscher, J.-C. How Do Teachers' Approaches to Geometric Work Relate to Geometry Students' Learning Difficulties? *Educ. Stud. Math.* **2011**, *77*, 129–147. [CrossRef]
13. Retnawati, H.; Arlinwibowo, J.; Sulistyaningsih, E. The Students' Difficulties in Completing Geometry Items of National Examination. *Int. J. New Trends Educ. Implic.* **2017**, *8*, 28–41.
14. Smith, R.R. Three Major Difficulties in the Learning of Demonstrative Geometry. *Math. Teach.* **1940**, *33*, 99–134. [CrossRef]
15. Puspitasari, N.; Afriansyah, E.A.; Nuraeni, R.; Madio, S.S.; Margana, A. What Are the Difficulties in Statistics and Probability? *J. Phys. Conf. Ser.* **2019**, *1402*, 077092. [CrossRef]
16. Garfield, J.; Ahlgren, A. Difficulties in Learning Basic Concepts in Probability and Statistics: Implications for Research. *J. Res. Math. Educ.* **1988**, *19*, 44–63. [CrossRef]
17. Levine, P.B.; Zimmerman, D.J. The Benefit of Additional High-School Math and Science Classes for Young Men and Women. *J. Bus. Econ. Stat.* **1995**, *13*, 137–149. [CrossRef]
18. Lee, J. Two Worlds of Private Tutoring: The Prevalence and Causes of after-School Mathematics Tutoring in Korea and the United States. *Teach. Coll. Rec.* **2007**, *109*, 1207–1234. [CrossRef]
19. Zhang, Y.; Cui, C.; He, Y.; Wang, L. Does Private Supplementary Tutoring Matter in Chinese Students' Learning of Mathematics: A Longitudinal Study. *ZDM Math. Educ.* **2022**, *54*, 737–747. [CrossRef]
20. Guill, K.; Bos, W. Effectiveness of Private Tutoring in Mathematics with Regard to Subjective and Objective Indicators of Academic Achievement Evidence from a German Secondary School Sample. *J. Educ. Res. Online* **2014**, *6*, 34–67.
21. Putnam, R.T. Teaching the "Hows" of Mathematics for Everyday Life: A Case Study of a Fifth-Grade Teacher. *Elem. Sch. J.* **1992**, *93*, 163–177. [CrossRef]
22. Kalchman, M. Using the Math in Everyday Life to Improve Student Learning. *Middle Sch. J.* **2011**, *43*, 24–31. [CrossRef]
23. Jansen, B.R.J.; Schmitz, E.A.; van der Maas, H.L.J. Affective and Motivational Factors Mediate the Relation between Math Skills and Use of Math in Everyday Life. *Front. Psychol.* **2016**, *7*, 513. [CrossRef]
24. Kang, S.; Shokeen, E.; Byrne, V.L.; Norooz, L.; Bonsignore, E.; Williams-Pierce, C.; Froehlich, J.E. ARMath: Augmenting Everyday Life with Math Learning. In Proceedings of the 2020 CHI Conference on Human Factors in Computing Systems, Honolulu, HI, USA, 25–30 April 2020; ACM: New York, NY, USA, 2020; pp. 1–15.
25. Li, H.-C.; Tsai, T. L. Education for Sustainable Development in Mathematics Education: What Could It Look Like? *Int. J. Math. Educ. Sci. Technol.* **2022**, *53*, 2532–2542. [CrossRef]
26. Joutsenlahti, J.; Perkkilä, P. Sustainability Development in Mathematics Education—A Case Study of What Kind of Meanings Do Prospective Class Teachers Find for the Mathematical Symbol "23"? *Sustainability* **2019**, *11*, 457. [CrossRef]
27. Moreno-Pino, F.M.; Jiménez-Fontana, R.; Domingo, J.M.C.; Goded, P.A. Training in Mathematics Education from a Sustainability Perspective: A Case Study of University Teachers' Views. *Educ. Sci.* **2022**, *12*, 199. [CrossRef]
28. Central Examination Board. Available online: https://cke.gov.pl/images/_EGZAMIN_OSMOKLASISTY/Informacje_o_wynikach/2022/sprawozdanie/EO_matematyka_sprawozdanie_2022.pdf (accessed on 23 January 2023).
29. Ferguson, G.A.; Takane, Y. *Statistical Analysis in Psychology and Education*; McGraw-Hill: New York, NY, USA, 1989.
30. Ha, C.; Ahmed, U.; Khasminsky, M.; Salib, M.; Andey, T. Correlative and Comparative Study Assessing Use of a Mock Examination in a Pharmaceutical Calculations Course. *Am. J. Pharm. Educ.* **2023**, *87*, ajpe8654. [CrossRef]

sustainability

MDPI

Article

Challenging Examples of the Wise Use of Computer Tools for the Sustainability of Knowledge and Developing Active and Innovative Methods in STEAM and Mathematics Education

Péter Körtesi [1], Zsolt Simonka [2], Zsuzsanna Katalin Szabo [3,*], Jan Guncaga [4] and Ramona Neag [3]

1 Faculty of Materials Science and Engineering, University of Miskolc, 3515 Miskolc, Hungary
2 Department of Mathematics & Actuarial Science, Faculty of Economic Informatics, University of Economics in Bratislava, Dolnozemska Cesta 1, 852 35 Bratislava, Slovakia
3 Faculty of Economics and Law, George Emil Palade University of Medicine, Pharmacy, Science, and Technology of Targu Mures, 540566 Târgu Mureș, Romania
4 Faculty of Education, Comenius University in Bratislava, Racianska 59, 813 34 Bratislava, Slovakia
* Correspondence: szabo.zs.katalin@gmail.com

Abstract: The rapid changes in information and communication technology (ICT), the increasing availability of processing power, and the complexity of mathematical software demand a radical rethinking of science, technology, engineering, arts, and mathematics (STEAM), as well as mathematics education. In the transition to technology-based classrooms, the constant use of educational software is a requirement for sustainable STEAM and mathematics education. This software supports a collaborative and actionable learning environment, develops 21st-century skills, and promotes the adoption of active and innovative methodologies. This paper focuses on learning and teaching mathematics and analyzes the role and utility of ICT tools in education as computer algebra systems (CAS) and dynamic geometry systems (DGS) in implementing active and innovative teaching methodologies related to sustainable STEAM education. Likewise, it highlights the necessity for learners to have extensive knowledge of mathematical theory, an essential asset to ensure the reliable and effective use of mathematical software. Through a practical experiment, this study aims to highlight that a mixed teaching method can significantly improve the sustainability of math knowledge. It provides various solid examples of CAS and DGS applications to emphasize its usage rooted in a mathematical background to enable learners to identify when the computer solution is unreliable. The study highlights that the proper use of CAS and DGS is an efficient method of deepening our understanding of mathematical notions and solving tasks in STEAM subjects and real-life applications. This paper's goal is to direct our attention to the proper and intelligent use of computer tools, especially symbolic calculators, such as CAS and DGS, without providing an in-depth analysis of the challenges of these technologies. The outcomes of the paper should offer educators and learners new elements of active strategies and innovative learning models that can be immediately applied in education.

Keywords: mathematics education; STEAM; ICT; wise use of symbolic calculators; computer algebra systems (CAS); dynamic geometry systems (DGS); graphing functions; didactics of mathematics; sustainable integration of technology; 21st-century skills

Citation: Körtesi, P.; Simonka, Z.; Szabo, Z.K.; Guncaga, J.; Neag, R. Challenging Examples of the Wise Use of Computer Tools for the Sustainability of Knowledge and Developing Active and Innovative Methods in STEAM and Mathematics Education. *Sustainability* **2022**, *14*, 12991. https://doi.org/10.3390/su142012991

Academic Editors: Marc A. Rosen, Jin Su Jeong and David González-Gómez

Received: 27 July 2022
Accepted: 28 September 2022
Published: 11 October 2022

1. Introduction

Education in the 21st century must be focused towards developing learners' creativity and innovativeness to teach them how to be more entrepreneurial and face the challenges of a rapidly changing world. These areas received limited attention in 20th-century education, which consisted of educational practices that are outdated need to be revised. As Yong recalled, "The system results in a population with similar skills in a narrow spectrum of talents" [1]. Education is the main driver of development, and it is still challenging to reform 20th-century education to provide the best teaching possible for a new generation.

Nevertheless, as Stephen Sterling argues, in a rapidly changing world, formal education "largely remains part of the problem of unsustainability" [2]. This anticipates the concept of 'Sustainable Education', which "offers the possibility of education that is appropriate and responsive to the new systemic conditions of uncertainty and complexity that reflected in the headlines every day; one that nurtures the increasingly important qualities of adaptability, creativity, self-reliance, hope and resilience in learners" [2].

Technological advances are encouraging changes in 21st-century teaching. The widespread use of ICT tools modifies the structure, format, and running of teaching and learning at all levels of school education. This learning process can be recorded and archived through teaching platforms and reused for further analysis to improve teaching and find optimal methods to solve problems during learning. Computer tools and mathematical software give new dimensions to teaching methodologies, opening up new innovative ways for introducing and applying pedagogical methods [3–5]. The rethinking of education in the digital era is today's main challenge. Policymakers, professionals, and the whole of society are engaged in finding answers to the most pressing questions such as: What is the role of education? How can education give learners the adequate skills to succeed in a rapidly changing digital world? How can new and traditional teaching methodologies work together in the 21st-century education system to develop skills that make young professionals succeed in the workforce? [6,7].

The COVID-19 pandemic accelerated the technologization of many industries and has brought significant educational changes [8], which were created under the intense time pressure of the "new normal." The different ways of integrating information and communication technologies (ICTs) in the curriculum exemplify how education has adapted to the "new normal" in all parts of the world [9,10]. According to Yong Zhao, "the pandemic has created a unique opportunity for educational changes that have been proposed before COVID-19 but were never fully realized" [11].

Education improved with technology, and this process is irreversible. However, not all students will have equal access to technology in the future; therefore, the problem of a digital divide persists [12]. Teaching in the 21st century needs to be different from the 20th- or 19th-century tutoring. The 21st-century population lives in a globalized, rapidly changing, chaotic, and less controllable world. Education is responsible for preparing the next generation to thrive in such an environment by equipping them with the necessary skills and abilities to understand problems, make decisions, learn from mistakes, and grow personally and professionally. Generation Alpha (those born after 2010) [13] already grow up in a Web 4.0 environment. They begin to use technology from their early childhood; therefore, acquiring new skills and broadening their knowledge through ICT constitutes a powerful tool that is easier for them to understand and adopt. Nevertheless, it is essential to highlight that computer tools can only be efficient when under human control.

It is widely accepted that integrating ICTs in the curriculum has many advantages: new innovative teaching and learning methods can be implemented; educators can employ a wide variety of learning styles to identify the best combination to meet the student's individual needs; in the technology-based classroom, learners are active participants and teachers are more equipped with different teaching tools [14]. According to [15], ICT tools offer a new means of communication, which is the basis of the educational process, allowing teachers to individually communicate with every student and students with each other, thus enabling a differentiated approach. Therefore, it can be stated that the introduction and usage of ICT in the curriculum revolutionizes education and helps transform the teaching-learning process toward the development of sustainable education [16–18]. However, there are large obstacles to achieving this: finding ways to incorporate ICT tools in the curriculum, teachers' preparedness, the development of digital skills, and students' access. Furthermore, it is very important to educate preservice teachers to acquire the relevant skills and competencies needed to teach generation alpha. According to [19], "educators are powerful change agents who can deliver the educational response needed to achieve the

Sustainable Development Goals (SDGs). Their knowledge and competencies are essential for restructuring educational processes and educational institutions towards sustainability".

For the sustainable development of STEAM (Science, Technology, Engineering, Arts and Mathematics) and mathematics education, emerging technologies, especially the use of symbolic calculators as computer tools (SCCT), present a great tool and opportunity to implement active and innovative teaching methodologies [20]. In this paper, under the SCCT framework, we will consider computer algebra systems (CAS) and dynamic geometry systems (DGS). The general term of computer algebra (CA) and separately CAS or DGS will be used when a difference must be emphasized. SCCT offers new directions in teaching and learning mathematics by offering a visualization of basic notions in mathematics and science and challenging users to solve complex mathematical problems. CA develops critical thinking and problem-solving skills by helping the user to understand and interpret unique theoretical examples [21,22]. Nevertheless, the successful utilization of CA and CAS in complex problems requires a solid theoretical understanding. According to Buchberger, "in the application of mathematics to itself, there lies an enormous driving force, which has reached a new dimension, especially through new mathematical software systems, and there is an unprecedented dynamism in mathematical research, education, and applications", and the "computational mathematics is one of the technologies, if not the key technology, of today's information society" [23].

In addition to the benefits of using math software, downsides and dangers must also be considered. It is essential to identify factors and define the criteria used to determine the effectiveness of these tools [24]. The paper of [25] reveals the side effects of education policies and practices and stresses the importance of considering them in the same way as medical products, which are required to disclose both their intended outcomes and known side effects.

The present paper points to the possible 'side effects' of CAS usage. In this respect, this paper warns that the simulations and results provided by SCCT must be looked at with a solid mathematical theory background to avoid any 'side effect,' the misleading interpretation. Only an extensive knowledge of mathematics can lead the user to the correct interpretation of the offered solutions. It is also crucial to know the barriers to the utilized SCCT in order to drive the development of critical thinking and problem-solving skills. At the same time, this can lead to incorrect decisions and solutions if the person who uses this method does not possess sufficient theoretical math knowledge. Moreover, teachers must have control over technology in this case. A theoretical background and knowledge are required to control and interpret the results obtained by CAS and DGS. The user has to decide whether to accept or reject the results provided by the computer and validate them against real-life conditions.

The present paper focuses on some unique aspects of teaching and learning mathematics. It analyzes the role and utility of using ICT tools and CAS/DGS toward implementing active and innovative teaching methodologies for sustainable STEAM education. Furthermore, it highlights the requirements for using these tools effectively. It provides concrete examples of CAS/DGS usage to highlight the importance of having a solid theoretical math background to detect when a computer solution is unreliable. This paper aims to demonstrate through a practical experiment that a mixed teaching method can significantly improve mathematical knowledge.

The paper's goal is to direct attention to the proper and intelligent use of these technologies and the importance and possibility of renewing the mathematics curriculum and teaching methods to follow rapid changes in the available ICT tools. The conclusions of this study provide educators and learners with new elements of active strategies and innovative learning models to be applied during the education process.

The paper is based on quantitative research led by the following questions: What are the most efficient, active, and innovative teaching and learning methods that ensure a higher level of mathematics learner comprehension? What is the most effective way of introducing CAS and DGS in mathematics teaching? What are the advantages of introducing CAS and

DGS in math education to develop 21st-century skills? Is using CAS/DGS in teaching and learning correlated with learners' mathematical knowledge and skills?

The paper also contains an integrative literature review to analyze the theoretical background and study recent research results. Furthermore, this literature review helps validate the accuracy of the obtained results in the context of the research questions.

The primary keywords used during this research are ICT use and CA use in math teaching, challenges, and failures, the effective use of CAS and DGS, mixed teaching methods, the effectiveness of math teaching, math curricula in the 21st-century, learning strategies and models, 21st-century skills, math computation and real-life confrontation, CAS/DGS use and the development of reliable knowledge in math, sustainable STEAM education, innovative and active teaching, and learning methods.

This article is organized as follows: Section 2 describes a practical experiment on the mathematics learning process with or without computer tools; Section 3 presents a short overview of computer tools used in STEAM education and why these tools are used; Section 4 introduces relevant methodological examples in graphing functions, with a comparative discussion of some possible failures and challenges in different CAS and DGS. Lastly, Section 5 presents our conclusions and further research possibilities.

2. Math Learning Process—With or without ICT Tools?—Practical Experiment

2.1. Materials and Methods

The goals of the experiment (presented below) were to study the utility of computer tools in math education, to analyze the need to redefine the didactical aspects of teaching in the 21st century, and to evaluate the necessity of identifying the most efficient, effective math teaching methods in the context of the new paradigm [26]. Our research question is: What are the most efficient, active, and innovative teaching and learning methods that ensure a higher level of mathematics learner comprehension? Is it the theoretical method—the classical teaching method (teacher-centered method/traditional method) or exclusively teaching through new technology, using computer tools—the innovative method (learner-child-centered method/laboratory method), or a combination of the two?

This section presents the main conclusions and methodological outcomes of the experiment.

2.1.1. Procedure and Sampling, Tools

For this extensive pedagogical research, the authors conducted a study at the University of Economics in Bratislava, a public institution of higher education in Slovakia. Forty-seven bachelor's level 1st year students participated in the study. The experiment focused on the most challenging teaching topic, as confirmed by students: graphing functions of one real variable and their transformations. To accomplish the primary research goal (to assess the impact of math software on the understanding of math concepts), a comparison between traditional teaching methods and teaching using ICT tools, the MS Excel environment, was used. A didactical software was created—Elem_F_Grapher (programmed and developed by Zsolt Simonka), Figure 1—which allows graphs of all elementary functions and their transformations to be interactively drawn (off-line function grapher). The program creates an internet-independent and syntax-free technological environment for graphing functions.

The MS Excel environment was chosen predominantly due to its accessibility for each student, whether at the university or home, and because all users had the basic skills required to use it [27,28].

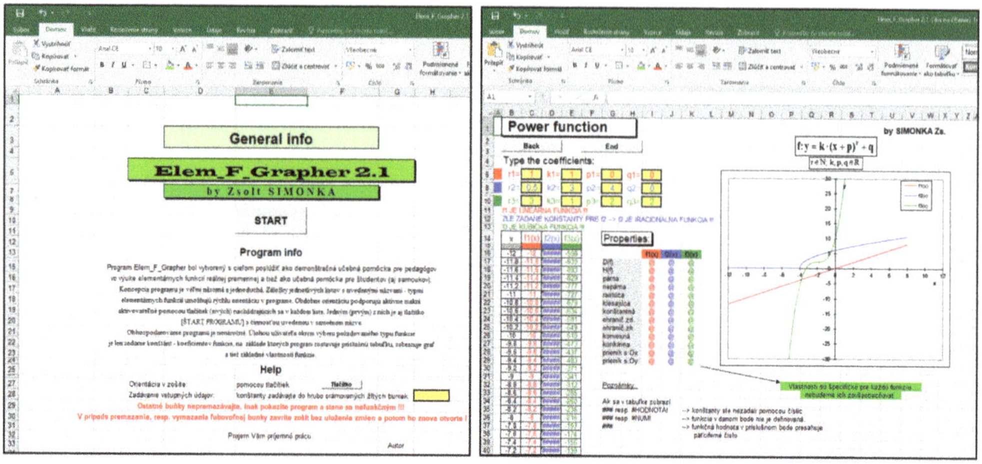

Figure 1. Elem_F_Grapher—the main window and the power functions.

2.1.2. Sample Description

The experiment itself was carried out in the first semester with a sample of 3 first-year student groups using the test–retest method. Non-probability sampling (non-random selection) was used to ensure that students with different high school backgrounds were equally represented in the test groups, guaranteeing the equivalence of groups. Test–retest is a method of repeating a test after a specific time with the same participants [29]. It is considered "One of the most suitable methods for researching the effectiveness of the educational process" [30]. The test–retest method is considered of high reliability [31].

The socio-demographic variables of the participants in relation to their university were as follows: the first group was called the control group (ContGr) and consisted of 16 students of the Faculty of Business Management. The second group was called experimental group 1 (ExpGr1) and consisted of 18 students of the Faculty of Business. The third group was called experimental group 2 (ExpGr2) and consisted of 13 students from the Faculty of Economics. The students were not acquainted with the goal of the experiment in order to not influence their behavior.

2.1.3. Procedure

A procedure was designed to help identify the most efficient teaching method between only traditional teaching, only computer tools, and a combination of traditional and computer tools as measured through the level of students' knowledge in graph functions before and after teaching.

In the first seminar, with the topic of the recapitulation of high school knowledge regarding functions, the knowledge level of all students in all three groups was tested by having them individually complete identical tests (Appendix A). Then, the completed tests from all students were collected and evaluated. The data obtained were named Test results 1. Next week, during the second seminar, different teaching methodologies were applied for each of the three groups. In the control group (ContGr), teaching was carried out using a classical method without the use of computer tools. Transformations of the linear, quadratic, power, square root, and rational graph functions were shown to students by manually building the graphs on the blackboard. At the end of the seminar, students were told that they needed to practice function graphing in preparation for the next seminar. In experimental group 1 (ExpGr1), teaching was carried out using computer tools only, without the classical method. A demo version of the program Elem_F_Grapher was distributed so that each student in group 1 could individually practice graphing functions on their computer in preparation for the next seminar. In experimental group 2

(ExpGr2), teaching was carried out using a combination of the classical teaching method applied to control and ICT tool-based teaching applied to group 1. A laptop/computer and video projector with a brief presentation of the program Elem_F_Grapher (used with experimental group 1) was used. Next, a demo version of the program was distributed for individual use. The experiment continued in the next math seminar, during the third week, where students from all three groups had to repeat the same test that was used in the first seminar (Appendix A). The completed test data (called Test results 2) were collected, analyzed, and evaluated. In each group, Test results 2 were separately compared with Test results 1 in order to identify the improvement in the level of students' knowledge of function graphing.

2.1.4. Data Analysis

The completed tests by the students in the first seminar form the data, called Test results 1, and show the level of the students' knowledge acquired in high school. The relationship between the Test results 1 and the mathematical entry exam results of the students participating in the experiment was checked. The calculated Pearson correlation coefficient, $r = 0.8782$, confirmed a high validity.

The second round of test, named Test results 2, contains the data used to evaluate the efficiency of the used teaching methods. The students' results and the data of Test results 1 and Test results 2 were separately compared in case of each group, measuring the level of knowledge progress of groups. A quantitative analysis is presented in Section 2.2.

2.2. *Results*

A quantitative analysis was conducted using the data of Test results 1 and Test results 2, and average group scores and average relative success rates were calculated. The maximum score that can be achieved by students is 16. Table 1 and Appendix B contains the average group scores noted by $\overline{x}$ and $\overline{x'}$.

Table 1. The average scores. Own calculation.

	TEST RESULTS 1	*TEST RESULTS 2*
Control Group (ContGr)	$\overline{x}\text{ContGr} = \frac{\sum_{\text{ContGr}} x_i}{n_{\text{ContGr}}} = 4.375$	$\overline{x'}\text{ContGr} = \frac{\sum_{\text{ContGr}} x'_i}{n_{\text{ContGr}}} = 10.0625$
Experimental Group 1 (ExpGr1)	$\overline{x}\text{ExpGr1} = \frac{\sum_{\text{ExpGr1}} x_i}{n_{\text{ExpGr1}}} = 7.0\overline{5}$	$\overline{x'}\text{ExpGr1} = \frac{\sum_{\text{ExpGr1}} x'_i}{n_{\text{ExpGr1}}} = 11.7\overline{2}$
Experimental Group 2 (ExpGr2)	$\overline{x}\text{ ExpGr2} = \frac{\sum_{\text{ExpGr2}} x_i}{n_{\text{ExpGr2}}} \doteq 5.154$	$x'\text{ ExpGr2} = \frac{\sum_{\text{ExpGr2}} x'_i}{n_{\text{ExpGr2}}} \doteq 11.46$

To determine the winning teaching methodology, the relative improvement in the level of knowledge was compared as measured via Test results 1 and 2. As each of the three groups had a different average score in Test results 1, an analysis needed to be conducted to assess the improvement based on relative changes in the average scores ($\Delta\overline{p}$) of Test results 1 and 2. The following formula was used:

$$\Delta\overline{p} = \frac{\overline{x'} - \overline{x}}{x_{max}} \cdot 100\%,$$

where $x_{max} = 16$ is the maximum achievable score of the math test (Appendix A). The results are shown in Figure 2.

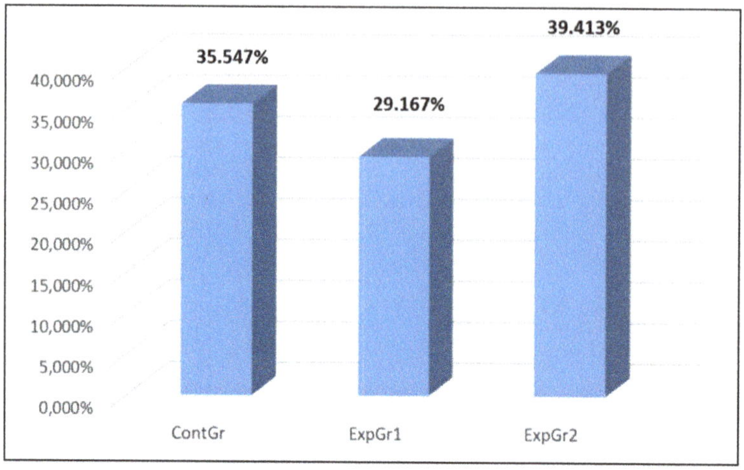

Figure 2. The change in average relative success rates in each group. Own calculations.

The data show that only providing access to ICT tools (a specialized math software in our case) for individual use is not sufficient for improving students' knowledge levels, as stated by [32]. Moreover, based on the above quantitative analysis, it was concluded that the most efficient teaching and learning method was a mix of traditional and ICT tools, as applied to experimental group 2. This supports the innovativeness and the sustainability of the transfer of math knowledge. For the teacher, the relative success according to individual tasks (in this case, the relative change in success according to individual tasks) of the didactic test is also important.

3. Short Overview of Computer Tools used in STEAM Education

3.1. Reasoning for the Use of CAS and DGS

Based on the above quantitative analysis (Section 2), it was concluded that the most successful method of learning mathematics, the process to achieve sustainable knowledge, is to use a mix of traditional teaching methods and computer tools.

The rapid changes in ICT, the increase in the available computer power, and the complexity of mathematical software accessible to students demand a radical re-thinking of how topics in the curriculum should be presented and the impact that they have on the teaching of mathematics in STEAM education [33–41].

There are two connected but distinct issues related to the increased availability of computers and software, which are both of considerable importance to the mathematics curriculum. The first issue is that new, innovative approaches to teaching and learning are made possible. The second issue is that enormously sophisticated mathematical software is now commonly available, allowing problems of such size and complexity to be tackled, problems that have only become part of research in recent years. Students require a personal knowledge of mathematics to be able to use mathematical software reliably and effectively. Students must not simply learn the relevant commands in the software package available to them. They must learn to discerningly use these packages, from a base of mathematical knowledge that will inform them when the computer solution may become unreliable.

According to Philip J. Davis: "In the hands of Newton and Leibnitz, calculus was a theory that involved geometrical figures. These formed a part of the reasoning. There followed thereupon a gradual decline of the image in mathematics in favor of the symbolic, and by the early 20th century, the image was all but dead. Why? Computer graphics has to some extent restored the image to its former prominence in mathematics and promises in the future to be an important but uneasy partner with the symbolic" [42].

CA System (such as Maple, Mathematica, Mathcad, Mathlab, Derive, Maxima, Reduce) is a software package "that facilitates symbolic mathematics. The core functionality is the manipulation of mathematical expressions in symbolic form" [43]. Dynamic Geometry Systems (DGS) (such as GeoGebra, Cabri, Cinderella, Geometer's SketchPad) "or interactive geometry software", "allows one to create and then manipulate geometric constructions, primarily in plane geometry" [43]. This software is based on programs using results of computer algebra and contains more or fewer possibilities for symbolic computations and graphing. CAS and DGS are compelling tools, but only to the extent that they are consciously used to correctly interpret the computer's "responses" and consciously find the methods that are most appropriate for a specific task or technical application. Questions need to be asked "correctly" and "translated" to the language of mathematics to solve real-life tasks using a computer program or other technical questions. Furthermore, the results obtained need to be interpreted and "reversed" into real-life language or technical applications. The principle of white box mathematics means that computer algebra tools help those who use them not only to assist them in solving tasks but also to develop them further, raise new questions and, ultimately, reach a higher level of understanding and application in mathematics [23].

Equally, the variety of available software could be an obstacle for teachers or students when deciding on the right software to use for different educational scenarios. The availability of software (commercial software, such as Maple [44], Mathematica, Geometer's SketchPad [45], or open-source software, such as Reduce, wxMaxima [46], GeoGebra [47]) might also be critical, especially for low-income countries. There are different ways to find more about computer tools, and this can be a good method for understanding the way that these tools function [34,36].

3.2. *Short Overview of the Most Relevant Software Packages*

The large variety of computer tools, CAS and DGS [48], can be analyzed and presented from different angles. It is essential to know details such as the operating systems and their support, the platforms these tools are available on, the language to implement them, their functionalities, and the main mathematical functions built in the software. Furthermore, it is important to know if the tools are commercial products. These tools make computations based on their programmed algorithms; therefore, the results must be analyzed in the context of mathematical or real-life problems in order to choose the solution that is suitable for the task in question. This analysis must rely on a solid mathematical background. Thus, the most crucial thing is to make the user aware that CAS data must be interpreted using mathematical knowledge.

The present paper contains examples of the comparative use of wxMaxima, GeoGebra, Maple, and Excel.

wxMaxima is an open source mathematics software, and it is released and distributed under the terms of the GNU General Public License (GPL). This allows everyone to modify and distribute it, as long as its license remains unmodified. In this article, the term "wxMaxima" is used more often, but the terms "Maxima" and "wxMaxima" can be used interchangeably. "wxMaxima is a document-based interface for the CAS Maxima ... provides menus and dialogs for many common maxima commands, autocompleting, inline plots, and simple simulations" [49]. A brief presentation on wxMaxima's strengths and limitations through the use of examples was recently published; see [50]. GeoGebra for interactive geometry and algebra is globally quite well-known among mathematics educators. It is an open-source mathematics software. In the article by Kovacs et al. [51], examples of GeoGebra's impact on different educational contexts are presented. Maple [52] is powerful math software that is easy to use and meets the requirements needed for STEAM education. It is one of the well-known commercial 3M mathematical software programs (Maple, Matlab, Mathematica); thus, one needs to purchase a license before using it. In [53], topics in education and different applications of Maple are presented. The well-known commercial spreadsheet program from Microsoft, MS Excel, was launched in 1985 and is widely available to students. When using MS Excel, the change in the content of one cell automatically leads to the recalculation of one or more cells based on a user-defined relationship. The user can activate its Tool packages, and the Solver package is successfully used in Operation research education for STEAM students. With the latest version of Excel 2019 and Excel365, it became the most flexible and most commonly used business application in the world due to its ability to adapt to almost any business process [54,55].

4. Challenging Examples in Graphing Functions: Comparative Discussion of Selected Failures in the Use of Different CAS and DGS

This section presents examples using symbolic calculators such as CAS and DGS, as well as MS Excel, highlighting the challenges of using these math software tools correctly and strategically. As the examples prove, these software tools support students' creativity in visualization, using innovative methods, and presenting graphical schemas to solve real-life problems and tasks. Besides their benefits, this section covers the limitations of such tools for both students and teachers, helping to develop a suitable methodology and find the right use case to adopt them.

4.1. MS Excel Environment

Examples of incorrect graphical displays of the MS Excel (in which the program Elem_F_Grapher was created) may appear accidentally, namely at the points of discontinuity in the case of rational functions (see Figure 3: Rational functions in Elem_F_Grapher). It is important to emphasize the need for the active participation of the teacher and their ability to explain these phenomena to students (the software errors/failures) in a suitable way.

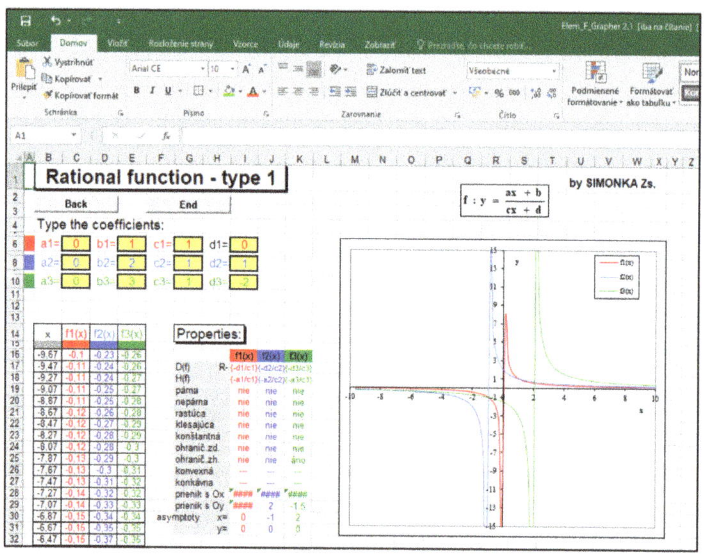

Figure 3. Rational functions in Elem_F_Grapher.

4.2. Computer Algebra Systems (CAS) and Dynamic Geometry Systems (DGS)

Failures may also appear when using CAS and DGS tools, and some examples are described in this section.

Example 1

Let us consider a frequent example in calculus—the computation of the following limit:

$$\lim_{x\to 1}\frac{x^2-1}{x^2-3x+2}$$

Usually, the students are advised to consider the factorization of the numerator and the denominator and conclude by simplifying the fraction, as shown below:

$$\frac{x^2-1}{x^2-3x+2} = \frac{(x-1)(x+1)}{(x-1)(x-2)} = \frac{x+1}{x-2}$$

Now, the computation of the limit is reduced to obtain the limit by the value of the simplified fraction for $x = 1$:

$$\lim_{x\to 1}\frac{x^2-1}{x^2-3x+2} = \lim_{x\to 1}\frac{x+1}{x-2} = -2$$

The question here is: are the two functions equal or not?

The use of CAS and DGS might help to find the right answer, but suddenly, as can be seen below, the computer gives "different" answers.

Let us consider the following functions, noted by f and g, and point out their domains:

$$f(x) = \frac{x^2-1}{x^2-3x+2},\ Dom(f) =]-\infty,1[\cup]1,2[\cup]2,\infty[$$
$$g(x) = \frac{x+1}{x-2},\ Dom(g) =]-\infty,2[\cup]2,\infty[$$

First, the GeoGebra program is used to look for the answer (Figures 4 and 5).

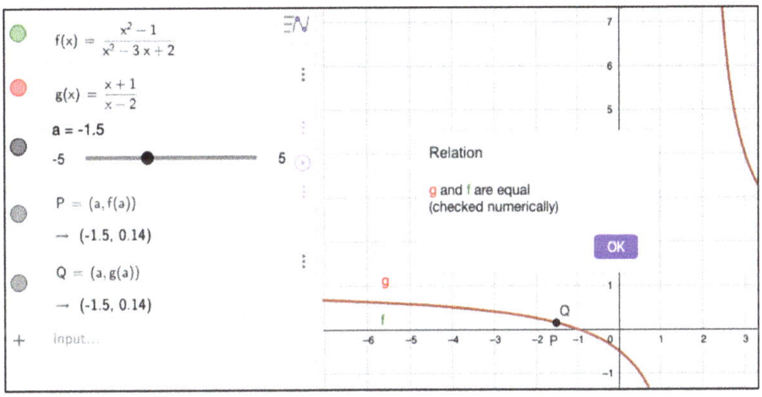

Figure 4. GeoGebra answer 1: "g and f are equal".

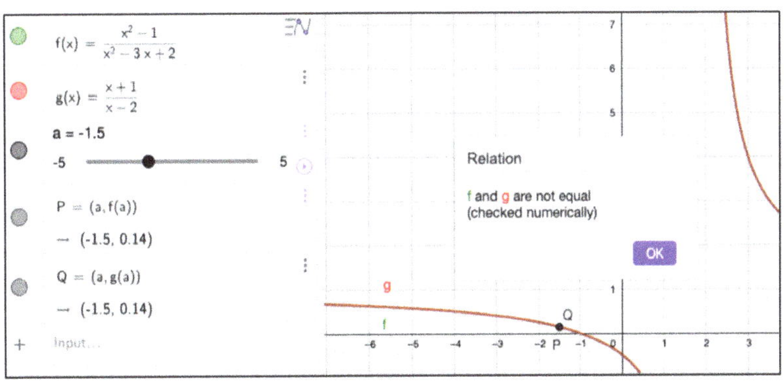

Figure 5. GeoGebra answer 2: "f and g are not equal".

Note here that the two "relations" from above differ only in the order; thus, normally, the answers should be the same. In addition, the mathematical reasoning is making a clear distinction between the two functions, as, by definition, the equality of two functions is true if their domains and codomains are the same, and the two functions have equal values for all the elements of the domain. In the above example the first condition fails, the domains differ. Most likely, the built-in command "relation between two objects" used by GeoGebra does not coincide with the content of the mathematical notion. Things might be more confusing in the following example with two logarithmic functions.

Example 2

Let us consider the following functions:

$$f(x) = \ln(x-1) + \ln(x+1),\ Dom(f) =]1, \infty[$$
$$g(x) = \ln(x^2-1),\ Dom(g) =]-\infty, -1[\cup]1, \infty[$$

The GeoGebra gives a strange answer here, "the two function are equal" (Figures 6 and 7) no matter of the order, but their graphs are represented differently of course (see Figures 8 and 9), reflecting the difference between the domains.

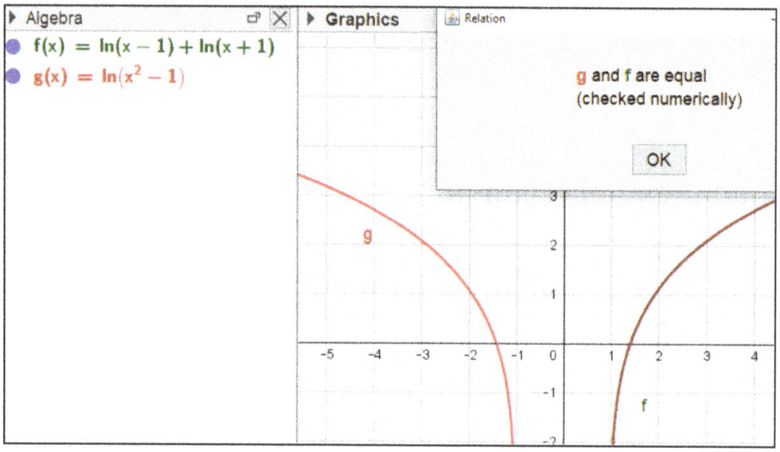

Figure 6. GeoGebra answer 1 for logarithms: "*g* and *f* are equal".

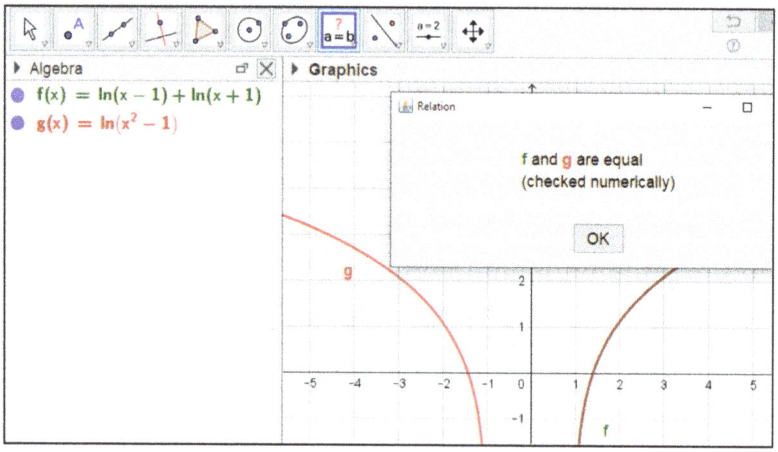

Figure 7. GeoGebra answer 2 for logarithms: "*f* and *g* are equal".

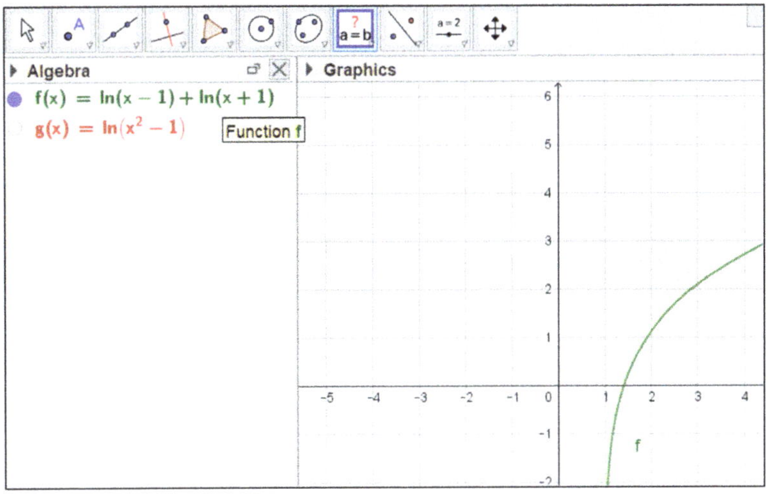

Figure 8. GeoGebra graph for function *f*.

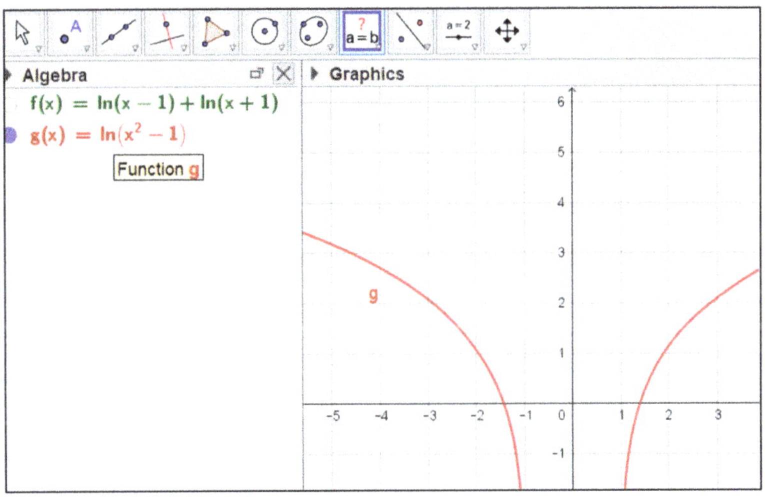

Figure 9. GeoGebra graph for function *g*.

In another type of failure, the GeoGebra answers that the two functions are equal, and this is true only at the intersection of the two domains for the interval $]1,\infty[$. From a mathematical point of view, the two functions are different.

When the same GeoGebra draws a graph, the difference between the two functions can be visualized, as seen in the subsequent two figures (Figures 8 and 9).

This difference is evident again from a mathematical point of view when considering the difference between the two domains. Even this example could offer an excellent opportunity for the teacher to argue for the importance of emphasizing the difference between the domains in case of the well-known algebraic relation, true only at the intersection of the two domains. The intersection of the two domains is in fact: $D_f \cap D_g =]1,\infty[$. Thus, the equality: $\ln(x-1)+\ln(x+1)=\ln(x^2-1)$ holds only for the interval $]1,\infty[$, the intersection of the two domains. The function $f(x)=\ln(x-1)+\ln(x+1)$ has the domain $D_f =]1,\infty[$, while the function $g(x)=\ln(x^2-1)$ has the domain $D_g =]-\infty,-1[\cup]1,\infty[$, thus $D_f \neq D_g$. The intersection of the two domains is $D_f \cap D_g =]1,\infty[$; thus, the above equality holds only for the interval $]1, \infty[$. When checking the results with other computer tools, and graphing the above two functions, one obtains the same answers. Below, Figures 10–12 show the respective graphs plotted in Maple and wxMaxima.

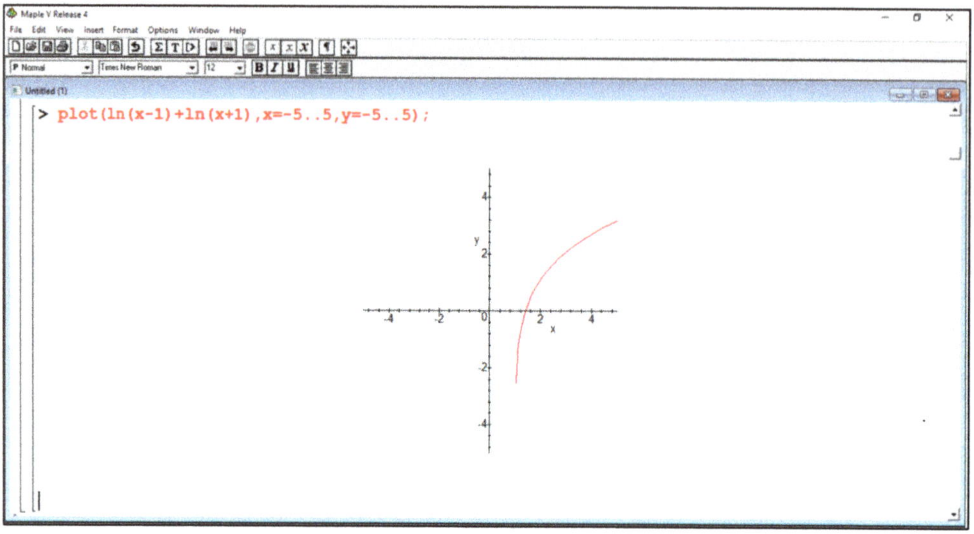

Figure 10. Maple graph for $f(x) = \ln(x-1) + \ln(x+1)$.

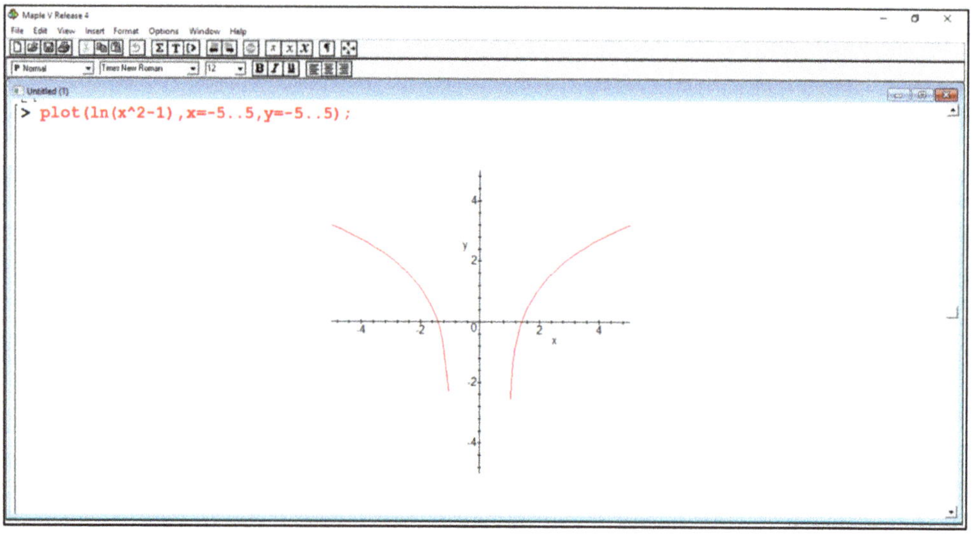

Figure 11. Maple graph for $g(x) = \ln(x^2 - 1)$.

Example 3

The following example shows that some other confusing facts may appear in a similar case. If the same logic is repeated for another two functions, the same problem appears but in a different manner. Let us consider the following functions:

$$f(x) = \sqrt{x-1} \cdot \sqrt{x-1},\ Dom(f) = [1, \infty[g(x) = \sqrt{x^2-1},\ Dom(g) =]-\infty, -1] \cup [1, \infty[$$

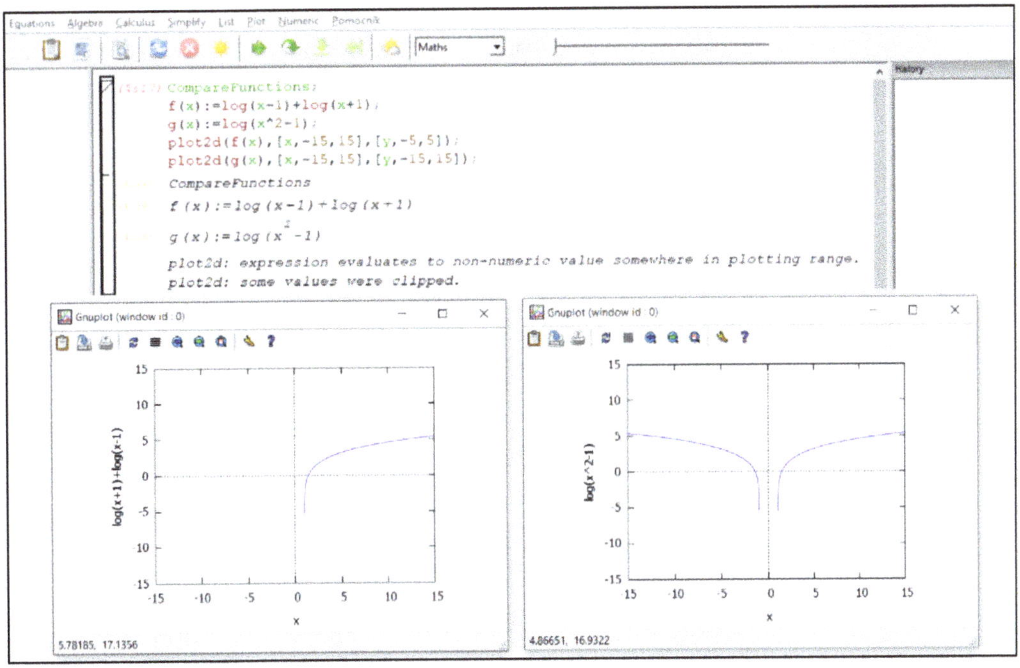

Figure 12. wxMaxima graphs for $f(x) = \ln(x-1) + \ln(x+1)$ and $g(x) = \ln(x^2-1)$.

The GeoGebra gives a strange answer again here—"the two functions are equal" (Figures 13 and 14)—no matter the order. However, when illustrated, their graphs are different, reflecting the difference between their domains (Figures 15 and 16).

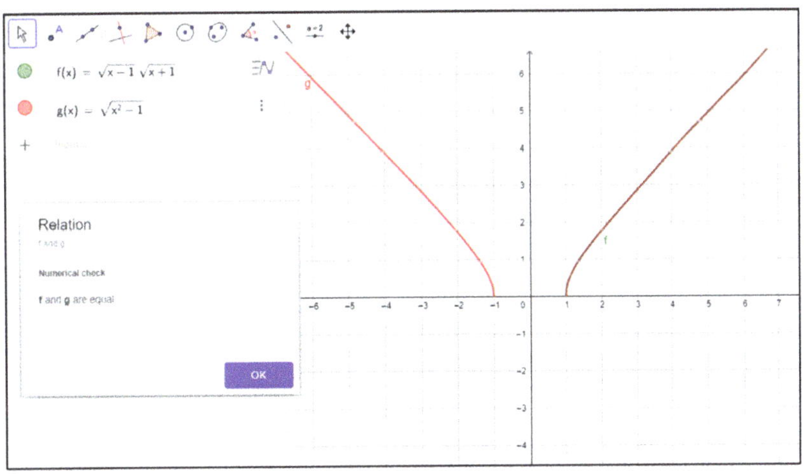

Figure 13. GeoGebra answer 1 for irrational functions "*f* and *g* are equal".

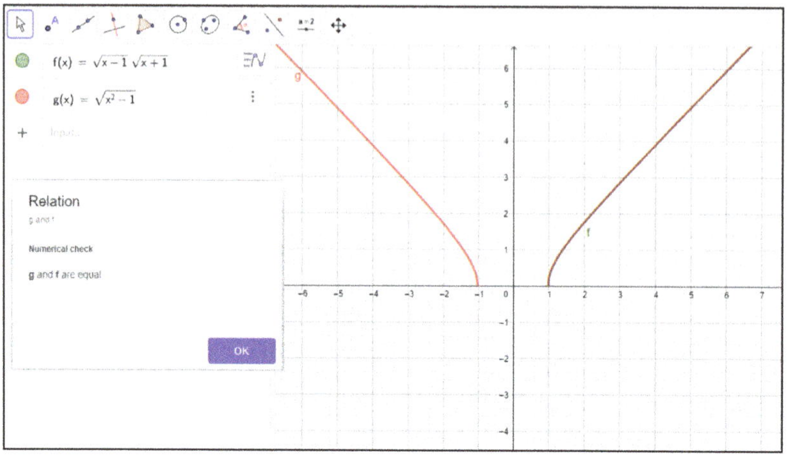

Figure 14. GeoGebra answer 2 for irrational functions "g and f are equal".

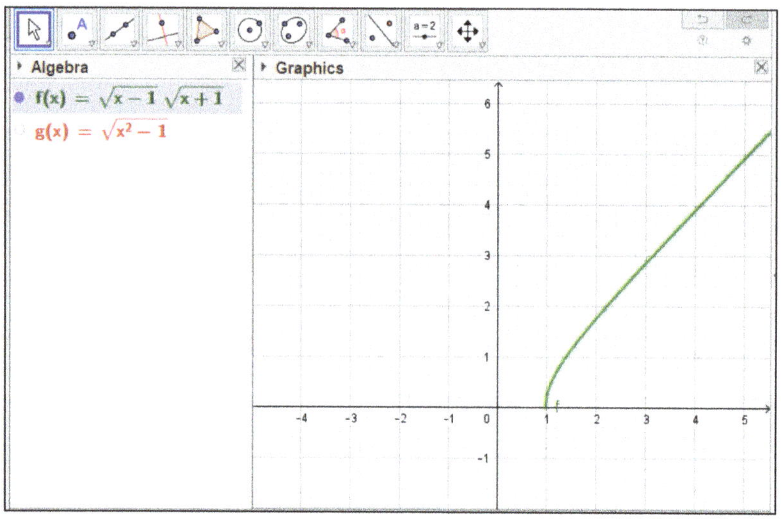

Figure 15. GeoGebra graph of function $f(x) = \sqrt{x-1} \cdot \sqrt{x+1}$.

The GeoGebra graphs of the two functions reveal a similar difference as seen in the previous case:

Once again, it is crucial to emphasize the validity of the equality of the two irrational algebraic expressions. The algebraic equality $\sqrt{x-1} \cdot \sqrt{x+1} = \sqrt{x^2-1}$ is only true at the intersection of the two domains. The function $f(x) = \sqrt{x-1} \cdot \sqrt{x+1}$ has the domain $D_f = [1, \infty[$, while the function $g(x) = \sqrt{x^2-1}$ the domain $D_g =]-\infty, -1] \cup [1, \infty[$, thus $D_f \neq D_g$. The intersection of the two domains is $D_f \cap D_g = D_f = [1, \infty[$; thus, the above equality holds only for the interval $[1, \infty[$.

Therefore, it is possible to verify this by plotting the two functions in Maple; however, a new challenge appears here, the two graphs not only differ in their domain, but Maple seems to tackle the function f in a different way to how GeoGebra does (Figures 17 and 18).

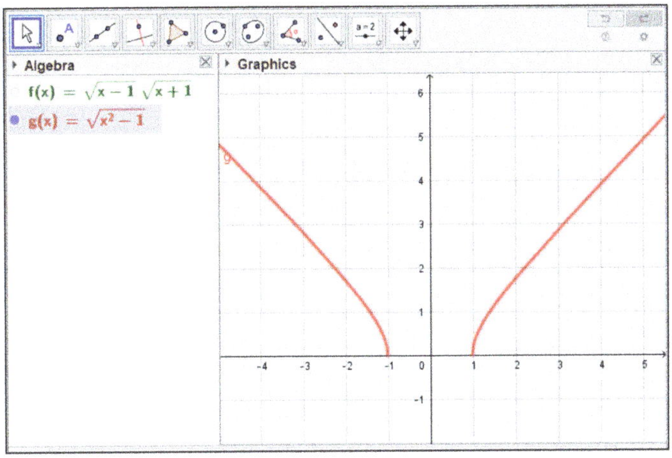

Figure 16. GeoGebra graph of function $g(x) = \sqrt{x^2 - 1}$.

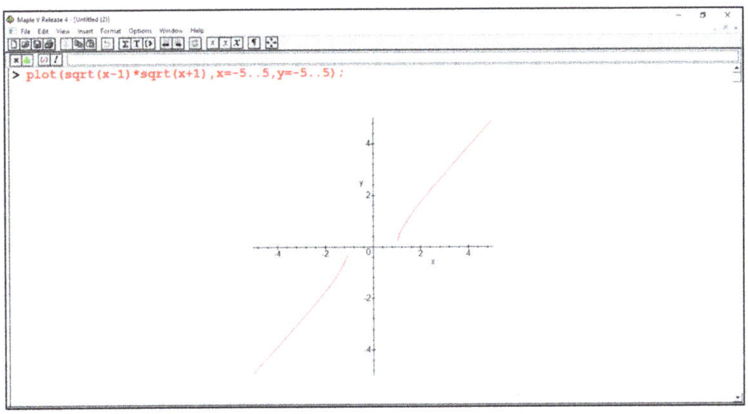

Figure 17. Maple graph of function $f(x) = \sqrt{x-1} \cdot \sqrt{x+1}$.

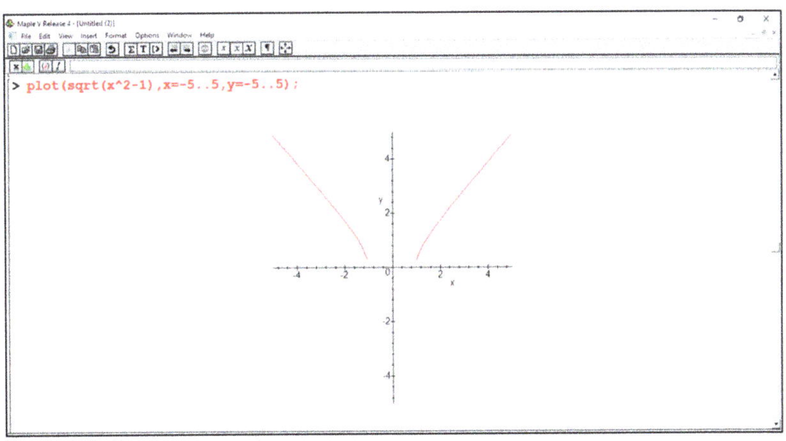

Figure 18. Maple graph of function $g(x) = \sqrt{x^2 - 1}$.

What happened here? Maple graphing uses complex functions too, and this explains the difference of the two graphs, obtained first by GeoGebra only working with real functions (Figures 15 and 16). However, Maple can also handle complex functions (Figures 17 and 18), and the difference is clear when comparing Figures 15 and 17.

Remark 1. In the last two Maple graphs, the graph of the functions is apparently not touching the x-axes. The reason for this is that, in $x = -1$ and $x = 1$, the graphs have almost vertical tangents. The teacher could ask students to compute the derivatives and challenge them to analyze the case.

The wxMaxima graphing gives similar results to Maple, as shown in Figure 19.

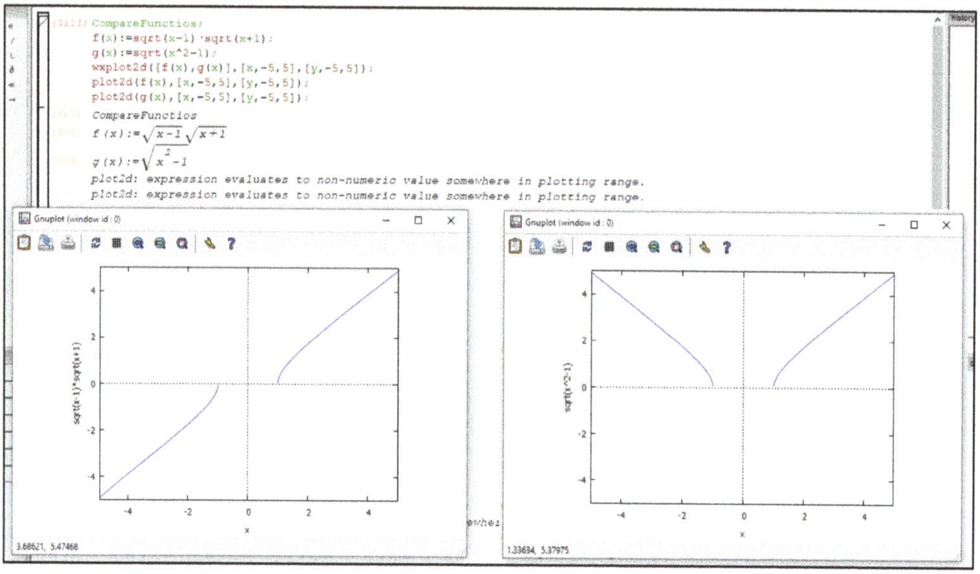

Figure 19. wxMaxima graphs for $f(x) = \sqrt{x-1} \cdot \sqrt{x+1}$ and $g(x) = \sqrt{x^2 - 1}$.

Example 4

The phenomenon met in Example 3 can also be seen by the students in the case of the following well-known function. Let us consider the functions: $f(x) = \sqrt{x^2 + 2x + 1}$ and $g(x) = \sqrt{x+1} \cdot \sqrt{x+1}$, where $D_f =]-\infty, \infty[$ and $D_g = [-1, \infty[$.

A similar challenge can be formulated for the students when graphing these two functions (Figures 20–22).

The differences are rooted in the way GeoGebra and wxMaxima/Maple are using real and complex numbers. GeoGebra only makes computations restricted to real numbers, while the computation with wxMaxima and Maple are using complex numbers by definition.

The above two examples show that the right CAS program for the given educational environment is needed. For secondary schools, the use of GeoGebra covers most of the content of their curricula. In contrast, for higher education (STEAM education), where complex functions are included in curricula, it is helpful to provide a detailed explanation of the differences between the graphs obtained using GeoGebra and Maple/wxMaxima.

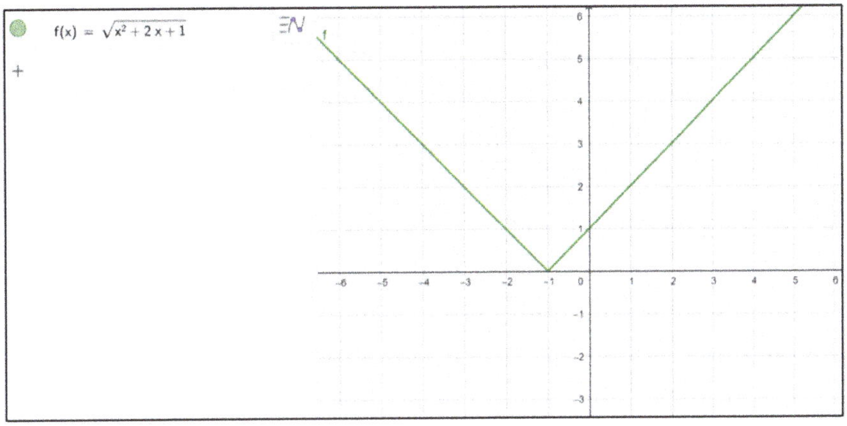

Figure 20. GeoGebra graph of function $f(x) = \sqrt{x^2 + 2x + 1}$.

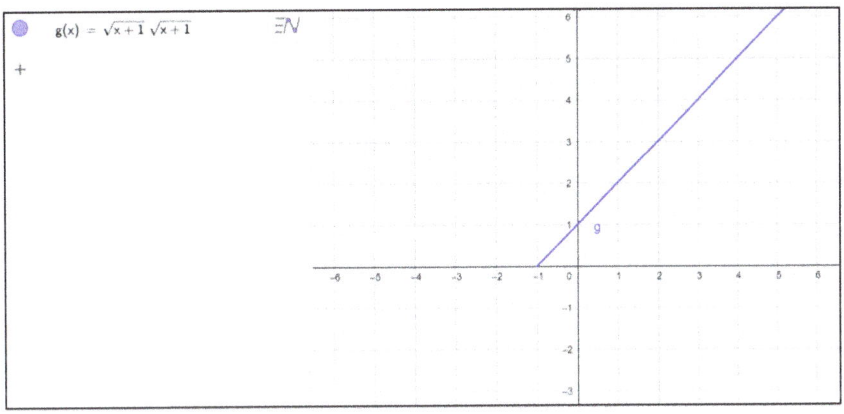

Figure 21. GeoGebra graph of function $g(x) = \sqrt{x+1} \cdot \sqrt{x+1}$.

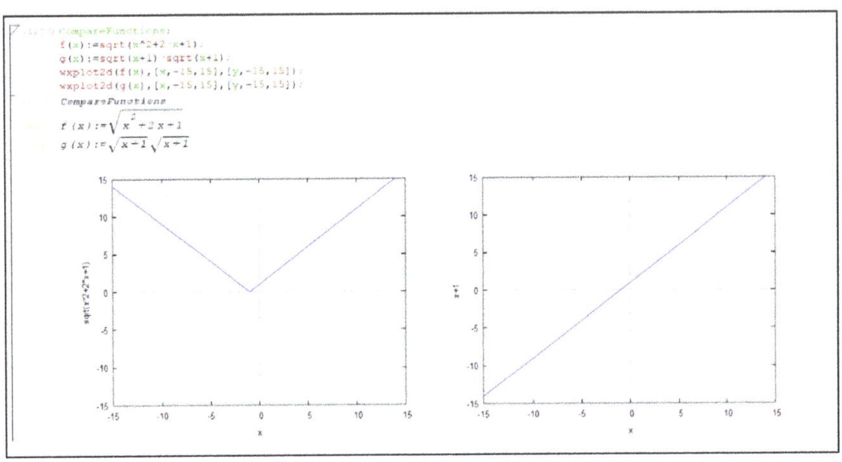

Figure 22. wxMaxima graphs for $f(x) = \sqrt{x^2 + 2x + 1}$ and $g(x) = \sqrt{x+1} \cdot \sqrt{x+1}$.

5. Conclusions

Technology is prevalent in all areas of life, and the possibilities offered by Web 4.0 are gradually becoming a reality. It is widely accepted that the integration of information and communication technologies (ICTs) and the wide variety of digital tools available are causing changes in traditional education and accelerating its transformation. Education must keep up with this transformation; thus, its technologization is both essential and inevitable.

During the experiment (Section 2), the increase in group knowledge level was measured through participants' individual progress. Based on the findings, the most significant improvement in knowledge, as shown in Figure 2, occurred when the mixed method was used (the case of experimental group 2).

Thus, the practical experiment shows that the winning methodology in teaching mathematics is a mix of traditional methods and the usage of computer tools/mathematical software. The experiment highlighted that giving only the computer tools for individual use is insufficient for improving students' sustainable knowledge level, as also stated by [33].

The transition to technology-based classrooms and the constant use of educational software is a prerequisite for sustainable STEAM and mathematics education. This enables a collaborative learning environment, and the teacher must be able to use digital technologies to foster and enhance learners' activities. Moreover, students can use digital technologies as part of collaborative assignments, improving communication, creativity, critical thinking, and collaborative knowledge-sharing [14]. According to [56], in the digital era, problem-solving using learners' cognition is the only skill required for artificial intelligence (AI). These transition challenges demand expanding the already known and applied pedagogical methods and adopting new, active, and innovative methodologies.

As the presented examples in Section 4 show, using CAS and DGS, the visualization of functions becomes easier. A deeper understanding of concepts can also be achieved [57–59] using the "black box mathematics" [60] method, i.e., only focusing on the advantages of clickable mathematics and the so-called principle of "white box mathematics" [60] notions [61,62]. This powerful tool can be used to "walk around" the concepts to be learned and understood several times, seeking a more precise outline and a deeper understanding of them. This helps to comprehend the limits and possible contradictions of applying these concepts in theoretical and real-world practical problems.

The examples chosen in the paper are considered the most efficient examples for students according to authors' teaching experience. Students' attention can easily be directed toward the possible or real-life mistakes that can appear in software usage, motivating them to find their own examples, and thus ultimately mastering these computer tools.

The outcomes of Section 4 offer educators and learners new ideas and elements for innovative learning models that can be immediately applied in math teaching.

CAS and DGS are essential tools for teaching and learning mathematics, developing students' math knowledge and performance, and increasing their analytical and critical thinking. It is a powerful and innovative tool for developing 21st-century skills. The CAS and DGS tools are useful not only in the education/teaching process (as observed during online teaching in the pandemic period), but they proved to be highly motivating and engaging for students. Among more skilled and qualified students, using these tools proved to be more stimulating, as they are more motivated to push the boundaries and constraints of the programs. The presented examples in this paper are significant for all users. They demonstrate the need for an enhanced, mathematical background knowledge in this case, in addition to adequate technical and user knowledge.

6. Limitations

The research was limited to one area of math, the graphing of functions, since it is one of the most critical and problematic issues faced by 1st year university students. From a critical point of view, the challenges were analyzed in this area to highlight the need for the proper and intelligent use of math software.

7. Future Research

The examples presented in Section 4 of this paper could serve as a starting point to deepen our understanding of the analysis of functions: the understanding of key notions and problems related to the computation of the limits of functions, studying continuous functions or points of discontinuity of functions, and studying graphing functions with two variables.

Author Contributions: Conceptualization, P.K., Z.S., Z.K.S., J.G., R.N.; methodology, P.K., Z.S., Z.K.S., J.G., R.N.; software, P.K., Z.S.,; validation, P.K., Z.S., Z.K.S.; formal analysis, Z.K.S., R.N.; investigation, P.K., Z.S., Z.K.S., J.G., R.N.; resources, P.K., Z.S., Z.K.S., J.G., R.N.; data curation, J.G., R.N.; writing—original draft preparation, P.K., Z.S., Z.K.S., J.G., R.N.; writing—review and editing, P.K., Z.K.S.; visualization, Z.K.S., Z.S., J.G.; supervision, P.K.; project administration, Z.K.S.; funding acquisition, P.K., Z.S., Z.K.S., J.G., R.N. All authors have read and agreed to the published version of the manuscript.

Funding: This research received no external funding.

Institutional Review Board Statement: Not applicable.

Informed Consent Statement: Not applicable.

Data Availability Statement: Not applicable.

Acknowledgments: Authors are grateful to the scientific and methodological support received from: grant KEGA 026UK-4/2022 "The concept of Constructionism and Augmented Reality in STEM Education (CEPENSAR), methodology research group in the Department of Applied Mathematical Studies (DAMS) of George Emil Palade University of Medicine, Pharmacy, Science, and Technology of Targu Mures. The authors of the paper wish to acknowledge the valuable remarks and support of partners of the CEEPUS Network CIII-HU-0028-15-2122.

Conflicts of Interest: The authors declare no conflict of interest.

Appendix A

Math test for 1st year study students.

MATH TEST

Sketch the graphs of elementary functions $f_{01}, f_{02}, f_{03}, f_{04}, f_{05}, f_{06}$ separately and based on them draw graphs of the functions $f_{11}, f_{21}, f_{22}, f_{23}, f_{31}, f_{32}, f_{41}, f_{51}, f_{61}, f_{62}$ (the same type of functions draw into the one picture).

1. $f_{01} : y = x$
   $f_{11} : y = 2x - 1$
2. $f_{02} : y = x^2$
   $f_{21} : y = 2x^2 + 1$
   $f_{22} : y = (x+3)^2 + 2$
   $f_{23} : y = x^2 + 6x + 11$
3. $f_{03} : y = x^3$
   $f_{31} : y = 3x^3 - 1$
   $f_{32} : y = (x-2)^3 + 3$
4. $f_{04} : y = \sqrt{x}$
   $f_{41} : y - \sqrt{x+1} - 2$
5. $f_{05} : y = \sqrt[3]{x}$
   $f_{51} : y = \sqrt[3]{x+2} - 1$
6. $f_{06} : y = \frac{1}{x}$
   $f_{61} : y = 1 + \frac{3}{x-1}$
   $f_{62} : y = \frac{x+2}{x-1}$

Appendix B

The average score of groups based on data Test results 1 and Test results 2.

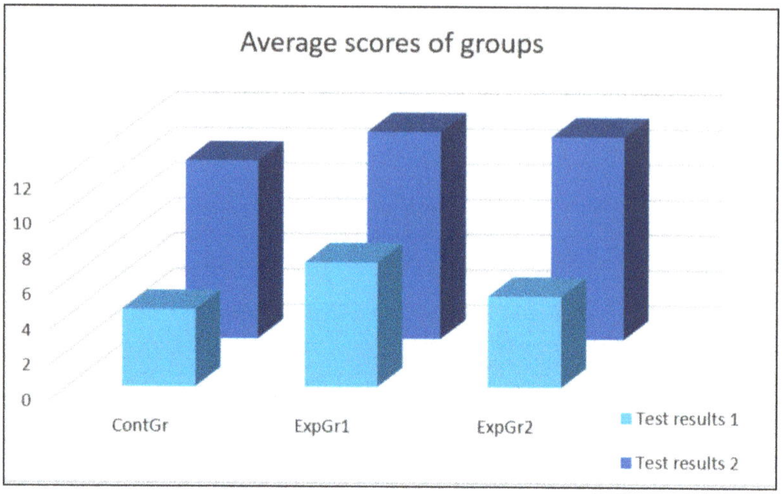

References

1. Yong, Z. *Who's Afraid of the Big Bad Dragon?* Jossey-Bass—A Wiley Brand: San Francisco, CA, USA, 2014.
2. Sterling, S. Sustainable Education—Towards a Deep Learning Response to Unsustainability. *Policy Pract. Dev. Educ. Rev.* **2008**, *6*, 63–68. Available online: https://www.developmenteducationreview.com/sites/default/files/article-pdfs/sterling%20pers%201.pdf (accessed on 10 November 2021).
3. Aldon, G.; Trgalová, J. (Eds.) *Technology in Mathematics Teaching*; Springer: Cham, Switzerland, 2019. [CrossRef]
4. Monaghan, J.; Trouche, L. *Mathematics Teachers and Digital Tools*; Springer: Cham, Switzerland, 2016; Volume 110. [CrossRef]
5. Ochkov, F.V.; Bogomolova, P.E. Teaching Mathematics with Mathematical Software. *J. Humanist. Math.* **2015**, *5*, 265–285. [CrossRef]
6. Juan, C.C. The technology acceptance model in the Collaborative learning of Web 2.0, Web 3.0 and Web 4.0: Higher Education case study. *WSEAS Trans. Environ. Dev.* **2020**, *16*, 680–688. [CrossRef]
7. Miri, B. Transition from traditional to ICT-enhanced learning environments in undergraduate chemistry courses. *Comput. Educ.* **2007**, *48*, 30–43. [CrossRef]
8. Alabdulaziz, M. COVID-19 and the use of digital technology in mathematics education. *Educ. Inf. Technol.* **2021**, *26*, 7609–7633. [CrossRef]
9. Carroll, N.; Conboy, K. Normalising the "new normal": Changing tech-driven work practices under pandemic time pressure. *Int. J. Inf. Manag.* **2020**, *55*, 102186. [CrossRef]
10. Pacheco, J.A. The "new normal" in education. *Prospects* **2021**, *51*, 3–14. [CrossRef]
11. Zhao, Y.; Watterston, J. The changes we need: Education post COVID-19. *J. Educ. Chang.* **2021**, *22*, 3–12. [CrossRef]
12. OECD. *OECD Digital Education Outlook 2021: Pushing the Frontiers with Artificial Intelligence, Blockchain and Robots*; OECD Publishing: Paris, France, 2021. [CrossRef]
13. Jha, A.K. *Understanding Generation Alpha*; OSFPreprints: Charlottesville, VA, USA, 2020; pp. 1–12. [CrossRef]
14. Punie, Y. *European Framework for the Digital Competence of Educators: DigCompEdu*; Redecker, C., Ed.; Publications Office of the European Union: Luxembourg, 2017. [CrossRef]
15. Klement, M.; Dostál, J.; Kubrický, J.; Bártek, K. *ICT versus teachers: Adoration or resistence? [ICT Nástroje a Učitelé: Adorace, či Rezistence?]*; Univerzita Palackého v Olomouci: Olomouc, Czech Republic, 2017. [CrossRef]
16. Lavicza, Z.; Weinhandl, R.; Prodromou, T.; Anđić, B.; Lieban, D.; Hohenwarter, M.; Fenyvesi, K.; Brownell, C.; Diego-Mantecón, J.M. Developing and Evaluating Educational Innovations for STEAM Education in Rapidly Changing Digital Technology Environments. *Sustainability* **2022**, *14*, 723. [CrossRef]
17. Drijvers, P. Digital Technology in Mathematics Education: Why It Works (or Doesn't). 2013. Available online: https://files.eric.ed.gov/fulltext/EJ1054923.pdf (accessed on 1 May 2022).
18. Hillmayra, D.; Ziernwalda, L.; Reinholda, F.; Hoferb, I.S.; Reiss, M.K. The potential of digital tools to enhance mathematics and science learning in secondary schools: A context-specific meta-analysis. *Comput. Educ.* **2020**, *153*, 103897. [CrossRef]
19. UNESCO. *Education for Sustainable Development Goals. Learning Objectives*; UNESCO Digital Library: Paris, France, 2022. Available online: https://unesdoc.unesco.org/ark:/48223/pf0000247444 (accessed on 18 April 2022).
20. Artigue, M. Learning Mathematics in a CAS Environment: The Genesis of a Reflection about Instrumentation and the Dialectics between Technical and Conceptual Work. *Int. J. Comput. Math. Learn.* **2002**, *7*, 245–274. [CrossRef]
21. Guin, D.; Ruthven, K.; Trouche, L. (Eds.) *The Didactical Challenge of Symbolic Calculators*, 1st ed.; Springer: New York, NY, USA, 2005. [CrossRef]

22. Hutkemri, Z.; Sharifah, N.A. The Effectiveness of the GeoGebra Software: The Intermediary Role of Procedural Knowledge on Students' Conceptual Knowledge and Their Achievement in Mathematics. *EURASIA J. Math. Sci. Technol. Educ.* **2017**, *13*, 2155–2180. [CrossRef]
23. Buchberger, B. Computer Algebra: The End of Mathematics? *ACM SIGSAM Bull.* **2002**, *36*, 3–9. [CrossRef]
24. Devlin, K. Teaching mathematics as a way of thinking—Not calculating. *Eest. Haridusteaduste Ajak.* **2021**, *9*, 33–59. [CrossRef]
25. Zhao, Y. *What Works May Hurt—Side Effects in Education*; Teachers College Press: New York, NY, USA, 2018.
26. Hervey, S. What Is Effective Teaching of Mathematics? 2013. Available online: https://www.generationready.com/our-services/mathematics-2/ (accessed on 2 December 2021).
27. Körtesi, P.; Desislava, G. Development of Mathematical Skills with Commonly Used Computer Software. In Proceedings of the MultiScience—XXIX microCAD International Multidisciplinary Scientific Conference, Miskolc, Hungary, 9–10 April 2015; pp. 1–9. [CrossRef]
28. Mucha, V.; Páleš, M. Visualization and presentation of actuarial demography data with dynamic elements using Microsoft Excel 2013. In Proceedings of the 10th International Scientific Conference, Bratislava, Slovakia, 12–13 November 2015.
29. Lapitka, M. *Tvorba a Použitie Didaktických Testov (The Creation and the Usage of the Didactical Tests)*; Štátny Pedagogický Ústav. (The State Pedagogical Institute): Bratislava, Slovakia, 1996.
30. Kulik, J.A.; Bangert, R.L.; Williams, G.W. Effects of computer-based teaching on secondary school students. *J. Educ. Psychol.* **1983**, *75*, 19–26. [CrossRef]
31. Middleton, F. Reliability vs. Validity in Research | Difference, Types and Examples. 2019. Available online: https://www.scribbr.com/methodology/reliability-vs-validity/ (accessed on 6 September 2022).
32. Felician, D.; Torres, L.; Santín, D. One Laptop per Child? Using Production Frontiers for Evaluating the Escuela2.0 Program in Spain. *Mathematics* **2021**, *9*, 2600. [CrossRef]
33. Trouche, L. Integrating Tools as an Ordinary Component of the Curriculum in Mathematics Education. In *Tools and Mathematics*; Springer: Cham, Switzerland, 2016; Volume 110, pp. 267–303. [CrossRef]
34. Lavicza, Z. Examining the use of Computer Algebra Systems in university-level mathematics teaching. *J. Comput. Math. Sci. Teach.* **2009**, *28*, 99–111. Available online: https://www.learntechlib.org/primary/p/30303/ (accessed on 14 July 2022).
35. Simonka, Z. Implementation of ICT in teacher training in the world and in Slovakia. *Ekon. Inform.* **2008**, *6*. Available online: https://fhi.euba.sk/en/51-science-and-research/scientific-journals/economics-and-informatics (accessed on 16 February 2022). ISSN 1336-3514.
36. Buteau, C.; Jarvis, D.; Lavicza, Z. On the Integration of Computer Algebra Systems (CAS) by Canadian Mathematicians: Results of a National Survey. *Can. J. Sci. Math. Technol. Educ.* **2014**, *14*, 35–57. [CrossRef]
37. Fenyvesi, K.; Koskimaa, R.; Lavicza, Z. Experiential Education of Mathematics: Art and Games for Digital Natives. *Kasvatus* **2015**, *9*, 107–134. Available online: http://www.kasvatus-ja-aika.fi/dokumentit/fenyvesial_0804151248.pdf (accessed on 14 July 2022).
38. Adelabu, F.; Makgato, M.; Ramaligela, M. The Importance of Dynamic Geometry Computer Software on Learners' Performance in Geometry. *Electron. J. e-Learn.* **2019**, *17*, 52–63. Available online: https://files.eric.ed.gov/fulltext/EJ1216699.pdf (accessed on 10 July 2022).
39. İbili, E. The Use of Dynamic Geometry Software from a Pedagogical Perspective: Current Status and Future Prospects. *J. Comput. Educ. Res.* **2019**, *7*, 337–355. [CrossRef]
40. Lavicza, Z.; Prodromou, T.; Fenyvesi, K.; Hohenwarter, M.; Juhos, I.; Koren, B. Integrating STEM-related Technologies into Mathematics Education at Large Scale. *Int. J. Technol. Math. Educ.* **2020**, *27*, 3–12. [CrossRef]
41. Rutherford, T.; Allison, S.L.; Arena, S.L.; Schenke, K. Impact on mathematics self-beliefs from a mastery-based mathematics software. *J. Res. Technol. Educ.* **2020**, *52*, 79–94. [CrossRef]
42. Davis, J.P. *The Decline, Fall, and Current Resurgence of Visual Geometry. History of Mathematics & Teaching of Mathematics*; University of Miskolc: Miskolc, Hungary, 2004. Available online: https://www.uni-miskolc.hu/hmtm/abstracts.html (accessed on 5 March 2022).
43. Braman, J.; Vincenti, G.; Trajkovski, G. *Handbook of Research on Computational Arts and Creative Informatics*; Information Science Reference (an imprint of IGI Global): Hershey, PA, USA, 2009. [CrossRef]
44. Petrášková, V.; Rosa, P. Using Maple Cloud Computing in Financial Education of Pre-service Teachers. *Math. Comput. Sci.* **2019**, *13*, 237–245. [CrossRef]
45. Nanteni, G.; Leong, K.E. The Effect of Dynamic Geometry Software Geometer's Sketchpad on Students' Achievement in Topic Circle among Form Two Students. *Malays. Online J. Educ. Technol.* **2020**, *8*, 58–68. [CrossRef]
46. Fedriani, E.; Moyano, R. Using Maxima in the Mathematics Classroom. *Int. J. Technol. Math. Educ.* **2011**, *18*, 171–176. Available online: https://www.semanticscholar.org/paper/Using-Maxima-in-the-Mathematics-Classroom-Fedriani-Moyano/ead206114e9f66d93e82249ee6855fbc486eb2a2 (accessed on 5 March 2022).
47. Sagesse, M.U.; Maniraho, J.F.; Michael, T. GeoGebra integration and effectiveness in the teaching and learning of mathematics in secondary schools: A review of literature. *Afr. J. Educ. Stud. Math. Sci.* **2020**, *16*, 1–13. [CrossRef]

48. Hohenwarter, J.; Hohenwarter, M.; Lavicza, Z. Evaluating Difficulty Levels of Dynamic Geometry Software Tools to Enhance Teachers' Professional Development. *Int. J. Technol. Math. Educ.* **2010**, *17*, 127–134. Available online: https://www.semanticscholar.org/paper/Evaluating-Difficulty-Levels-of-Dynamic-Geometry-to-Hohenwarter-Hohenwarter/093dc16d839f35151701ea3594d765ab59131a7c (accessed on 5 March 2022).
49. Maxima. 2022. Available online: https://wxmaxima-developers.github.io/wxmaxima/ (accessed on 14 July 2022).
50. Karjanto, N.; Husain, H. Not Another Computer Algebra System: Highlighting wxMaxima in Calculus. *Mathematics* **2021**, *9*, 1317. [CrossRef]
51. Kovacs, Z.; Recio, T.; Velez, M.P. GeoGebra Discovery in Context. *Electron. Proc. Theor. Comput. Sci.* **2021**, *352*, 141–147. [CrossRef]
52. Maple. 2022. Available online: https://www.maplesoft.com/ (accessed on 5 March 2022).
53. Robert, M.; Jürgen, G.; Ilias, S.K. *Maple in Mathematics Education and Research*; Springer: Cham, Switzerland, 2021; Volume 1414. [CrossRef]
54. Dostál, J. MS Excel pro učitele. In *MS Excel for Teachers*; Palacký University of Olomouc: Olomouc, Czech Republic, 2009.
55. Microsoft-Excel. 2022. Available online: https://www.britannica.com/technology/Microsoft-Excel (accessed on 12 April 2022).
56. UNESCO. Human Learning in the Digital Era. 2019. Available online: https://unesdoc.unesco.org/ark:/48223/pf0000367761.locale=en (accessed on 2 May 2022).
57. Wester, J.M. *Computer Algebra Systems: A Practical Guide*, 1st ed.; Wiley: Hoboken, NJ, USA, 2000.
58. Chan, K.; Leung, S. Dynamic Geometry Software Improves Mathematical Achievement: Systematic Review and Meta-Analysis. *J. Educ. Comput. Res.* **2014**, *51*, 311–325. [CrossRef]
59. Tola, B.B.; France, M. The effect of GeoGebra on STEM students learning trigonometric functions. *Cogent Educ.* **2022**, *9*, 1–18. [CrossRef]
60. Buchberger, B. Should students learn integration rules? *ACM SIGSAM Bull.* **1990**, *24*, 10–17. [CrossRef]
61. Cunska, A.; Savicka, I. Use of ICT teaching-learning methods make school math blossom. *Procedia-Soc. Behav. Sci.* **2012**, *69*, 1481–1488. [CrossRef]
62. Simonka, Z. Graphing of elementary functions and their transformations using ICT (projects and studies in the world). In Proceedings of the APLIMAT 2008—7th International Conference on Applied Mathematics, Department of Mathematics, Faculty of Mechanical Engineering STU, Bratislava, Slovakia, 5–8 February 2008.

sustainability

MDPI

Article

Sustainable Learning, Cognitive Gains, and Improved Attitudes in College Algebra Flipped Classrooms

Natanael Karjanto [1,*] and Maxima Joyosa Acelajado [2]

[1] Department of Mathematics, University College, Natural Science Campus, Sungkyunkwan University, Suwon 16419, Korea
[2] Department of Science Education, College of Education, De La Salle University, 2401 Taft Ave, Malate, Manila 1004, Philippines
* Correspondence: natanael@skku.edu

Abstract: To respond to global issues positively, education systems in higher education institutions play a significant role in empowering learners as well as promoting sustainable development goals. By implementing curricula that cultivate cross-cutting and transversal key competencies for sustainability, such as critical thinking, problem-solving, and collaboration, we prepare our pupils to become sustainability citizens, who not only sustain learning throughout their lives in various circumstances and across different disciplines but also engage constructively and responsibly toward any future world's challenges through their dispositions, strategies, and skills. One such sustainable teaching methodology is known as the flipped classroom, an active-learning, student-centered, flexible, and multidimensional pedagogy. Our objective is to investigate the effect of such pedagogy on learners' academic achievement and their attitude toward mathematics using both quantitative and qualitative methods. We cultivated sustainable learning in mathematics education for college freshmen ($n = 55$) by exposing them to both the conventional teaching method (CTM) and flipped classroom pedagogy (FCP). By splitting them into control and experimental groups alternately ($n_1 = 24$, $n_2 = 31$) and by selecting the four most challenging topics in college algebra, we measured their cognitive gains quantitatively via a sequence of pre- and post-tests. The topics are factorization, rational expressions, radical operations, and applied problems. Both groups improved academically over time across all these four topics with statistically very significant outcomes ($p < 0.001$). Although they were not always statistically significant ($p > 0.05$) in some topics, the post-test results suggest that generally, the FCP trumps the CTM in cognitive gains, except for the first topic on factorization, where the opposite is true with a very statistically significant mean difference ($p < 0.001$). By examining non-cognitive gains qualitatively, we analyzed the students' feedback on the FCP and their responses to a perception inventory. The finding suggests a favorable response toward the FCP with primary improvements in the attitudes toward mathematics and increased levels of cooperation among students. Since these students are so happy to have control of their own learning, they were more relaxed, motivated, confident, active, and responsible in learning under the FCP. We are confident that although this study is relatively small in scale, it will yield incremental and long-lasting effects not only for the learners themselves but also for other role-takers in education sectors who aspire in nurturing sustainable long-life learning and achieving sustainable development goals successfully.

Keywords: flipped classroom; cognitive gains; attitude toward mathematics; college algebra; mathematics education; sustainable learning

Citation: Karjanto, N.; Acelajado, M.J. Sustainable Learning, Cognitive Gains, and Improved Attitudes in College Algebra Flipped Classrooms. *Sustainability* **2022**, *14*, 12500. https://doi.org/10.3390/su141912500

Academic Editors: Jin Su Jeong and David González-Gómez

Received: 30 July 2022
Accepted: 17 September 2022
Published: 30 September 2022

1. Introduction

One of the 17 United Nations' Sustainable Development Goals (SDGs) for achieving a better and more sustainable future for all beings is "quality education" (Goal 4, herein SDG4). The two facets of SDG4 ensure inclusive and equitable quality education and promote lifelong learning opportunities for all [1]. Although the primary targets of SDG4 are children and youths in their primary and secondary educations, the principle can be extended into higher education institutions (HEIs) as well. HEIs play an essential role in promoting sustainability and advocating the SDGs by educating and training prospective teachers, policymakers, future leaders, entrepreneurs, and other professionals [2].

In particular, through their cutting-edge research and continuously improving their curricula, HEIs advance sustainability competencies among their pupils. The specialized agency UNESCO also supports SDG4 through education for sustainable development (ESD) and global citizenship education (GCED). In turn, this education will enhance necessary cross-cutting key competencies for sustainability that are relevant not only to SDG4 but also to all SDGs. Among these crucial key competencies are critical thinking, integrated problem-solving, and collaboration competencies [3].

Keeping this association in mind, having a solid foundation in science, technology, engineering, art, and mathematics (STEAM) education is very essential in order to develop sustainable citizens since these subjects inherently cultivate critical thinking and problem-solving skills, the very qualities which are aligned with ESD and GCED in SDG4 [4]. These skills can be used throughout life irrespective of whatever professions the students might choose to pursue. The very basics of all these disciplines, namely mathematics as well as the way we teach and how students learn about it (mathematics education), should be of particular interest to many educators around the world who aspire to instill lifelong learning among their students, and at the same time, nourish them with the ESD and GCED transversal competencies.

Thanks to the progress in information technology, many novel pedagogical methodologies have emerged and received more attention during the past two decades, particularly those that emphasized student-centered learning activities. These include but are not limited to flipped classroom, blended learning, problem and/or project-based learning, inquiry-based learning, collaborative learning, and inclusive education, including its framework of universal learning design, among others. The body of literature confirms that these educational pedagogies demonstrated excellent educational outcomes and provided platforms for supportive learning environments that align with the SDGs [5,6]. Focusing on a particular pedagogy and observing how it affects learners' cognitive gains and improves attitudes toward a particular subject or learning in more general would be an essential step in cultivating transferable competencies for sustainability.

In this article, we discuss how flipped classrooms in a college algebra module could improve students' cognitive gains as well as their attitudes toward mathematics. Unless specified differently, throughout this paper, cognitive gains refer to the positive gains in knowledge and comprehension, which can be translated through academic performance. Although non-cognitive gains encompass a wide range of abilities such as communication, teamwork, perseverance, conscientiousness, and motivation, among others, what we refer to is related to the latter, i.e., students' attitudes toward mathematics. Furthermore, we will refer to the two pedagogical approaches used in this study as the FCP and the CTM, which refer to the flipped classroom pedagogy and conventional teaching method, respectively.

With the nature of present-day students and their exposure to a variety of technological tools, nowadays teaching requires a degree of flexibility in addressing diverse learning styles, a wide range of level capability, and handling large classes. A breed of confident and competent problem-solvers who are eager to learn new things on their own and the improvement of students' achievement in and attitudes toward mathematics are very much desired not only to survive in this modern society but also to develop sustainable citizens. Hence, there is a pressing need for advancing from teacher-centered, passive-receptive learning of the CTM into a student-centered, active-learning FCP. Some evidence suggests that the FCP could be a sustainable active-learning pedagogy when learning is disrupted such as during the recent COVID-19 pandemic [7].

This study investigated the impact of the FCP on students' academic achievements in a mathematics course among freshmen majoring in mathematics education at a private, Catholic coeducational research university in the Philippines. Additionally, we are addressing the following research questions:

- Between the experimental and control groups and among the four most challenging topics in college algebra, which group obtains better cognitive gains? Which topics would be learned better using the FCP?
- For non-cognitive gains, what are students' reactions to the use of the FCP? What are their attitudes toward mathematics after learning the course via the FCP?

The article is organized as follows. After this introduction, Section 2 provides a literature review on sustainable learning in education (SLE) and how the FCP has been successfully implemented in a wide range of disciplines from mathematics to ESD. It then offers a conceptual framework for this study that SDG4 covers both ESD and SLE, and the FCP is one possible option for SLE. Section 3 covers research objectives and methodology, including the participants and applied measurement. Section 4 outlines the result of our experiments and discusses what these findings mean. Finally, Section 5 concludes our discussion and provides further recommendations.

2. Literature Review

2.1. Sustainable Learning in Mathematics Education

Ben-Eliyahu (2021) clarified a potentially baffling concept between "learning sustainable development" and "sustainable learning in education (SLE)" [8]. The designated terminology for the former as "sustainable learning" might be easily mistaken for the latter or vice versa. On the one hand, a better terminology for learning sustainable development would be "sustainability learning" or "ESD", which refers to an approach to education that emphasize the importance of humans living in harmony with nature, either by specifically teaching sustainability principles or by integrating them into a curriculum by including key issues of sustainable development, as we have mentioned briefly in the introduction [9–12].

On the other hand, SLE refers to "learning that lasts"; it can be achieved through well-structured and responsive teaching that matters for all learners [13]. It consists of two facets of learning that can be likened to two sides of a coin. On the one side of the coin, the learning that takes place in formal education has lasting value to learners into the future. On the other side of the coin and at the same time, this sustainable type of learning will also encourage pupils to continue their education journey by embracing lifelong learning themselves. Indeed, SLE interlaces marvelously with SDG4; see [1]. Furthermore, as suggested by Hays and Reinders (2020), any type of sustainable learning and education should instill in students the skills and dispositions to thrive in a complicated and challenging world as well as the desire to contribute positively in creating the world a better place to live [14].

Among one of the seven recommendations for creating sustainability education at HEIs, Moore (2005) proposed establishing a space for pedagogical transformation. The endorsement encompasses not only improving the interaction between students and instructors but also promoting student-centered, reflective, critical, transformative, and experiential learning [15]. This advocate is underpinned by a recent article on the learning environment in the context of SDG4, where it is confirmed that an educational environment that accommodates sustainable learning is one that encouraged active role from both the pupils and educators, such as class participation, critical thinking, nurturing curiosity, and cultivating creativity, among others [6].

Similar to renewable and sustainable energy, SLE preserves the learning process throughout one's life as the scene of the world is changing. It endows learners with skills and strategies to rejuvenate themselves through inquiry, self-assessment, and evaluation of their environment and social systems. SLE encompasses four aspects of self-regulated learning models that exhibit an analogy with sustainable nature: renewing and relearning; independent and collaborative learning; active learning; transferability [8,16–20]. As educators, in addition to grooming our students with future-focused experiences and skills, we also ought to nurture their confidence and refine their awareness of achieving positive changes.

On a larger scale, HEIs play a key role in sustainability; again like the two sides of a coin, by promoting both ESD as well as SLE. On the one side of the coin, by facilitating and designing curricula that center around sustainability, i.e., ESD, HEIs contribute crucially to creating a sustainability mindset among their members, including faculty, administrative personnel, and, in particular, the student body, where a new generation of future leaders would emerge [21–24]. On the other side of the coin, by improving their didactics and pedagogy, i.e., SLE, HEIs prepare current and prospective learners with transversal competencies that are necessary for tackling not only personal, a smaller-level, challenging circumstances but also global, a larger-level, economic, social, and environmental challenges [25–30].

Although the aforementioned cited works administer the field of education in the general context, the same principles are also certainly appropriate for mathematics education. Renert (2011) attempted to address an inquiry about how to reconcile the urgent need to act for a sustainable future with the current practices of mathematics education by presenting a model of possible responses to sustainability in mathematics education [31]. The proposed model adapts two existing stage models of approaches to sustainability to the context of mathematics education, i.e., Sterling's (2001) and Edwards' (2010) models of educational responses and organizational approaches to sustainability, respectively [25,32]. The three types of the educational response of accommodation, reformation, and transformation mean education *about*, *for*, and *as* sustainability, respectively. The former two fit well with ESD, whereas the final type aligns with SLE.

With an exception of Summer's (2020) exploration of how a sustainable primary mathematics education ought to be implemented [33], other works that intersect between sustainability and mathematics education usually concentrate around ESD instead of SLE, e.g., [34–39]. By suggesting concrete pedagogical initiatives to tackle primary students' challenges in learning mathematics, Summer (2020) demonstrated that quality education and competent teachers not only decrease their learning difficulties but also equip children with essential mathematical and critical thinking skills [33]. These skills, in turn, will eventually provide a solid foundation for acquiring transversal competencies when they grow up, join the workforce, and assess sustainability principles.

2.2. Flipped Classroom Pedagogy

To better comprehend the FCP, we need to agree with what we mean by the traditional teaching and learning approach, dubbed in this article the CTM. In a CTM situation, the students listen to the teacher's lecture-discussion on the day's lesson and they are given homework or assignment to measure their understanding of the lesson. The homework may be discussed in class or simply submitted to the teacher. Teachers are the instruments by which knowledge is communicated. The CTM is primarily teacher-centered where all students are taught the same materials at the same time. The CTM emphasizes direct instruction, predominantly lectures, and a fixed seatwork so that students learn through listening and observation. Classroom instruction is often solely based on textbooks, lectures, and individual written assignments.

An FCP, which is otherwise known as an inverted classroom [40], flipped teaching, flipped learning, or the Thayer method [41,42], is a combination of viewing video recordings and reading module materials related to the lesson anywhere and anytime before class and applying what has been learned during the face-to-face time [43]. Class time is used more interactively for group discussion, discovery activities, experiments, and class presentations where the teacher's role is to facilitate and assist the students in their quest for further understanding of the lesson. As a result, the students become active learners rather than plain receptacles of information [44].

The FCP is a form of blended learning in which students learn new contents from various modes, such as by viewing video lectures online, browsing websites, reading textbooks, and viewing PowerPoint-like (or Beamer) lecture slide presentations that are either provided by the teacher or the result of their own search mechanisms, usually at home before the class time. What used to be the homework/assignment is now done in class with teachers offering more personalized guidance and interaction with students, instead of lecturing full-time. During the class, the students apply the knowledge they have learned by solving problems, doing practical work, and collaborating with their peers.

Both classroom and online learning should provide materials that would enable the students to practice and interact with others and still deliver favorable outcomes. Doing the homework to keep them engaged for a deeper understanding of concepts and mastery of skills is done during face-to-face class time. The students work on their own phase and work level in a result-based environment while using online content that serves as the preliminary source of learning and this may be used repeatedly in class or as review materials. The content acquisition comes ahead and concept engagement takes place in class with the students doing interactive activities. By inquiring the teachers on materials that they could not grasp or more challenging topics, positive response and feedback from the teacher could assist both adept and slow learners.

There are several implications in the context of educational theory when implementing the FCP. First, the knowledge becomes personal since each learner might differ in interpretations and thus possesses a distinctive point of view when following a flexible and multi-dimensional pedagogy such as the FCP [45]. Second, we allow students to construct knowledge by themselves when they experience different things, rather than passively absorbing it in the case of the CTM [46,47]. Third, and as a consequence, learning becomes an active process. Learners construct meaning in their understanding not only through active engagement with their environment but also by establishing meaningful connections between prior and new knowledge [48,49]. These aspects constitute a constructivist approach in education theory and we could confirm from the following literature study that the FCP does belong to this approach [50,51].

According to Strayer (2007), an FCP is a more active, student-centered style of teaching through the use of group projects, discovery activities, experiments, and class presentations that are implemented during classroom time with information-rich, lecture-based direct instruction being used during an out-of-class time, usually delivered through online videos that students view before arriving in class [52]. In short, this model aims to move the easier parts of teaching and learning into independent practice ahead of learning the more difficult concepts, which are taught face-to-face [53].

Flipped teaching is a pedagogical approach in which direct instruction is moved from the group learning space to the individual learning space, and the resulting group space is transformed into a dynamic, interactive learning environment where the teacher guides students as they apply concepts and engage creatively in the subject matter. The FCP has benefited students who missed classes since they may use the online materials to review and reinforce lessons and it gives them the opportunity to radically rethink how they should use class time effectively [54,55].

In the FCP, instruction is delivered online and outside the classroom through video, podcasts, or the online learning environment. Homework is moved into the classroom. In this approach, students can take in the information at their own pace and discuss it with the teacher and peers. This creates time in the classroom for collaborative work by the students and more room for a differential approach and remediation by the teacher. Fulton (2012) enumerated the following advantages of the flipped classroom: (1) students move at their own pace; (2) doing "homework" in class gives teachers better insight into students' difficulties and learning styles; (3) teachers can more easily customize and update the curriculum and provide it to students anytime; (4) classroom time can be used more effectively and creatively; (5) teachers using the method report seeing increased levels of student achievement, interest, and engagement; (6) learning theory supports the new approaches; (7) the use of technology is flexible and appropriate for the 21st-century learning [56].

Moreover, Herreid and Schiller (2012) argued that with the FCP: (1) there is more time to spend with students on authentic research; (2) students get more time working with scientific equipment that is only available in the classroom; (3) students who missed class can watch the lectures while on the road; (4) the method promotes thinking inside and outside of the classroom; (5) students are more actively involved in the learning process [57]. The survey conducted by Bishop and Verleger (2013) revealed that most studies on the FCP used single-group study designs to explore student perceptions and reports are somewhat mixed but generally positive overall. Some students tend to prefer in-person lectures to video lectures, but prefer interactive classroom activities over lectures. Anecdotal evidence suggests that student learning is improved for the FCP compared to the CTM [58].

Hantla (2014) indicated that the flipped classroom is a new iteration of an old way of teaching that enables instructors to do more during face-to-face classroom time than is otherwise possible. The FCP provides an incentive to students to come to class prepared and assess their understanding, focuses on higher-level cognitive activities under the guidance of the teacher, and provides students adequate time to carry out their assignments and get on-the-spot feedback about their work. The FCP is carried out in a learning environment with the support of educational technology where students learn through activity-oriented activities [59].

When it comes to teaching mathematics at all levels of education, the body of published literature does not lack examples. The FCP has been successfully implemented in teaching mathematics at the primary [60–63] and secondary levels [64–70]. The FCP also encompasses a broad mathematical subjects, such as college algebra [71,72], precalculus [56], calculus [73–78], vector calculus [79], linear algebra [80–83], statistics [84], and actuarial science [85]. Generally, both students' cognitive gains in terms of understanding and academic performance, as well as students' positive attitudes in terms of enjoyment and confidence are affirmative; see also [86–90].

Certainly, the successful—and fruitless—attempts of the FCP are not only limited to mathematics education. Many works provide a narrative on pedagogy in other fields, including both STEM and non-STEM fields. In particular, in what follows we provide some evidence in the learning about sustainability and sustainable developments. Buil-Fabregá et al. (2019) demonstrated that the FCP has successfully assisted students in improving their transversal competencies and being more conscious of sustainable development requirements [91]. Rodríguez-Chueca et al. (2020) measured the efficiency of the FCP and challenged-based learning to facilitate learning of sustainability principles and circular economy and discovered that the former is more satisfactory than the latter [92]. Howell (2021) revealed positive student perceptions when they were exposed to the FCP when enrolling in education for sustainable development courses [93].

2.3. Conceptual Framework

Based on the literature review of both SLE as well as the superiority of the FCP over the CTM, we propose a conceptual framework to examine that our study on flipped classrooms in mathematics education can also be expanded to other topics outside mathematics, duplicated, and even improved for better learning outcomes as well as with more positive psychological well-being. Figure 1 displays a conceptual framework for this study.

Our study is based upon a conceptual framework that sustainability and education influence each other and contribute to each other's development. As we reviewed in Section 2.1, SLE in general and sustainable learning in mathematics education in particular require not only strategic HEI curricula that empower learners but also flexible and multidimensional teaching approaches that equip learners with transferable skills to renew themselves when facing personal and world challenges [8,14]. Among the highlighted pedagogies that promote SLE are project-based learning and the FCP. Our focus is the latter. Conversely, the body of published literature reviewed in Section 2.2 has demonstrated that the FCP was not only viewed very favorably among students but also compels them to commit to sustainability principles and practice sustainable development when they join the labor market, as demonstrated in several recent studies [91–93]. Hence, as illustrated in Figure 1, ESD and SLE could function both ways.

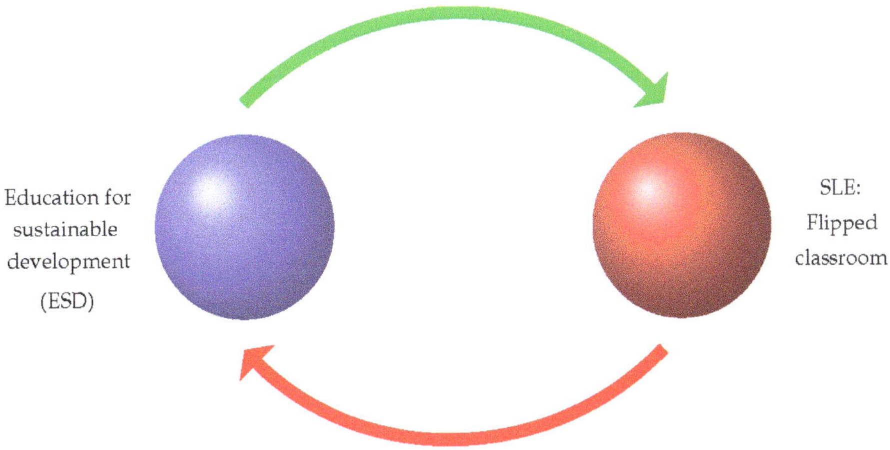

Figure 1. A conceptual framework for sustainable development in mathematics education employing SLE and FCP. HEIs play an active role in promoting SDG4 by equipping learners with sustainable learning in (mathematics) education to cope with future challenging circumstances. In particular, the FCP offers quality education and SLE through its flexible teaching yet effective learning approach (top green arrow). Conversely, the FCP can be implemented to cultivate learners in contributing to SDGs via ESD with the aim of embracing and practicing sustainability principles (bottom red arrow).

3. Research Material and Methodology

3.1. Participant

Two classes of 24 and 31 freshmen mathematics education majors at De La Salle University, Manila, the Philippines, enrolled in a three-credit course of College Algebra (course code: TCHALGE) during the first term of the academic year 2014/2015 (September–December 2014), served as respondents of this study. The classes were referred to as Group 1 and Group 2, respectively. The selection methodology was based on the convenience technique since these two classes were simply assigned to and taught by one of us (the second author). The age range of the participants is typically between 18 and 19 years old. The percentages of the female and male students are around 60% and 40%, respectively.

3.2. Measurement

A quasi-experimental design with switching replication was implemented for both groups. The group exposed to the FCP is considered the experimental group, whereas the group exposed to the CTM is regarded as the control group. For the first and second topics, Group 2 was considered the experimental group, while for the third and fourth topics, Group 1 was designated as the experimental group. Both groups were also administered pre- and post-tests on all four topics considered in this study. The two groups were alternately exposed to the FCP and the CTM in the delivery of topics in college algebra which are identified as relatively hard by students who took the subject during preceding terms. These are factoring/factorization, rational expressions, operations on radicals, and solving applied problems.

On the one hand, the experimental group was provided with a list of all websites, videos, lecture notes, PowerPoint presentation slides, and course modules that they need to watch and read as well as the corresponding assessment materials that they need to respond while viewing the videos or listening to the lectures before class to test their understanding of the topics. During the face-to-face class time, they were given more interactive activities which were done either individually or in groups with the teacher acting as a facilitator. On the other hand, the control group was taught identical topics and was provided with the same interactive activities in class after which the students were given the assignment/homework for submission during the next meeting.

In our empirical study, we have employed both quantitative and qualitative methods to investigate the effect of the FCP on students' cognitive and non-cognitive gains. The former was investigated through test results on four different topics that were considered the most challenging ones applied to both the control and experimental groups. A validated, multiple-choice type, teacher-made pre-test and post-test on these particular topics were administered to gauge and compare their academic achievements in each topic. This validation was conducted by other instructors who possess expertise in the field, have taught the subject for at least five years, and obtained distinguished results on students' feedback on teaching evaluation.

The levels of difficulty in both the pre-test and post-test were parallel and identical. Each test consists of 25 items for each topic and the students must complete it within 120 minutes. The pre-test was administered before the start of the experiment, whereas the post-test was conducted during the final exam period. All scores were converted into percentages. To determine if any statistically significant difference exists between the pre-test mean scores, pre-test and post-test mean scores, and the post-test mean scores of the two respondent groups, we conducted a sequence of Student's t-tests for dependent and independent samples. To assess learners' non-cognitive gains after they experienced learning the module using the FCP, we requested them to write their opinion on the FCP in student journals as well as to respond to an FCP perception inventory.

Table 1 shows the classroom activities for both groups during the pre-test period. Both groups experience both the FCP and the CTM but with different topics. The first group covers the first two topics in the CTM and the other two topics in the FCP, while the second group is the opposite, first the FCP, and then the CTM. For the post-test activity, the students dedicate the remaining time to journal writing and the completion of the FCP perception inventory.

Table 1. Group activities and their associated topics in college algebra conducted during the pre-test period. The boldface letter **G** refers to the group, which can be 1 or 2, whereas **F** and **T** refer to the classroom approaches of flipped and traditional, respectively.

Topic and Content	Group (G1 or G2) and Classroom Approach (F or T)	
	G1T	**G2F**
Factorization: Common Monomial Factor, Factors of General Trinomials, Difference of Two Squares, Perfect Square Trinomial, Sum/Difference of Two Cubes	Students listened to the teacher's regular classroom lecture-discussion aided by some PowerPoint presentations, printed references, solved sample problems, and applications. Homework/assignment was given to the students for submission during the next meeting.	Students viewed downloaded and/or teacher-provided video and PowerPoint presentations, read modules and textbooks, and completed worksheets and other assessment materials prior to class. Interactive activities were done by the students in the classroom.
Rational Expressions: Operations on Rational Expressions, Simplifying Complex Rational Expressions		
	G1F	**G2T**
Operations on Radicals: Simplifying Radicals, Rationalizing Denominators	Students viewed, downloaded, and/or teacher-provided video and PowerPoint presentations, read modules and textbooks, and completed worksheets and other assessment materials prior to class. Interactive activities were done by the students in the classroom.	Students listened to the teacher's regular classroom lecture-discussion aided by some PowerPoint presentations, printed references, solved sample problems, and applications. Homework/assignment was given to the students for submission during the next meeting.
Solving Applied Problems: Routine and Nonroutine Problems (Number, Age, Investment, and Mensuration)		

4. Result and Discussion

4.1. Cognitive Gains

To achieve the objectives of the study, the pre-test and post-test scores were subjected to statistical treatment. Moreover, the students' journal report and their responses to the perceptions inventory were analyzed. Below are the tables showing the descriptive statistics for the gathered data and the corresponding discussion.

As reflected in Table 2, the pre-test mean scores of Group 1 and Group 2 are far below the passing score of 60% in all topics under consideration. Very clearly, the students had very low mean scores in the topics of radical operations and solving applied problems. The small values of the respective standard deviations indicate little variation among the scores; that is, the groups are more or less homogeneous. The non-statistically significant difference between the computed t-values of the pre-test means confirmed the comparability of the two groups at the beginning of the experiment in so far as performance in all the topics under consideration is concerned. This indicates that any changes in their achievement can be attributed to the utilized teaching approach.

Table 2. Descriptive statistics comparing the pre-test mean scores of the two groups (df = 53, $t_{crit} = 2.0057$). All results were not statistically significant ($p > 0.05$).

Topic	Group 1 ($n = 24$)		Group 2 ($n = 31$)		Statistics	
	$\bar{x}$	s	$\bar{x}$	s	$\lvert t\rvert$-Value	p-Value
Factorization	45.50	11.77	45.87	7.67	0.1408	0.8886
Rational expressions	44.71	9.72	45.52	6.27	0.3746	0.7095
Radical operations	33.21	10.18	34.35	10.17	0.4121	0.6819
Applied problems	38.96	9.29	39.23	10.46	0.0996	0.9210

The pre-test and post-test mean values of each group in all topics as shown in Table 3 indicate that learning took place using any of the two approaches—flipped or traditional. In teaching factorization and rational expressions, the use of the CTM for Group 1 accounted for respective mean gains of 39.58 and 27.58 as compared with the respective mean gains of 27.52 and 36.16 which can be attributed to the use of the FCP for Group 2. Moreover, the use of the CTM in teaching radical operations and solving applied problems accounted for mean gains of 28.68 and 30.21, respectively, as compared to the respective mean gains of 32.62 and 36.42 of the students exposed to the FCP. Obviously, based on the mean gains, using the FCP is better than the CTM when it comes to the topics of rational expressions, radical operations, and solving applied problems. The highest mean gain is evident in the use of the CTM in factorization (39.58), followed by the use of the FCP in solving applied problems (36.42), rational expressions (36.16), and radical operations (32.62). Thus, the FCP proved to be a better teaching approach in three out of the four identified difficult topics in this study in terms of academic achievement gains. Statistically significant cognitive gains are also evident as indicated by the calculated t-values, which are all greater than the tabular t-values. This indicates that each approach has a significant positive effect on learning, but the use of the FCP generally accounts for greater mean gain in the identified difficult topics in college algebra.

As observed in Table 4, the result of the t-test for independent samples applied to the post-test mean scores indicate a statistically significant difference between the FCP and CTM in the delivery of factorization and rational expressions, but no significant difference in radical operations and solving applied problems. This implies that the CTM appears to be a better teaching approach for factorization while the FCP might be a better delivery method for rational expressions.

Table 3. Descriptive statistics between the pre-test and post-test mean scores of the respondents. Group 1 adopted the FCP for the third and fourth topics, with df = 23 and t_{crit} = 2.0687. Group 2 adopted the FCP for the first and second topics, with df = 30 and t_{crit} = 2.0423. All results were statistically very significant ($p < 0.001$).

Topic	Group 1 (n = 24)						Group 2 (n = 31)					
	Pre-Test		Post-Test		Statistics		Pre-Test		Post-Test		Statistics	
	$\bar{x}$	s	$\bar{x}$	s	$\|t\|$	p	$\bar{x}$	s	$\bar{x}$	s	$\|t\|$	p
Factorization	45.50	11.77	85.08	11.72	3878.03	0.000	45.87	7.67	73.39	10.42	55.72	0.000
Rational expressions	44.71	9.72	72.29	12.28	52.78	0.000	45.52	6.27	81.68	11.92	35.63	0.000
Radical operations	33.21	10.18	65.83	10.58	399.51	0.000	34.35	10.17	63.03	14.65	35.64	0.000
Applied problems	38.96	9.29	75.38	14.86	32.03	0.000	39.23	10.46	69.44	14.12	45.96	0.000

Table 4. Descriptive statistics comparing the post-test mean scores of the two groups (df = 53, t_{crit} = 2.0057). Only the mean scores for the first two topics appear to be statistically significant and the FCP produced higher mean scores except for the first topic on factorization.

Topic	Group 1 (n = 24)		Group 2 (n = 31)		Statistics	
	$\bar{x}$	s	$\bar{x}$	s	$\|t\|$-Value	p-Value
Factorization	85.08	11.72	73.39 *	10.42	3.9076	0.0000 *
Rational expressions	72.29	12.28	81.68 *	11.92	2.8595	0.0061 *
Radical operations	65.83 *	10.58	63.03	14.65	0.7897	0.4332
Applied problems	75.38 *	14.86	69.44	14.12	1.5123	0.1364

* post-test mean score under the FCP; * significant at $\alpha = 0.05$ level.

It can be noted that the students registered a higher mean score in factorization under the CTM while the use of the FCP accounts for the higher mean scores in rational expressions, radical operations, and solving applied problems. However, non-statistically significant mean differences in the respondents' post-test scores in the last two topics imply that learning can be achieved whether delivery of these topics was done using the FCP (where they view the lessons online, through other modes, and do their supposed homework in the classroom) or the CTM (where they have their lessons in the classroom and really do their homework at home).

There could be several factors that might influence this result. The first one is related to the topic and content investigated in this study. The second factor relates to the quality of lectures in both pedagogies. The third factor is associated with the variation in students' academic performance when they encounter various college algebra topics in different learning environments, i.e., the FCP vs. CTM. Our finding suggests that learners achieved better cognitive gains on the first topic of factorization when it was delivered using the CTM instead of the FCP. This topic may be relatively less challenging than the other three topics, and for some of the contents, the students may have seen them in primary or secondary mathematics. For example, finding the greatest common factor between monomials relates to basic number theory on prime numbers in primary school mathematics. Another example is factoring general trinomials, which requires a direct explanation of how to find two integers whose product is one monomial and whose sum is another monomial for easier understanding. As might be the case in the latter, although viewing video recordings might explain the procedure, an absence of immediate feedback might hinder further progress and comprehension, and thus the CTM might be better suited for these concepts than the FCP.

The second factor might be related to the quality of the lectures in both traditional and flipped settings. From the learners' perspective, it does not reserve a possibility that the teaching delivery during direct instruction has better quality and the recorded video recording for this particular topic of factorization. As argued by Krantz (2015), teaching mathematics by lecturing directly—offline mode, face-to-face—and hence predominantly the CTM, is still a powerful teaching device that has stood the test of time for more than three millennia provided that the instructor does it very well [94]. However, we need to be cautious in directly swallowing this argument since Krantz's argument was also seriously disputed by a meta-analysis study from several STEM disciplines by Freeman et al. (2014) who discovered that student academic achievements are significantly better when some kind of active learning is implemented [95].

For the third factor, we observed that the pre-test and post-test results from Group 1 on the first topic factorization yield sample standard deviation values of around 11.75, i.e., 11.77 and 11.72 for the former and latter, respectively. These not-so-low values suggest that the CTM only improved the students' scores and translated the mean to a higher score but it did not really close the difference gap between the more academically and less academically prepared students. In other words, the CTM has successfully prevented the widening gap between the various spectrum of students' academic strengths. By looking at the results for Group 2, the standard deviation values increase from 7.67 to 10.42 for the pre-test and post-test results, respectively. This nearly 3% increase in standard deviation could influence a lower post-test mean score compared to Group 1, and thus the mean difference for post-test results appears to be statistically significant. This could suggest that the FCP fails to prevent gap widening or enclosing the existing gap in academic achievement for this particular topic.

A similar outcome was observed for the third topic of radical operations, but then the pedagogy was reversed. Although the mean difference for the post-test scores did not appear to be statistically significant, we could argue that the FCP has had relative success in preventing the widening gap in academic performance among students in Group 1 for this particular topic. The situation, however, was slightly different for the fourth topic of applied problems. The increases in sample standard deviations from the pre-test to post-test results were nearly 6% and 4% for Groups 1 and 2, respectively. This finding might suggest that the mean difference for the post-test scores did not appear to be statistically significant for this topic even though the result from the FCP in Group 1 was better than the one from the CTM in Group 2. A similar argument could be applied to the second topic of rational expressions where both groups experienced an increase in standard deviation values. But since the mean score for Group 2 is much higher than Group 1 (nearly 9%), the increase in standard deviations was still compensated and the mean difference for the post-test scores appeared to be statistically significant. It would be interesting to investigate whether other studies yield comparable outcomes when a similar methodology was applied to different cohorts of students.

4.2. Non-Cognitive Gains: Attitudes Toward Mathematics

In addition to cognitive gains of better academic performance, the students also acquired non-cognitive gains after experiencing the FCP. By soliciting their opinions regarding the use of the FCP through a flipped classroom perception inventory, we collect further insight regarding their learning characteristics. Table 5 summarizes the percentages of those who agreed on each item in the perception inventory.

From this outcome, we observe that students' responses are generally favorable when it comes to learning using the FCP. All the ten items yield either more than 90% or nearly 90% of students' ratings, where the highest percentage (94%) belongs to the seventh item that the FCP promotes a positive attitude toward mathematics. The high percentages of students in favor of each item in the inventory indicate that they welcome this type of pedagogy in learning mathematics, particularly college algebra. In addition to improving positive attitudes toward mathematics, providing the opportunity to understand mathematical concepts in a better way, and offering various alternatives in dealing with mathematical problems, the non-cognitive benefit of the FCP also goes beyond mathematics learning per se.

Table 5. The percentage of students who agreed on each item in the FCP perception inventory. The rating is either more than 90% or very close to 90%.

No.	Item	Rating (%)
1	Motivates students to study	92
2	Develops students' confidence to solve problems	92
3	Offers a variety of alternatives in understanding the lesson	93
4	Improves students' creative and critical thinking ability	92
5	Provides students the opportunity to understand mathematical concepts better	92
6	Allows students to participate actively in the learning process and progress independently	91
7	Promotes positive attitudes toward mathematics	94
8	Strengthens students' retention of subject matter	88
9	Promotes better cooperation between teacher and students and among students	92
10	Offers powerful ways of dealing with mathematical problems	89

In the essence of SLE, the students in our study confirmed that the FCP has motivated them to study, improved their critical thinking, enhanced their knowledge retention rate, and promoted better communication with the teachers and among themselves. Overall, the FCP has assisted them not only in acquiring non-cognitive gains but also in appreciating some aspects of SLE. Note that although some of them may forget the details of certain mathematical concepts or theorems in this particular course, the acquired qualities and positive attitudes will be transferable to another advanced or different course, even to a totally different setting, whether in their profession as educators or elsewhere; see [8].

Other non-cognitive gains from the use of the FCP indicated by the students in their journals are as follows: a higher level of participation in active learning sessions, a stronger sense of being connected to the teacher and with other students, a positive perception of the course, and an enjoyable learning environment. Moreover, they acquired from wide resources of learning tools at varying speeds in their chosen learning environment and were able to have a continuing review as the need arises in as much as the materials were accessible even for absentees. Hence, the element of flexibility. Finally, they also like the idea that their teacher had more time in guiding them during activity sessions.

Empirical results confirmed that the use of the FCP enhances students' intellectual experiences and learning outcomes. Additionally, although some students in the CTM setting were more satisfied with the clarity of instruction provided by the teacher and the facility of getting feedback immediately after the lecture, they felt more strongly in terms of gaining a greater appreciation of college algebra concepts through the FCP.

Some unedited comments culled from the students' journals are given as follows:

"Learning math has never been like this before because this time I did not have to sit in class very long doing nothing but listen and be bored.";

"I enjoyed learning math because I can switch from viewing videos to reading the PowerPoint presentation sent by my teacher to simply reading the textbook, trying to get ready for our class activities.";

"Whenever I have questions, I kept on viewing the online lesson wherever I am so long as I have my laptop with me and I have access to internet.";

"While viewing/reading the lesson, I had to list down my questions and ask my teacher and my classmates for clarification.";

"I could have learned more math had this approach been introduced earlier.";

"I felt helpless whenever I had questions, but nobody was there to answer immediately.";

"Sometimes, I cannot access the lesson because there was no internet facility.";

"There should be a larger viewing center at the library where we can stay whenever no internet is available at home.";

> *"Sometimes I was tempted to skip my class when I thought I already understand the lesson very well and my answers to the assessment materials were almost alright.";*
>
> *"I enjoyed the extra time given me to do the viewing of the lesson over and over again until it becomes clear to me.";*
>
> *"I got a strong bonding session with my classmates and my teacher during the face-to-face encounter.";*
>
> *"As I read my lessons alone, I became independent, patient and confident in learning math.";*
>
> *"It seems that I became more interested in math.";*
>
> *"During class time I had a fruitful exchange of ideas with my peers as well as with our teacher."*

From these comments, we observe that there is a mixed reaction from the students regarding their experience with the FCP. For the negative aspect of the situation, we could categorize it into at least three factions: the students' side, the teachers' part, and the intermediary party that links the learners with their facilitators. For the first group, some students consider that studying on their own is a mammoth challenge while for others, they could understand easily. The former might feel helpless whenever they have some questions that need immediate assistance, but nobody was there to provide prompt help and feedback. For the latter, they might be tempted to skip the face-to-face sessions with their teachers. For both sides of the spectrum, we observe that the students might be at risk, and thus educators should interfere promptly and determine better strategies to facilitate their sustainable learning.

The challenge in the intermediary party that links students with their teachers is mostly related to technical issues, such as Internet connectivity and computer facilities. Several students brought up these points in the aforementioned comments, and their learning might be hindered as well. This leads us to address burdens on teachers' shoulders. Although they are not mentioned in the students' comments, the previous two challenges would commit educators to additional tasks in assisting their learners. They might need to prepare additional teaching materials for the students who were lacking with Internet access, they might need to schedule additional activities for the students who struggle understanding basic materials, or they might adjust the teaching speed and material coverage to accommodate those who were left behind.

Despite these downsides, we recognize that more students wrote about their constructive experience with the FCP. They enjoyed learning mathematics and became more interested in delving deeper into mathematics. In addition to acting more independently when studying on their own, they also developed patience and gained more confidence in learning mathematics. Since they underwent different learning experiences, they did not feel bored anymore when studying mathematics. Clearly, many of these students have an improved, better attitude toward mathematics; we hope that their learning will last since some of them will eventually become educators themselves. This type of empowering non-cognitive gains is transferable not only to this particular module but also to other skills they might learn in the future. To a great extent, by actively renewing and relearning through self-assessment and inquiry, they are on the right track to adopting SLE; see [8].

5. Conclusions and Recommendation

We admit several limitations in this study. In addition to a relatively small number of participants, we did not collect any details such as their precise age and/or birthday as well as gender information. Another concern relates to the outcome of the experiment. Although we initially hypothesized that the FCP would always yield better learners' cognitive gains for all the four topics in college algebra considered in this study, it turned out that the first topic on factorization yielded better students' cognitive gains under the CTM. A further investigation on the root of this cause might be an interesting topic and a potentially open problem that is worth pursuing further, as we recommend in the final paragraph of this section.

To conclude, we have considered an attempt in improving essential competencies for sustainability, particularly in the context of mathematics teaching and learning in higher education. These ESD competencies are not only crucial but also cross-cut for achieving other goals in SDGs in addition to attaining SDG4. The focus of our study is cultivating sustainable learning among learners in mathematics education through active learning and the student-centered FCP. Learning formats under this pedagogy allow learners not only to renew and rebuild knowledge but also to cope with challenging and complex situations. In particular, we have selected students majoring in mathematics education who were enrolled in a college algebra module at a private university in the Philippines. Since the majority of these students will become educators themselves, they need to acquire the habit of lifelong learning early in their careers, thus adopting sustainable learning for themselves and cultivating it to their potential pupils.

Furthermore, since four topics were considered pretty challenging for many students enrolled in this course, our quantitative study then concentrated on the assessment results of these four topics. For both the control (CTM) and experimental (FCP) groups, the cognitive gains across the four topics in college algebra appeared to be statistically significant, as indicated by improved mean scores in the post-test. However, by comparing data of the post-test results, we observed that the experimental Group 1 achieved better cognitive gains under the FCP for the third and fourth topics of radical operations and applied problems, respectively, although the mean difference with the control Group 2 under the CTM did not appear to be statistically significant. The outcome for the second topic materialized as we have anticipated, Group 2 under the FCP gained better cognitively than Group 1 under CTM. However, for the first topic, the outcome is rather surprising where cognitive gains under the CTM trumped those of the FCP. In both instances, the mean differences appeared to be statistically significant. This seemingly rather puzzling outcome could potentially invite further discussion, debate, and other follow-up studies related to this topic.

Regarding non-cognitive gains, we qualitatively analyzed the result of the FCP perception inventory as well as student journals that they wrote regarding their feedback on the FCP. For the former, the rating for each of the ten items is overwhelmingly positive, ranging from 88% to 94%. In particular, promoting a positive attitude toward mathematics gained the highest rate of 94%. For the latter, we observed that the reaction from the students is mixed although they also generally wrote positive things about the FCP. Some of the hindrances are related to technical issues, such as Internet connectivity. However, many students were happier to have control of their own learning, more confident, learned more, enjoyed learning, and developed more interest in mathematics. Certainly, these positive traits could lead to more active and intentional learning, not only in mathematics modules but also in other courses. Any incremental skills and strategies that they acquired in this course could also potentially be transferred beyond the classroom setting, into different contexts and domains. Hence, these qualitative findings suggest that the FCP is a possible excellent teaching approach for cultivating SLE among mathematics learners.

Upon conducting this study, on the one hand, we would not recommend entirely abandoning the CTM and its predominantly lecturing method in teaching mathematics at all levels of education. Lectures and the CTM, after all, have their place in transmitting and imparting mathematical knowledge to the learners albeit their nature tends to be teacher-centered and students' role is passive-receptive. Coupled with the fact that many teachers do not lecture very well when they teach, or rather, they do not teach very well when they lecture, it is imperative to equip mathematics educators—whether they are teachers at the primary and secondary levels, or instructors and professors at the tertiary level—with sufficient training in teaching techniques, including delivering lectures. Marrying excellent lecturing with active learning approaches, such as the FCP, may have a tremendous positive impact on students' learning. When learners' curiosity is aroused and they experience a sparkle of light in mathematical understanding, they will take the next step of actively learning beyond the course syllabus, renewing their learning, whether independently or collaboratively, and transferring it to different domains. Indeed, they are on the right track toward lifelong learning and SLE.

On the other hand, we would like to recommend practitioners in education to continue embracing and implementing student-centered, active learning approaches in their classrooms, such as the FCP. Although there are some cases where the FCP fails in enhancing students' learning and arousing their interest in learning, the majority of studies in the body of published literature consistently support positive outcomes of the FCP, along with other active and sustainable learning methodologies, such as problem-based and/or project-based learning. Very often, when implementing the FCP, some instructors were distracted from the main objectives of the pedagogy itself by focusing too much on preparing video recordings that the students are supposed to view before they come to class and hence they come unprepared to the class. However, we need to emphasize that the main point of the FCP is what happens when learners are inside the class, doing and solving problems, collaborating with their peers, and interacting with the instructor. When learners take charge and are actively involved in their own learning, they are one step closer to embracing sustainable learning for their own life.

To end this recommendation, we would like to invite other researchers to delve deeper into some possible causes regarding the effectiveness of the FCP vs. the CTM when it comes to cognitive gains. Since our results suggest different outcomes for different algebraic topics, we would like to see a similar study be replicated in the future, preferably on a wider scale of subjects and objects. Any potential outcomes would be interesting for further discussion, debate, and collaboration.

Author Contributions: Conceptualization, M.J.A.; methodology, M.J.A.; software, M.J.A. and N.K.; validation, N.K. and M.J.A.; formal analysis, M.J.A. and N.K.; investigation, M.J.A.; resources, M.J.A.; data curation, M.J.A.; writing—original draft preparation, M.J.A.; writing—review and editing, N.K.; visualization, N.K.; supervision, M.J.A.; project administration, M.J.A.; funding acquisition, N.K. All authors have read and agreed to the published version of the manuscript.

Funding: This research was supported by the National Research Foundation (NRF) of Korea and funded by the Korean Ministry of Science, Information, Communications, and Technology (MSICT) through Grant No. NRF-2022-R1F1A1-059817 under the scheme of Broadening Opportunities Grants—General Research Program in Basic Science and Engineering.

Institutional Review Board Statement: Not applicable.

Informed Consent Statement: Not applicable.

Data Availability Statement: Data available upon request.

Acknowledgments: The authors acknowledged all three anonymous reviewers whose comments and remarks have improved the article significantly.

Conflicts of Interest: The authors declare no conflict of interest.

References

1. Department of Economic and Social Affairs, United Nations. Sustainable Development Goals. 2015. Available online: https://www.un.org/development/desa/disabilities/envision2030.html (accessed on 30 July 2022).
2. Department of Economic and Social Affairs, UN. The HESI Working Group. Higher Education Sustainability Initiative, Assessments of Higher Education's Progress towards the UN Sustainable Development Goals, Volume 2: For Higher Education Institutions Participating in Assessments. 2021. Available online: https://sdgs.un.org/ (accessed on 30 July 2022).
3. UNESCO. Education for Sustainable Development Goals: Learning Objectives. 2017. Available online: https://unesdoc.unesco.org/ark:/48223/pf0000247444 (accessed on 30 July 2022).
4. Rogovaya, O.S.; Larchenkova, L.; Gavronskaya, Y. Critical thinking in STEM (science, technology, engineering, and mathematics). Utopía y Praxis Latinoamericana: Revista Internacional de Filosofía Iberoamericana y Teoría Social. *Lat. Am. Utop. Praxis Int. J. -Ibero-Am. Philos. Soc. Theory* **2019**, *24*, 32–41.
5. Kioupi, V.; Voulvoulis, N. Education for sustainable development: A systemic framework for connecting the SDGs to educational outcomes. *Sustainability* **2019**, *11*, 6104. [CrossRef]

6. Galán-Casado, D.; Moraleda, A.; Martínez-Martí, M.L.; Pérez-Nieto, M.Á. Sustainable environments in education: Results on the effects of the new environments in learning processes of university students. *Sustainability* **2020**, *12*, 2668. [CrossRef]
7. Collado-Valero, J.; Rodríguez-Infante, G.; Romero-González, M.; Gamboa-Ternero, S.; Navarro-Soria, I.; Lavigne-Cerván, R. Flipped classroom: Active methodology for sustainable learning in higher education during social distancing due to COVID-19. *Sustainability* **2021**, *13*, 5336. [CrossRef]
8. Ben-Eliyahu, A. Sustainable learning in education. *Sustainability* **2021**, *13*, 4250. [CrossRef]
9. Venkataraman, B. Education for sustainable development. *Environ. Sci. Policy Sustain. Dev.* **2009**, *51*, 8–10. [CrossRef]
10. Hopkins, C.; McKeown, R. Education for sustainable development: An international perspective. In *Education and Sustainability: Responding to the Global Challenge*; Tilbury, D., Stevenson, R.B., Fien, J., Schreuder, D., Eds.; International Union for Conservation for Nature and Natural Resources–Commission on Education and Communication: Gland, Switzerland; Cambdridge, UK, 2002; pp. 13–24.
11. Boeve-de, Pauw, J.; Gericke, N.; Olsson, D.; Berglund, T. The effectiveness of education for sustainable development. *Sustainability* **2015**, *7*, 15693–15717. [CrossRef]
12. Huckle, J.; Wals, A.E. The UN decade of education for sustainable development: Business as usual in the end. *Environ. Educ. Res.* **2015**, *21*, 491–505. [CrossRef]
13. Graham, L.; Berman, J.; Bellert, A. *Sustainable Learning*; Cambridge University Press: Cambridge, UK, 2015.
14. Hays, J.; Reinders, H. Sustainable learning and education: A curriculum for the future. *Int. Rev. Educ.* **2020**, *66*, 29–52. [CrossRef]
15. Moore, J. Seven recommendations for creating sustainability education at the university level: A guide for change agents. *Int. J. Sustain. High. Educ.* **2005**, *6*, 326–339. [CrossRef]
16. Zimmerman, B.J. Self-regulated learning and academic achievement: An overview. *Educ. Psychol.* **1990**, *25*, 3–17. [CrossRef]
17. Pintrich, P.R. Understanding self-regulated learning. *New Dir. Teach. Learn.* **1995**, *1995*, 3–12. [CrossRef]
18. Boekaerts, M. Self-regulated learning: Where we are today. *Int. J. Educ. Res.* **1999**, *31*, 445–457. [CrossRef]
19. Puustinen, M.; Pulkkinen, L. Models of self-regulated learning: A review. *Scand. J. Educ. Res.* **2001**, *45*, 269–286. [CrossRef]
20. Zimmerman, B.J.; Schunk, D.H. Self-regulated learning and performance: An introduction and an overview. In *Handbook of Self-Regulation of Learning and Performance*; Schunk, D.H., Zimmerman, B., Eds.; Routledge: New York, NY, USA, 2011; pp. 15–26.
21. Cortese, A.D. The critical role of higher education in creating a sustainable future. *Plan. High. Educ.* **2003**, *31*, 15–22.
22. Jones, P.; Selby, D.; Sterling, S. (Eds.) *Sustainability Education: Perspectives and Practice Across Higher Education*; Routledge: Oxfordshire, UK, 2010.
23. Žalėnienė, I.; Pereira, P. Higher education for sustainability: A global perspective. *Geogr. Sustain.* **2021**, *2*, 99–106. [CrossRef]
24. Veidemane, A. Education for sustainable development in higher education rankings: Challenges and opportunities for developing internationally comparable indicators. *Sustainability* **2022**, *14*, 5102. [CrossRef]
25. Sterling, S.; Orr, D. *Sustainable Education: Re-visioning Learning and Change*; Green Books for the Schumacher Society: Totnes, UK, 2001.
26. Van den Branden, K. Sustainable education: Basic principles and strategic recommendations. *Sch. Eff. Sch. Improv.* **2012**, *23*, 285–304. [CrossRef]
27. Fullan, M. *The New Meaning of Educational Change*, 5th ed.; Teachers College Press: New York, NY, USA, 2015.
28. Peris-Ortiz, M.; Lindahl, J.M.M. (Eds.) *Sustainable Learning in Higher Education: Developing Competencies for the Global Marketplace*; Springer: Berlin/Heidelberg, Germany, 2015.
29. Sá, M.J.; Serpa, S. Transversal competences: Their importance and learning processes by higher education students. *Educ. Sci.* **2018**, *8*, 126. [CrossRef]
30. Membrillo-Hernández, J.; Lara-Prieto, V.; Caratozzolo, P. Sustainability: A public policy, a concept, or a competence? Efforts on the implementation of sustainability as a transversal competence throughout higher education programs. *Sustainability* **2021**, *13*, 13989. [CrossRef]
31. Renert, M. Mathematics for life: Sustainable mathematics education. *Learn. Math.* **2011**, *31*, 20–26.
32. Edwards, M. *Organizational Transformation for Sustainability: An Integral Metatheory*; Routledge: New York, NY, USA, 2010.
33. Summer, A. A sustainable way of teaching basic mathematics. *Discourse Commun. Sustain. Educ.* **2020**, *11*, 106–120. [CrossRef]
34. Hamilton, J.; Pfaff, T.J. Sustainability education: The what and how for mathematics. *PRIMUS* **2014**, *24*, 61–80. [CrossRef]
35. Barwell, R. Some thoughts on a mathematics education for environmental sustainability. In *The Philosophy of Mathematics Education Today*; ICME-13 Monographs; Ernest, P., Ed.; Springer: Cham, Switzerland, 2018; pp. 145–160.
36. Widiati, I.; Juandi, D. Philosophy of mathematics education for sustainable development. *J. Physics: Conf. Ser.* **2019**, *1157*, 022128. [CrossRef]
37. Li, H.C.; Tsai, T.L. Philosophy of education for sustainable development in mathematics education: Have we got one? *J. Math. Teach. Res. J.* **2020**, *12*, 136–140.
38. Li, H.C.; Tsai, T.L. Education for sustainable development in mathematics education: What could it look like? *Int. J. Math. Educ. Sci. Technol.* **2021**, *53*, 2532–2542. [CrossRef]
39. Moreno-Pino, F.M.; Jiménez-Fontana, R.; Domingo, J.M.C.; Goded, P.A. Training in mathematics education from a sustainability perspective: A case study of university teachers' views. *Educ. Sci.* **2022**, *12*, 199. [CrossRef]
40. Lage, M.J.; Platt, G.J.; Treglia, M. Inverting the classroom: A gateway to creating an inclusive learning environment. *J. Econ. Educ.* **2000**, *31*, 30–43. [CrossRef]

41. Connors, E. The Thayer method: Student active learning with positive results. *J. Math. Sci. Collab. Explor.* **2000**, *4*, 101–117. [CrossRef]
42. Shell, A.E. The Thayer method of instruction at the United States Military Academy: A modest history and a modern personal account. *Probl. Resour. Issues Math. Undergrad. Stud.* **2002**, *12*, 27–38. [CrossRef]
43. Overmyer, J. Flipped Classrooms 101. Principal (September/October), 2012; pp. 46–47. Available online: https://www.google.com.hk/url?sa=t&rct=j&q=&esrc=s&source=web&cd=&cad=rja&uact=8&ved=2ahUKEwji-MWNg7z6AhUAm1YBHaX7Bp4QFnoECAgQAQ&url=https%3A%2F%2Fwww.naesp.org%2Fsites%2Fdefault%2Ffiles%2FOvermyer_SO12.pdf&usg=AOvVaw2WBDogZHmzorH_a5e1SHMP (accessed on 30 July 2022).
44. King, A. From sage on the stage to guide on the side. *Coll. Teach.* **1993**, *41*, 30–35. [CrossRef]
45. Fox, R. Constructivism examined. *Oxf. Rev. Educ.* **2001**, *27*, 23–35. [CrossRef]
46. Arends, R.I. *Resource Handbook. Learning to Teach*, 4th ed.; McGraw-Hill: Boston, MA, USA, 1998.
47. Phillips, D.C. The good, the bad, and the ugly: The many faces of constructivism. *Educ. Res.* **1995**, *24*, 5–12. [CrossRef]
48. Good, T.L.; Brophy, J.E. *Looking in Classrooms*; HarperCollins College Publishers: New York, NY, USA, 1994.
49. Cooperstein, S.E.; Kocevar-Weidinger, E. Beyond active learning: A constructivist approach to learning. *Ref. Serv. Rev.* **2004**, *32*, 141–148. [CrossRef]
50. Elliott, S.N.; Kratochwill, T.R.; Littlefield, C.J.; Travers, J. *Educational Psychology: Effective Teaching, Effective Learning*, 3rd ed.; McGraw-Hill: Boston, MA, USA, 2000.
51. Vygotsky, L.S. *Mind in Society: The Development of Higher Psychological Processes*; Harvard University Press: Cambridge, MA, USA, 1978.
52. Strayer, J. The Effects of the Classroom Flip on the Learning Environment: A Comparison of Learning Activity in a Traditional Classroom and a Flip Classroom That Used an Intelligent Tutoring System. Ph.D. Thesis, The Ohio State University, Columbus, OH, USA, 2007 . Available online: https://etd.ohiolink.edu/apexprod/rws_etd/send_file/send?accession=osu1189523914&disposition=inline (accessed on 30 July 2022).
53. Bretzmann, J. *Flipping 2.0: Practical Strategies for Flipping Your Class*; Bretzmann Group: New Berlin, WI, USA, 2013.
54. Bergmann, J.; Sams, A. *Flip Your Classroom: Talk To Every Student In Every Class Every Day*; International Society for Technology in Education, Eugene, OR, USA, 2012.
55. Tucker, B. The flipped classroom. *Educ. Next* **2012**, *12*, 82–83. Available online: http://www.msuedtechsandbox.com/MAETELy2-2015/wp-content/uploads/2015/07/the_flipped_classroom_article_2.pdf (accessed on 30 July 2022).
56. Fulton, K. Upside down and inside out: Flip your classroom to improve student learning. *Learn. Lead. Technol.* **2012**, *39*, 12–17.
57. Herreid, C.F.; Schiller, N.A. Case studies and the flipped classroom. *J. Coll. Sci. Teach.* **2013**, *42*, 62–66. Available online: https://www.jstor.org/stable/pdf/43631584.pdf (accessed on 30 July 2022).
58. Bishop, J.; Verleger, M.A. The flipped classroom: A survey of the research. In Proceedings of the The 120th ASEE Annual Conference & Exposition Conference Proceedings, Atlanta, GA, USA, 23–26 June 2013; pp. 23–39.
59. Hantla, B.F. The Effects of Flipping the Classroom on Specific Aspects of Critical Thinking in a Christian College: A Quasi-Experimental, Mixed-Methods Study. Ph.D. Thesis, Southeastern Baptist Theological Seminary, Wake Forest, NC, USA, 2014. Available online: https://www.proquest.com/docview/1547356249?pq-origsite=gscholar&fromopenview=true (accessed on 30 July 2022).
60. Bergmann, J.; Sams, A. *Flipped Learning for Elementary Instruction*; International Society for Technology in Education: Eugene, OR, USA, 2016.
61. Alrouqi, F. Using Flipped Classrooms to Teach Mathematics to Elementary Students in Saudi Arabia. Ph.D. Thesis, University of South Florida, Tampa, FL, USA, 2019. Available online: https://www.proquest.com/docview/2321034820?pq-origsite=gscholar&fromopenview=true (accessed on 30 July 2022).
62. Panahi, M.; Jafarkhani, F.; Bozorg, Z.J.; Nikkho, L. Reviewing learning environments: Effect of flipped classroom on learning levels of mathematics in primary schools. In *Proceedings of the ICERI2019 12th International Conference of Education, Research and Innovation, Seville, Spain, 11–13 November 2019*; International Academy of Technology, Education and Development (IATED): Valencia, Spain, 2019; pp. 8561–8566.
63. Khalel, I.A.; Al-tmaran, O.S.; Hashmi, A.E. Employing the flipped classroom strategy in primary mathematics classes. *Int. J. Res. Educ. Sci.* **2021**, *4*, 497–529. [CrossRef]
64. Saunders, J.M. The Flipped Classroom: Its Effect on Student Academic Achievement and Critical Thinking Skills in High School Mathematics. Ph.D. Thesis, Liberty University, Lynchburg, VA, USA, 2014. Available online: https://www.proquest.com/docview/1639087375?pq-origsite=gscholar&fromopenview=true (accessed on 30 July 2022).
65. Bhagat, K.K.; Chang, C.-N.; Chang, C.-Y. The impact of the flipped classroom on mathematics concept learning in high school. *J. Educ. Technol. Soc.* **2016**, *19*, 134–142.
66. Makinde, S.O.; Yusuf, M.O. The flipped classroom: Its effects on students' performance and retention in secondary school mathematics classroom. *Int. J. Innov. Technol. Integr. Educ.* **2017**, *1*, 117–126.
67. Lo, C.K.; Hew, K.F. Using "first principles of instruction" to design secondary school mathematics flipped classroom: The findings of two exploratory studies. *J. Educ. Technol. Soc.* **2017**, *20*, 222–236.
68. Lo, C.K.; Hew, K.F. A critical review of flipped classroom challenges in K-12 education: Possible solutions and recommendations for future research. *Res. Pract. Technol. Enhanc. Learn.* **2017**, *12*, 1–22. [CrossRef]

69. Weinhandl, R.; Lavicza, Z.; Süss-Stepancik, E. Technology-enhanced flipped mathematics education in secondary schools: A synopsis of theory and practice. *K-12 STEM Educ.* **2018**, *4*, 377–389. [CrossRef]
70. Wei, X.; Cheng, I.; Chen, N.S.; Yang, X.; Liu, Y.; Dong, Y.; Zhai, X. Effect of the flipped classroom on the mathematics performance of middle school students. *Educ. Technol. Res. Dev.* **2020**, *68*, 1461–1484. [CrossRef]
71. Overmyer, G.R. The Flipped Classroom Model for College Algebra: Effects on Student Achievement. Ph.D. Thesis, Colorado State University, Fort Collins, CO, USA, 2014. Available online: https://www.proquest.com/docview/1615100148?pq-origsite=gscholar&fromopenview=true (accessed on 30 July 2022).
72. Van Sickle, J. Adventures in flipping college algebra. *PRIMUS* **2015**, *25*, 600–613. [CrossRef]
73. Jungić, V.; Kaur, H.; Mulholl, J.; Xin, C. On flipping the classroom in large first year calculus courses. *Int. J. Math. Educ. Sci. Technol.* **2015**, *46*, 508–520. [CrossRef]
74. Maciejewski, W. Flipping the calculus classroom: An evaluative study. *Teach. Math. Its Appl. Int. J. IMA* **2016**, *35*, 187–201. [CrossRef]
75. McGivney-Burelle, J.; Xue, F. Flipping calculus. *PRIMUS* **2013**, *23*, 477–486. [CrossRef]
76. Sahin, A.; Cavlazoglu, B.; Zeytuncu, Y.E. Flipping a college calculus course: A case study. *J. Educ. Technol. Soc.* **2015**, *18*, 142–152.
77. Sonnert, G.; Sadler, P.M.; Sadler, S.M.; Bressoud, D.M. The impact of instructor pedagogy on college calculus students' attitude toward mathematics. *Int. J. Math. Educ. Sci. Technol.* **2015**, *46*, 370–387. [CrossRef]
78. Karjanto, N.; Simon, L. English-medium instruction Calculus in Confucian-Heritage Culture: Flipping the class or overriding the culture? *Stud. Educ. Eval.* **2019**, *63*, 122–135. [CrossRef]
79. Ziegelmeier, L.B.; Topaz, C.M. Flipped calculus: A study of student performance and perceptions. *PRIMUS* **2015**, *25*, 847–860. [CrossRef]
80. Talbert, R. Inverting the linear algebra classroom. *PRIMUS* **2014**, *24*, 361–374. [CrossRef]
81. Murphy, J.; Chang, J.M.; Suaray, K. Student performance and attitudes in a collaborative and flipped linear algebra course. *Int. J. Math. Educ. Sci. Technol.* **2016**, *47*, 653–673. [CrossRef]
82. Novak, J.; Kensington-Miller, B.; Evans, T. Flip or flop? Students' perspectives of a flipped lecture in mathematics. *Int. J. Math. Educ. Sci. Technol.* **2017**, *48*, 647–658. [CrossRef]
83. Karjanto, N.; Lee, S.G. Flipped classroom in Introductory Linear Algebra by utilizing Computer Algebra System *SageMath* and a free electronic book. *arXiv* **2017**, arXiv:1705.00739.
84. Wilson, S.G. The flipped class: A method to address the challenges of an undergraduate statistics course. *Teach. Psychol.* **2013**, *40*, 193–199. [CrossRef]
85. Butt, A. Student views on the use of a flipped classroom approach: Evidence from Australia. *Bus. Educ. Accredit.* **2014**, *6*, 33–43.
86. Poffenberger, T.; Norton, D. Factors in the formation of attitudes toward mathematics. *J. Educ. Res.* **1959**, *52*, 171–176. [CrossRef]
87. Hodges, C.B.; Kim, C. Improving college students' attitudes toward mathematics. *TechTrends* **2013**, *57*, 59–66. [CrossRef]
88. Lo, C.K.; Hew, K.F.; Chen, G. Toward a set of design principles for mathematics flipped classrooms: A synthesis of research in mathematics education. *Educ. Res. Rev.* **2017**, *22*, 50–73. [CrossRef]
89. Karjanto, N. Attitude toward mathematics among the students at Nazarbayev University Foundation Year Programme. *Int. J. Math. Educ. Sci. Technol.* **2017**, *48*, 849–863. [CrossRef]
90. Turra, H.; Carrasco, V.; González, C.; Soval, V.; Yáñez, S. Flipped classroom experiences and their impact on engineering students' attitudes towards university-level mathematics. *High. Educ. Pedagog.* **2019**, *4*, 136–155. [CrossRef]
91. Buil-Fabrega, M.; Martínez, Casanovas, M.; Ruiz-Munzón, N.; Filho, W.L. Flipped classroom as an active learning methodology in sustainable development curricula. *Sustainability* **2019**, *11*, 4577. [CrossRef]
92. Rodríguez-Chueca, J.; Molina-García, A.; García-Ar, A.C.; Pérez, J.; Rodríguez, E. Understanding sustainability and the circular economy through flipped classroom and challenge-based learning: An innovative experience in engineering education in Spain. *Environ. Educ. Res.* **2020**, *26*, 238–252. [CrossRef]
93. Howell, R.A. Engaging students in education for sustainable development: The benefits of active learning, reflective practices and flipped classroom pedagogies. *J. Clean. Prod.* **2021**, *325*, 129318. [CrossRef]
94. Krantz, S.G. *How to Teach Mathematics*, 3rd ed.; American Mathematical Society: Providence, RI, USA, 2015.
95. Freeman, S.; Eddy, S.L.; McDonough, M.; Smith, M.K.; Okoroafor, N.; Jordt, H.; Wenderoth, M.P. Active learning increases student performance in science, engineering, and mathematics. *Proc. Natl. Acad. Sci. USA* **2014**, *111*, 8410–8415. [CrossRef]

sustainability

MDPI

Article

Technology-Based Pedagogy for Mathematics Education in South Africa: Sustainable Development of Mathematics Education Post COVID-19

Jayaluxmi Naidoo

Mathematics and Computer Science Education Cluster, School of Education, College of Humanities, University of KwaZulu-Natal, Durban 4041, South Africa; naidooj2@ukzn.ac.za

Abstract: The COVID-19 pandemic has transformed life globally. So too, teaching and learning in higher education have transformed to include technology-based pedagogy and online learning to curb the spread of the contagious Coronavirus. In the era of the 21st century, technology-based pedagogy is important for supporting teaching and learning. In mathematics higher education contexts, coupled with embracing abstract concepts in mathematics, there are different notions of what it means to include technology-based pedagogy during the COVID-19 pandemic. This qualitative study explored 38 postgraduate students' experiences and perceptions of using technology-based pedagogy during the COVID-19 pandemic. These participants were mathematics education students and practicing mathematics teachers at schools in South Africa. The theory of virtual communities of practice framed this study. Participants were invited to two interactive virtual workshops using various technology-based pedagogy during the COVID-19 pandemic. Subsequently, these participants were invited to participate in online interviews focusing on their experiences and perceptions of technology-based pedagogy for learning mathematics education. This study reveals the challenges and strengths of using technology-based pedagogy for learning mathematics education during COVID-19. These results are relevant when considering perceptions, experiences, and implications of technology-based pedagogy for sustainable mathematics education during and post-COVID-19.

Keywords: COVID-19; higher education; Hovercam; mathematics education; Moodle; online learning; technology-based pedagogy; virtual communities of practice; WhatsApp

Citation: Naidoo, J. Technology-Based Pedagogy for Mathematics Education in South Africa: Sustainable Development of Mathematics Education Post COVID-19. *Sustainability* **2022**, *14*, 10735. https://doi.org/10.3390/su141710735

Academic Editor: Linda Hagedorn

Received: 11 July 2022
Accepted: 22 August 2022
Published: 29 August 2022

1. Introduction

Currently, technology is increasingly being used to teach innovatively. In addition, technology is significantly used by society for different daily activities [1]. Additionally, in the 21st century, in the Fourth Industrial Revolution (4IR) era, there are several discussions on how current educational environments need to be transformed to support and incorporate technology-based pedagogy. Moreover, developments in technology-based pedagogy have provided lecturers with different technology-based tools and interactive online systems to change their education environment. Institutions on a global level are encouraging online and distance learning, and technology-based pedagogy is key in facilitating online and distance learning. In mathematics education contexts, coupled with embracing the effective understanding of abstract mathematics concepts, are issues of what it means to integrate technology-based pedagogy in mathematics educational environments in the context of COVID-19 and beyond.

In this study, technology-based pedagogy provided lecturers and students with different technology-based tools, devices, and online platforms to discuss content, clarify concepts and enhance knowledge to promote teaching and learning using the Internet (The Internet is a connected online system using computer networks which are made up of public, government, private, business and academic computer systems linked by wireless, electronic, and optical networking technologies). Additionally, within this study, the technology-based devices used were desktop computers, laptops, tablets, portable document cameras (Hovercam (The Hovercam is a portable document camera. The Hovercam is connected to the lecturer's computer. This allows the lecturer to teach in real time and display notes and other information on the computer as the lecturer teaches. The Hovercam is able to magnify resources, pictures and images so that they are easier and clearer for the student to see. Lectures can also be recorded using the Hovercam)), and cell phones. These technology-based devices supported technology-based pedagogy by using images, pictures, videos, text, and audio [2]. Technology-based tools in this study included PowerPoint presentations, websites, online resources, and software programmes. Thus, technology-based pedagogy validates online platforms and supports the integration of technology-based tools and devices [3]. Consequently, online platforms assist lecturers when using technology-based resources and tools to upload learning materials online. For example, the Zoom (Zoom is a software app programme which permits one to communicate virtually with peers, family, colleagues, lecturers, tutors, students and friends when communication through face-to-face means is not possible) and WhatsApp (WhatsApp is a free messenger application (app) programme. WhatsApp allows individuals to use the Internet to send and receive calls, images, pictures, messages and videos. The participating university used WhatsApp unofficially as an online platform for communication between students, tutors and lecturers. This was used prominently during the COVID-19 pandemic) online platforms are combined software solutions that support technology-based pedagogy.

This study sought to answer one fundamental research question: What are mathematics education students' experiences and perceptions of using technology-based pedagogy during the COVID-19 pandemic? This pandemic resulted in lockdowns worldwide, which has changed how we live our lives. In the 21st century and contemplating the conditions of the COVID-19 pandemic, Higher Education Institutions (HEIs) are using technology-based tools, devices, and online platforms to support technology-based pedagogy. Using technology-based tools, devices, and online platforms to teach and learn is an effective strategy for preventing the spread of the contagious COVID-19 virus [4]. So too, using technology-based pedagogy and online learning are seen as viable pedagogies for sustainable education post-COVID-19.

2. Literature Review

2.1. Teaching and Learning Mathematics Education

Teaching in the 21st century in the Fourth Industrial Revolution (4IR) incorporates progressive strategies in which technology is embedded within people and society [5]. Although we live in a technologically progressive world, many students still cannot use technology-based tools and devices competently [6]. In addition, some students do not have the essential technology-based tools and devices to succeed with technology-based pedagogy. South African students are from different socio-economic backgrounds and have diverse access to technology-based tools and devices. Challenges are complex because students need to constantly acquire new competencies to succeed in a technology-based world [7].

Thus, educational environments need to adopt different strategies to prepare students to succeed in this transforming educational landscape [8]. If technology-based pedagogy is going to be used, teacher knowledge and engagement is important to ensure that technology-based devices, platforms, and tools are used successfully to sustain education [9]. Professional teacher development is important for transforming pedagogy, evolving educational environments, using technology-based tools, and advancing learning outcomes [10]. Additionally, mathematics teacher education in the 4IR era entails that teachers are sufficiently developed to successfully use and integrate technology-based devices and tools in their educational environments [11]. However, due to the COVID-19 pandemic, HEIs were forced to abruptly transform and use technology-based pedagogy without preparing their lecturers or students sufficiently for this abrupt transformation. This is concerning for students who live in unequal social environments and do not have the essential technology-based tools, devices, and resources to join equally in technology-based educational environments [12].

2.2. Technology-Based Pedagogy for Mathematics Education

Technology-based tools and devices are prominent in educational environments, and technology-based pedagogy is progressively being used in educational environments in new ways [13,14]. While the use of graphical calculators, computers, and mathematics software programmes, for example GeoGebra, for mathematics teaching and learning has amplified [15], much research has focused on using technology for communication and presenting information in mathematics educational environments [16]. Within the era of COVID-19, research has been done on online courses, resources, and platforms [17–19]. Research has revealed some lecturers' and students' experiences regarding the opportunities and challenges of using technology for mathematics teaching and learning [18]. This research study aims to add to these discussions to promote sustainable mathematics education post-COVID-19.

In addition, this study focuses on online platforms and resources that may be used with technology-based devices, for example, desktop computers, laptops, tablets, or mobile phones. These technology-based devices function with images, text, videos, pictures, and audio files to scaffold teaching and learning [2]. Research [20] maintains that several online websites and online educational applications are available that can scaffold students' mathematics learning and achievement. In the era of the 4IR during the COVID-19 pandemic, HEIs were compelled to use technology-based tools and devices for distance, online, and technology-based pedagogy. Thus, technology-based pedagogy and online learning are a premise for promoting sustainable education post-COVID-19.

Research [21] has revealed that the use of technology-based tools and devices for mathematics teaching has enhanced students' performance. Similarly, technology-based pedagogy for mathematics education may support successful and struggling students [22]. However, within the context of distance and online teaching and learning, for the lecturer to select a suitable technology-based tool to traverse the multiple available technology-based tools is overwhelming. The notion is to choose technology-based tools that are most effective for successful technology-based teaching that scaffolds students' achievement and learning. It is, therefore, the responsibility of the teacher/lecturer to advance technology-based educational environments [23].

Research shows that technology-based tools and devices are changing the role of the lecturer. Thus, we need to know more about how best to assist lecturers during this change [13], more so post-COVID-19.

2.3. Virtual Communities of Practice Theory

The theory of Communities of Practice is a social theory of learning, with the community as the central unit of analysis [24]. Communities of Practice (CoP) focus on the experiences of individuals in the CoP [25]. This theory maintains that the CoP is fashioned by individuals who share the learning process with other community members. Virtual Communities of Practice involve a technology component and online platforms, making virtual communities of practice different from the traditional communities of practice [26]. Through collaboration using technology-based tools, devices, online platforms, online resources, and the Internet, the traditional CoPs advance to Virtual Communities of Practice (VCoPs). A Virtual Community of Practice may be described as a community of professionals who share ideas, and meaning-making, thus adding to their knowledge base and improving their professional practice [27].

In this study, as the VCoP under focus worked together virtually, they shared knowledge and developed a community that supported learning and development [28]. The connection between VCoP theory and this study is clarified: VCoP has a shared interest; in this study, the shared interest is the participants' perceptions of technology-based pedagogy for mathematics education during the COVID-19 pandemic. Within VCoP, community members collaborate in activities and critically reflect and discuss the content under discussion [29]. In this study, members of the VCoP used technology-based tools, devices, resources, and the Internet to collaborate and learn together. They used the Zoom online platform, WhatsApp, and the Hovercam to share ideas, learn together, and clarify misunderstandings. Members of a VCoP are professionals; in this study, the members of the VCoP were postgraduate students who were also mathematics teachers. Thus, this study focused on using technology-based pedagogy for mathematics education during the COVID-19 pandemic and was suitably framed by the VCoP theory.

3. Materials and Methods

3.1. General Background

This qualitative, interpretive study was located at one South African Higher Education institution during the COVID-19 pandemic. Qualitative methods were selected for this study since this was the most appropriate method for providing exploratory data focusing on the perceptions and experiences of the participants [30]. All participants provided their consent to participate in the study. The university's ethics committee approved ethical clearance and gatekeeper access. The study included two interactive online workshops and individual online interviews, which were semi-structured with postgraduate mathematics education students. These participants were also mathematics teachers in South African schools.

3.2. Participants

For this study, thirty-eight postgraduate students were invited to participate. Participants were given an informed consent form. This form explained and described the research design, methods, and processes. Positive responses were received from thirty-three students (15 male and 18 female). Six participants who responded positively were randomly selected to participate in the pilot study.

3.3. Pilot Study

To improve the study's dependability, trustworthiness, conformability, and transferability, a pilot study was conducted. During the pilot study, some participants lost connectivity during the pilot workshop, and some were unsure about the questions asked during the interview. Experiencing these issues during the pilot study assisted with the planning for the main study. The main study was held during off-peak periods to ensure stable and fast internet connectivity. The interview questions were revised to ensure clarity. These revisions improved the credibility of the research design, methods, and instruments.

3.4. Main Study

For the main study, twenty-seven postgraduate mathematic students participated. Data was gathered from two online workshops and online interviews. While 27 participants were present during the workshops, only 15 participants were available to be interviewed due to family, work, or personal commitments. Thus, for the main study, seven male and eight female participants were interviewed. Pseudonyms were assigned to each participant. This ensured the participants' confidentiality. The pseudonyms used for this study are shown in Table 1.

3.5. Interactive Online Workshops

The two interactive online workshops focused on teaching mathematics education students using technology-based pedagogy in the context of the COVID-19 pandemic. The researcher (lecturer) was the facilitator for the two workshops. Each participant was given additional online notes, resources, website addresses, and PowerPoint presentations. The online workshops were conducted via the Zoom platform during the second semester (July–December) of the 2020 academic year. The first online workshop focused on controversial issues in mathematics education and problem-solving. The second online workshop focused on misconceptions in mathematics education and problem-solving. All participants were invited to online interviews at the end of the second workshop. The online interviews were focused on exploring mathematics education students' perceptions and experiences of using technology-based pedagogy during the COVID-19 pandemic.

Table 1. Interview Participants Gender and Pseudonyms.

Participant	Pseudonym	Gender	Participant	Pseudonym	Gender
1	Anand	Male	9	Ned	Male
2	Bongani	Male	10	Pretty	Female
3	Cameron	Male	11	Rakesh	Male
4	Glenda	Female	12	Sam	Male
5	Jane	Female	13	Susan	Female
6	Linda	Female	14	Thuli	Female
7	Lungi	Female	15	Tom	Male
8	Nancy	Female			

3.6. Online Semi-Structured Interviews

The interview aimed to examine mathematics education students' experiences and perceptions of technology-based pedagogy during the pandemic. Each interview lasted about 30–40 min. The interviews were conducted using online platforms, for example, Zoom and WhatsApp. The interviews were conducted at a time that each participant chose. To place the participants at ease, every online interview began with a few common questions. It then advanced with precise items focusing on the participants' perceptions and experiences of using technology-based pedagogy during the COVID-19 pandemic. The online interviews concentrated on the critical question that follows:

- What are mathematics education students' perceptions and experiences of using technology-based pedagogy during the COVID-19 pandemic?

The online interviews included the following sub-questions:

- What are your perceptions and experiences of using technology-based tools, devices, and pedagogy during the COVID-19 pandemic?
- What are your perceptions and experiences of using the WhatsApp platform during the COVID-19 pandemic?
- What are your perceptions and experiences of using the online Zoom platform during the COVID-19 pandemic?
- What are your perceptions and experiences of using the Learning Management System (Moodle) during the COVID-19 pandemic?
- What are your perceptions and experiences of the virtual communities of practice created during the COVID-19 pandemic?
- What are your perceptions and experiences of studying remotely/online during the COVID-19 pandemic?

Each interview was recorded and transcribed. Thereafter, each participant was emailed a copy of the transcript. Participants were asked to peruse the transcripts to ensure that their responses to the interview questions were captured correctly.

3.7. Data Analysis

Qualitative thematic and content analysis were used to analyse the content of the interviews. The data were analysed manually by coding and categorising the content of the interviews into themes. Data analysis was based on the theoretical framing for this study, i.e., Virtual Communities of Practice (VCoP). The coding and categorising of themes were founded on this framework. The codes and themes were linked to the shared interest, critical reflection, and collaborative technology-based activities of the mathematics professionals within the VCoP under study.

Data analysis started with open coding to examine and organise the data into codes; secondly, codes were classified and clustered into themes. Then, each participant's responses were compared to see if there were any similarities or differences. Moreover, to ensure the accuracy of the results, all transcripts were sent to the relevant participants to ensure that their responses were accurately captured and transcribed. Participants could also clarify their responses if responses were not clear.

4. Results

The results reveal that participants had positive and negative perceptions and experiences concerning using technology-based pedagogy during the COVID-19 pandemic. These perceptions and experiences were categorised into two sub-themes strengths and challenges of using technology-based pedagogy during the COVID-19 pandemic. The results for these sub-themes are presented as follows.

4.1. Strengths of Using Technology-Based Pedagogy

The participants explained that using technology-based pedagogy within the context of the pandemic revealed strengths. The participants commented on the strengths of WhatsApp as an online platform, useful technology-based tools, such as the Hovercam for online teaching and learning and the use of different online platforms that fostered a supportive online community of practice. Figure 1 represents the number of participants who indicated that there were strengths to using technology-based pedagogy.

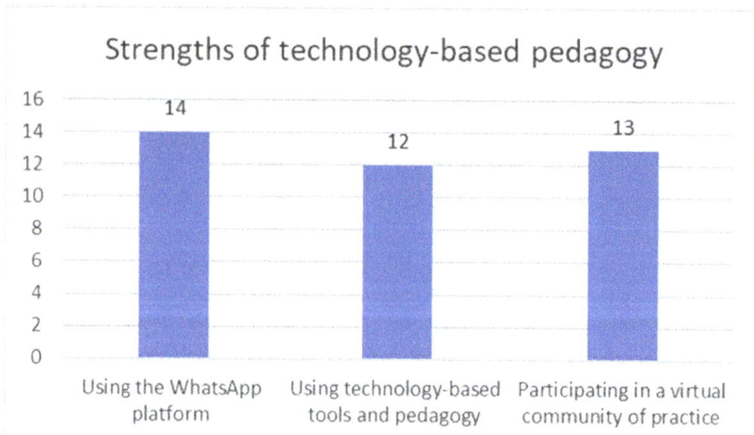

Figure 1. Number of Participants Who Indicated the Strengths of Using Technology-Based Pedagogy.

The participants' views that best convey these strengths are presented in Table 2.

From the interview transcripts, it was evident that the participants maintained that when using the different online platforms, technology-based tools and supportive online community of practice, there were strengths to using technology-based pedagogy during the COVID-19 pandemic. The results reveal participants' positive perceptions and experiences about using technology-based pedagogy.

4.2. Challenges of Using Technology-Based Pedagogy

The participants also explained that when using technology-based pedagogy, they experienced particular challenges. Participants expressed their unease with specific online platforms, the challenges of studying online during the COVID-19 pandemic, and the challenges they experienced with specific Learning Management Systems (LMS). Figure 2 represents the number of participants who indicated that there were challenges when using technology-based pedagogy.

Table 2. Transcript Excerpts of Participants Focusing on the Strengths of Using Technology-Based Pedagogy.

Question	Feedback
What are your perceptions and experiences of using the WhatsApp platform during the COVID-19 pandemic?	Anand: " ... WhatsApp was convenient and easy to use ... " Cameron: " ... we used WhatsApp and created a group ... we could share ideas and resources on WhatsApp ... the whole class formed our own online community WhatsApp group ... " Glenda: " ... I first try the work on my own, and then I ask for help on WhatsApp if I need it or check if my attempt is correct ... I take a picture of my work and send it to the group ... it is easy and quick ... " Lungi: " ... I am used to using WhatsApp ... it was easy to respond to questions ... with WhatsApp, we get responses from others quickly ... this help supported me ... " Pretty: " ... WhatsApp was quick, and I could send messages easily about my work ... sometimes if I had family problems, I could also ask for help on the group ... " Rakesh: " ... WhatsApp was easier to use than the university Learn platform ... I could get the lecture links for the Zoom lectures from my WhatsApp group ... we could share the assignment questions on the group ... I did not need to learn how to use WhatsApp since I normally use it with my friends and family ... " Sam: " ... we could share our teaching problems with each other ... we could share teaching resources and examples for different math sections ... the WhatsApp group was a very supportive group ... "
What are your perceptions and experiences of using technology-based tools, devices, and pedagogy during the COVID-19 pandemic?	Anand: " ... it was the first time we saw the Hovercam ... it was great to do the math problems step by step online ... " Bongani: " ... Hovercam made the math problem solving easy ... it was like being in the class with the lecturer ... even though we used videos and PowerPoints before so they were also interesting especially since we were working online from home ... " Jane: " ... the advanced technology tools we used in the lectures were great ... learnt a lot about even using my cell phone to teach and learn ... the lecturer showed us how to use the Hovercam, how to use teaching videos and how to make sure our PowerPoint presentations to benefit our learners ... it was like being in the class and following all the steps ... " Linda: " ... I never used the Hovercam before ... it was good to learn about the different devices like the sharing button on Zoom and the question banks on different websites ... the lecturer even showed us how to use Moodle to set up tests and have the computer mark them for us ... the Hovercam made math problem solving easy ... we could easily follow the problem-solving process step by step ... " Nancy: "I found this exciting ... these new gadgets we were learning about ... we could stop the lecturer at any time if we need more explanations ... the Hovercam and online videos even allowed for us to go back and forth ... " Sam: " ... I never knew about this device, the Hovercam ... it looks like something that will benefit my learners when I teach online ... something good to know for future teaching ... the explanations and step by step process was useful ... "

Table 2. *Cont.*

Question	Feedback
What are your perceptions and experiences of the virtual communities of practice created during the COVID-19 pandemic?	Bongani: " . . . the online class group helped me a lot . . . even when my family had COVID, I could talk to the class online . . . they supported me . . . they even told me about the COVID testing site near me . . . the drive-through place..." Ned: " . . . we worked well together as an online group . . . we supported each other with the math problems online . . . the lecturer was also part of this online group and assisted us immediately if we needed help or any further explanations or examples . . . we could also check up on each other at any time . . . " Rakesh: " . . . with the online group, we could use WhatsApp or Zoom since we also had an online discussion group on Zoom and Moodle . . . if someone was behind with their work, we could support them and encourage them . . . some people had a problem working from home, so we provided constant online support . . . . . . the online WhatsApp group also helped me understand and complete some math tasks . . . ". Susan: " . . . the online group supported me throughout the semester . . . it was good to know that the class was with me every step of the way . . . " Thuli: " . . . I had good help from the online group. The help was not only about the academic work but also with personal family issues . . . I knew I could count on the group at any time . . . " Tom: " . . . some of us had the same challenges with the tasks . . . we could work together with the problem-solving steps online . . . we helped each other . . . "

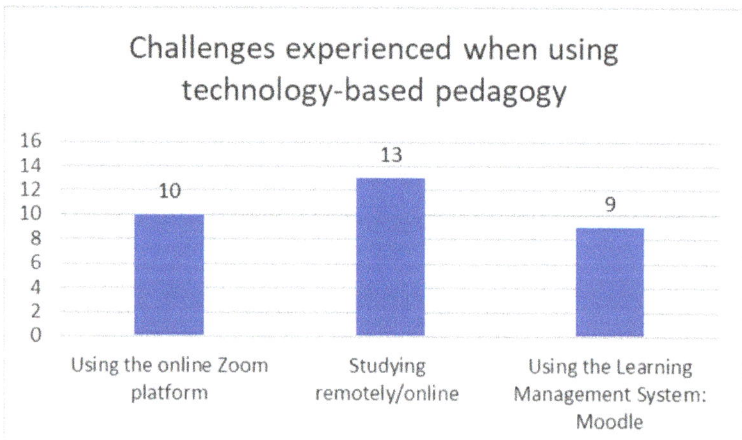

Figure 2. Number of Participants Who Indicated They Experienced Challenges When Using Technology-Based Pedagogy.

The participants' views that best convey these challenges are presented in Table 3.

From the interview transcripts, it was evident that when using the different online platforms, the use of the university's official Learning Management System (Moodle) and while studying remotely/online during the COVID-19 pandemic, participants experienced many challenges. The results of this study are discussed in more detail in the following section.

Table 3. Transcript Excerpts of Participants Focusing on the Challenges Experienced When Using Technology-Based Pedagogy.

Question	Feedback
What are your perceptions and experiences of using the online Zoom platform during the COVID-19 pandemic?	Bongani: " … didn't know anything about Zoom … first time experience … eish … I had to ask for help … had to learn about this … "Cameron: " … the workshops took too much data … Zoom is expensive … the resources were hard and took too long to download … very expensive too … takes a lot of data … " Jane: " … was challenging at first … I had to use trial and error with Zoom … had problems with my video … I think the university should teach us to use Zoom first … " Lungi: " … very expensive to buy a computer … didn't think I would need this when I registered … I struggled also with the expense for data and Internet … " Ned: " … Zoom is new to me … I needed to learn everything about how to use Zoom … I think it was unfair for the university to ask us to use Zoom without providing any training to us students … just when I learnt how to use Zoom … the Zoom buttons for the system were changed … I had to learn this again … " Thuli: " … online lectures … expensive … lots of money … I didn't know that I would be studying online when I registered in February … this was frustrating for my family and me … " Tom: " … the workshops were challenging to follow … my Internet was not stable … I had issues with connectivity … "
What are your perceptions and experiences of studying remotely/online during the COVID-19 pandemic?	Anand: " … working from home is distracting … need … space to work … attend...Zoom lectures … my car … " Lungi: " … my small children are with me at home … it is difficult to attend Zoom lectures with them around me … I found it a problem to focus on my studies while at home … " Nancy: " … my family is big … live together … because of COVID-19 everyone is at home … my house is not big … I often attend lectures in my kitchen … it is embarrassing … I cannot put my video on because of where I am sitting for lectures … my parents don't understand that I need to attend lectures … I also have other work at home … this was very stressful for me … I couldn't balance my home life while studying online … my husband also fought about this … I could not cope … but lucky I had my friends from class who understood what I was going through, and they supported me … " Pretty: " … as a mother and wife, I have lots of other responsibilities … because of COVID-19 … housework and cook … it was difficult to concentrate on my studies … I prefer studying campus … " Sam: … at my place, there is no privacy … I can't participate in my lectures because everyone at home is listening or talking during the lecture … I found it hard to ask questions during the lectures. I used the WhatsApp group to ask my questions … the online group supported me during the semester … " Tom: " … I am angry with the university … I paid for face to face lectures because I can't study on my own online … if I wanted to study online, I would have paid for online classes … I normally work in the university library because the network and connectivity where I live are not good … I don't even have a laptop … I thought I would use the university computers at the campus for my assignments … it is expensive working online from home … I do not have the money for this … "

Table 3. *Cont.*

Question	Feedback
What are your perceptions and experiences of using the Learning Management System (Moodle) during the COVID-19 pandemic?	Cameron: " ... Moodle was a big problem for me ... I came back to studying after a few years ... I forgot my password ... it took the university a long time to help me with my password problems ... I started attending lectures late in the semester ... I had a lot of work to catch up on ... if it were not for the class's online support, I would have been in big trouble completing my work ... it was very stressful for me ... " Glenda: " ... Moodle gave me trouble from the start ... I always took a long time to get onto Moodle ... the university also had something called Global Protect, which I needed to be on before logging onto Moodle ... this was difficult since the connectivity in my area was slow and unreliable ... the university must use a more user-friendly system for us students ... " Linda: " ... I used Moodle a long time ago for my first degree ... I know how to use it, but the university made changes, and we had to migrate to a new platform ... this was difficult, and then we could not see our modules ... I had to catch up because I started attending my lectures late ... I even did not know when we had the lectures ... lucky for me, we had the online group, which helped me a lot..." Lungi: " ... sometimes because most of the class were having problems with Moodle, if I used the discussion group, it took a long time for others to respond ... I then used WhatsApp to ask for help ... Moodle is only good for submitting our assessments ... it is not good for online discussions ... ." Thuli: " ... Moodle was ok ... I knew how to use it, but the class hardly looked at the questions on the discussion group, so I could not get help there ... it was a challenge for me to use Moodle to discuss the math questions that I had problems with ... " Rakesh: " ... hard ... took me time ... found I was replying to old discussion questions ... I felt like the Moodle discussion group wasted my time and effort ... the university must relook at this system ... it does not work for all of us ... "

5. Discussion

5.1. Strengths of Using Technology-Based Pedagogy during the COVID-19 Pandemic

Figure 1 represents the number of participants who explained the strengths of using technology-based pedagogy during the COVID-19 pandemic. From Table 2, it was evident that the participants' responses indicated that the WhatsApp platform, when used as a technology-based pedagogy, was efficient for obtaining instantaneous responses and provided a means of quick communication between peers and the lecturer. This view is supported by research [31,32], which indicates that students value WhatsApp for quick collaboration and communicating thoughts and ideas. In this study, WhatsApp communication included both social and academic messages. Based on the excerpts as reflected in Table 2, creating a supportive and collaborative online community was beneficial for the participants.

Moreover, the increase in the use of cell phones allows lecturers to engage and interact with their students on a personal level [32]. This allows students to interact with their lectures from any location, thus providing them with dynamic learning [33]. The participants in this study appreciated the online mathematics support they received when the mathematics concepts and topics were discussed. Additionally, the participants mentioned the value of the social support they received online from their community of peers and the lecturer. During the COVID-19 pandemic, an online community was initiated to provide a social support structure for the participants. The participants' perceptions and experiences of this support structure are presented in Table 2. This view is supported by research [32,34], which indicates that online social support is helpful and promotes the psychosocial quality of life.

Furthermore, the participants valued using the different technology-based tools and devices during the COVID-19 pandemic. The participants mentioned the Hovercam, an innovative technology-based device used during online teaching and learning. Research indicates that using innovative and dynamic digital devices is advantageous to online teaching and learning groups [33]. This study shows that online technology-based pedagogy encouraged and founded cooperative interactions within the VCoP that was created [32,35]. Based on the analysis of the interview data, it was evident that the participants received encouragement and motivation from members of the VCoP. Through shared interests and collaboration [32,36], the participants worked together virtually. The participants incorporated the notions of a VCoP by participating online, sharing ideas and resources, and supporting other VCoP members virtually.

Additionally, during the interviews, the participants indicated that the VCoP promoted effective collaboration and engagement [32,37]. In addition, the participants indicated in their interviews that when using the different online platforms, for example, WhatsApp and Zoom, the VCoP sustained and encouraged participants to acquire new knowledge. When the participants had challenges, they discussed these virtually with other community members. These interactions were apparent from the interview data. Thus, the virtual community of practice created for these participants provided an additional layer of academic, job-related, social, and mental support for the members of the VCoP.

5.2. Challenges Experienced When Using Technology-Based Pedagogy during the COVID-19 Pandemic

Figure 2 represents the number of participants who indicated that they experienced challenges when using technology-based pedagogy during the COVID-19 pandemic. Based on the interview excerpts in Table 3, it was evident that the participants were first exposed to Zoom during COVID-19. As a result, the participants indicated they required time to prepare and practice using Zoom. Consequently, the participants initiated collaborative online engagement to navigate this platform successfully. Additionally, certain Zoom functions were reviewed during the workshops, requiring the participants to learn new skills during the second workshop. Thus, valuable teaching time was used during the second workshop due to reskilling participants and addressing procedural queries. To use online platforms successfully, participants required access to tools, devices, and data. Some participants experienced some challenges, as shown in Figure 2. Participants were affected by unstable connectivity, issues with electricity shortages, and limited access to the Internet. These challenges limited participants from participating equally with online technology-based pedagogy. Similarly, research [32,38] suggests that students must have access to stable Internet connections to participate equally in online teaching and learning. However, many developing countries have facilities and infrastructure that are inadequate, which creates hurdles around access to the Internet and issues with unstable connectivity [32,39].

Participants experienced many unforeseen expenses in the 2020 academic year due to the purchase of laptops, computers, and data. Under normal circumstances, the research site offers the contact mode of teaching and learning, but like many HEI worldwide, this university initiated online teaching and learning due to COVID-19. Hence both students and lecturers needed to adapt quickly to this new way of teaching and learning.

Due to COVID-19, many people were forced to work from home. As a result, this created challenges for the participants since they constantly attended lectures with their families around them. As a result of studying at home, family duties were undertaken before the participants could start with academic work. It is evident from Table 3 that the participants had challenges when using Moodle. Previously, the participants did use the university's Learning Management System (Moodle). However, they did experience challenges with quick feedback and responses from their online community [32]. This online platform was also awkward and time-consuming to use. Based on the analysis of the online interviews, it was evident that the participants' challenges with access to data and devices affected their equal participation in online teaching and learning. The participants indicated that they needed access to technology-based tools, devices, and the Internet to engage equally and actively with online teaching and learning [38]. Similarly, research [32,40] maintains that to promote and develop notions of independent learning, it is important for students to access the necessary and essential resources for teaching and learning.

6. Conclusions

This study aimed to explore postgraduate students who were also mathematics teachers, perceptions, and experiences of using technology-based pedagogy during the COVID-19 pandemic. The results reveal that the participants had both positive and negative perceptions and experiences when using technology-based pedagogy. From the results and the discussion section, it was apparent that if the Virtual Communities of Practice's (VCoP) philosophies are incorporated within online teaching and learning environments, technology-based pedagogy may increase the advantages of meaningful online teaching and learning. The higher education environment for 21st century pedagogy has changed, and higher education environments enriched with technology are encouraged. Notions founded on the VCoP framework endorse interactive, supportive, and shared strategies. Along similar lines, in this study, traditional pedagogy was transformed to include technology-based pedagogy, and a collaborative and engaging technology-enriched VCoP was created. The VCoP that was created supported online teaching and learning. For this study, the participating students incorporated the notions of the VCoP. They interacted online, discussed, shared, and acquired new knowledge and resources.

This study concludes with recommendations for lecturers wanting to use technology-based tools, devices, and pedagogy post-COVID-19. Firstly, the difficulties of using technology-based pedagogy during COVID-19 are essential for lecturers to be aware of. To avoid these challenges, the lecturer must check with their students if they have the essential technology-based tools, devices, and Internet data. To further improve on the challenges mentioned in this study, lecturers should test the online resources to see if they are indeed accessible, economical to use, and user-friendly. One way to establish this is for lecturers to practice using online resources to understand better the amount of data used and the time required to download this information. This will assist the lecturer in informing the students about the data and time requirements for downloading these resources.

Secondly, when planning online lectures and workshops, lecturers must remember that their students have personal responsibilities. Thus, online lectures and workshops should be held at suitable times, and these lectures should be recorded and uploaded so that students can view these when they are available. Lecturers should make time available to allow for flexible online consultation if students require additional support.

Thirdly, the advantages of the VCoP created to support students when studying remotely/online during the COVID-19 pandemic were important. The participants in this study engaged with members of the VcoP and expressed the value of their interactive and supportive online discussions. Through this support and interactions, the participants were helped during online teaching and learning. Lecturers ought to recognise that students are sociable, and they need to communicate and socialise with other community members and provide and receive feedback and support. As was evident from the results of this study, the members of VCoP discussed academic work, family issues, and issues related to COVID-19, and the community members also supported each other with the job-related issues. It is important to understand that students need to communicate, collaborate, discuss, and interact with others during the pandemic to support them socially and emotionally. This supportive VCoP, as discussed in this study, provided an important resource that may also be implemented when using technology-based pedagogy post-COVID-19.

Finally, the results of this study reveal that for the successful use of technology-based pedagogy post-COVID-19, professional development workshops for lecturers are needed to support lecturers in effectively using technology-based pedagogy to reinforce and enhance their pedagogy in online environments. These professional development sessions would be advantageous to lecturers on a global level as we focus on sustaining education post-COVID-19. Due to the global importance of this study, more research on a larger scale involving other national and international higher education institutions could provide different perceptions and experiences of using technology-based pedagogy. These further research studies may also reveal additional perceptions and experiences of students in general and perceptions and experiences of mathematics students specifically on using technology-based pedagogy. The results, discussions, and implications for this study will enhance knowledge in the field. This enhanced knowledge will be advantageous to the global field as we integrate technology-based pedagogy to promote sustainable education post-COVID-19.

Funding: This research was partially funded by the National Research Foundation South Africa Grant Number: TTK170408226284, UID: 113952.

Institutional Review Board Statement: Not applicable.

Informed Consent Statement: Informed consent was obtained from all participants involved in the study.

Data Availability Statement: Data is available on request from the author: naidooj2@ukzn.ac.za.

Conflicts of Interest: The author declares that there is no conflict of interest.

References

1. Qurat-Ul, A.; Farah, S.; Muhammad, A.; Muhammad, A.I.; Muhammad, A.I.; Muhammad, M.Y. A review of technological tools in teaching and learning computer science. *Eurasia J. Math. Sci. Technol. Educ.* **2019**, *15*, 1–17. [CrossRef]
2. Peachey, N. *Digital Tools for Teachers*; PeacheyPublications: Kent, UK, 2017; p. 22.
3. Buzzard, C.; Crittenden, V.L.; Crittenden, W.F.; McCarty, P. The use of digital technologies in the classroom: A teaching and learning perspective. *J. Mark. Educ.* **2011**, *33*, 131–139. [CrossRef]
4. Murgatrotd, S. COVID-19 and Online Learning. 2020; *preprint*; pp. 1–3.
5. Schwab, K. *The Fourth Industrial Revolution: What It Means, How to Respond*; World Economic Forum: Geneva, Switzerland, 2016; pp. 1–8.
6. Bennett, S.; Maton, K.; Kervin, L. The 'digital natives' debate: A critical review of the evidence. *Br. J. Educ. Technol.* **2008**, *39*, 775–786. [CrossRef]
7. Ng'ambi, D.; Brown, C.; Bozalek, V.; Gachago, D.; Wood, D. Technology enhanced teaching and learning in South African higher education—A rearview of a 20 year journey. *Br. J. Educ. Technol.* **2016**, *47*, 843–858. [CrossRef]
8. Butler-Adam, J. The Fourth Industrial Revolution and education. *S. Afr. J. Sci.* **2018**, *114*, 1. [CrossRef]
9. Qing, L. Would we teach without technology? A professor's experience of teaching mathematics education incorporating the Internet. *Educ. Res.* **2003**, *45*, 61–77.
10. Darling-Hammond, L. Teacher education around the world: What can we learn from international practice? *Eur. J. Teach. Educ.* **2017**, *40*, 291–309. [CrossRef]
11. Naidoo, J.; Singh-Pillay, A. Teachers' perceptions of using the blended learning approach for stem-related subjects within the Fourth Industrial Revolution. *J. Balt. Sci. Educ.* **2020**, *19*, 583–593. [CrossRef]
12. Zilka, G.C.; Finkelstein, I.; Cohen, R.; Rahimi, I.D. Implications of the digital divide for the learning process during the COVID-19 crisis. *Rev. Eur. Stud.* **2021**, *13*, 57–71. [CrossRef]
13. Grand-Clement, S.; Devaux, A.; Belanger, J.; Manville, C. *Digital Learning: Education and Skills in the Digital Age*; RAND Corporation and Corsham Institute: Cambridge, UK, 2017; pp. 1–23.
14. Benning, I.; Linsell, C.; Ingram, N. Using technology in mathematics: Professional development for teachers. In *Making Waves, Opening Spaces*; Hunter, J., Perger, P., Darragh, L., Eds.; MERGA: Auckland, New Zealand, 2018; pp. 146–153. Available online: https://files.eric.ed.gov/fulltext/ED592505.pdf (accessed on 10 July 2022).
15. Cullen, C.J.; Hertel, J.T.; Nickels, M. The roles of technology in mathematics education. *Educ. Forum* **2020**, *84*, 166–178. [CrossRef]
16. Chinnappan, M.; Thomas, M.O.J. Technology in mathematics learning and teaching. *Math. Educ. Res. J.* **2012**, *12*, 173–176. [CrossRef]
17. Sacristán, A.I. Digital technologies in mathematics classrooms: Barriers, lessons and focus on teachers. In Proceedings of the 39th Annual Meeting of the North American Chapter of the International Group for the Psychology of Mathematics Education, Indianapolis, IN, USA, 5–8 October 2017; pp. 90–99. Available online: https://files.eric.ed.gov/fulltext/ED581397.pdf (accessed on 10 July 2022).

18. Zulu, J.; Nalube, P.P.; Changwe, R.; Mbewe, S. The challenges and opportunities of using Zoom App in the teaching and learning of mathematics in Higher Education Institutions (HEIS) during COVID-19 pandemic: Lecturers' and students' perspective. *Int. J. Res. Innov. Appl. Sci.* **2021**, *6*, 92–101. Available online: https://www.researchgate.net/publication/354193246 (accessed on 10 July 2022).
19. Alabdulaziz, M.S. COVID-19 and the use of digital technology in mathematics education. *Educ. Inf. Technol.* **2021**, *26*, 7609–7633. Available online: https://link.springer.com/content/pdf/10.1007/s10639-021-10602-3.pdf (accessed on 10 July 2022). [CrossRef] [PubMed]
20. Pope, S.; Mayorga, P. Enriching Mathematics in the Primary Curriculum. In *Exploring the Primary Curriculum*; Sage: London, UK, 2019.
21. Cheung, A.C.K.; Slavin, R.E. The effectiveness of educational technology applications for enhancing mathematics achievement in K-12 classrooms: A meta-analysis. *Educ. Res. Rev.* **2013**, *9*, 88–113. [CrossRef]
22. Bruce, C.D. Technology in the mathematics classroom: Harnessing the learning potential of interactive whiteboards. What works? *Res. Pract.* **2012**, *38*, 1–4.
23. Montrieux, H.; Vanderlinde, R.; Schellens, T.; De Marez, L. Teaching and learning with mobile technology: A qualitative explorative study about the introduction of tablet devices in secondary education. *PLoS ONE* **2015**, *10*, e0144008. [CrossRef]
24. Graven, M.; Lerman, S. Book Review: Wenger, E. Communities of practice: Learning, meaning and identity. *J. Math. Teach. Educ.* **2003**, *6*, 185–194. [CrossRef]
25. Wenger, E. Communities of Practice a Brief Introduction. 2009, pp. 1–5. Available online: https://neillthew.typepad.com/files/communities-of-practice-1.pdf (accessed on 10 July 2022).
26. Dubé, L.; Bourhis, A.; Jacob, R. Towards a typology of virtual communities of practice. *Interdiscip. J. Inf. Knowl. Manag.* **2006**, *1*, 69–93. [CrossRef]
27. Al-ghamdi, H.A.K.; Al-ghamdi, A.A.K. The Role of Virtual Communities of Practice in Knowledge Management Using Web 2.0. *Procedia Comput. Sci.* **2015**, *65*, 406–411. [CrossRef]
28. Schieffer, L. The benefits and barriers of virtual collaboration among online adjuncts. *J. Instr. Res.* **2016**, *5*, 109–124. [CrossRef]
29. Osterholt, D.A.; Barratt, K. Ideas for practice: A collaborative look to the classroom. *J. Dev. Educ.* **2010**, *34*, 26–35.
30. Mohajan, H.K. Qualitative research methodology in social sciences and related subjects. *J. Econ. Dev. Environ. People* **2018**, *7*, 23–48. Available online: https://mpra.ub.uni-muenchen.de/85654/1/MPRA_paper_85654.pdf (accessed on 10 July 2022). [CrossRef]
31. Barhoumi, C. The effectiveness of WhatsApp mobile learning activities guided by activity theory on students' knowledge management. *Contemp. Educ. Technol.* **2015**, *6*, 221–238. [CrossRef]
32. Naidoo, J. Postgraduate mathematics education students' experiences of using digital platforms for learning within the COVID-19 pandemic era. *Pythagoras* **2020**, *41*, a568. [CrossRef]
33. Cook, C.W.; Sonnenberg, C. Technology and online education: Models for change. *Contemp. Issues Educ. Res.* **2014**, *7*, 171–188. [CrossRef]
34. Dietrich, C. Online social support: An effective means of mediating stress. *Inq. J. /Stud. Pulse* **2010**, *2*, 1–5.
35. Mlotshwa, N.; Chigona, A. Using Moodle to enhance Mathematics learning in Grade 10 classrooms in South Africa. In Proceedings of the E-Learn 2018 Conference, Las Vegas, NV, USA, 15–18 October 2018.
36. Wenger, E. Communities of Practice. Learning as a social system. *Syst. Think.* **1998**, *9*, 1–10. [CrossRef]
37. Langley, A.; Patel, H.; Houghton, R.J. Fostering a Community of Practice for Industrial Processes. In *Dynamics of Long-Life Assets*; Grösser, S., Reyes-Lecuona, A., Granholm, G., Eds.; Springer: Cham, Switzerland, 2017; pp. 151–168.
38. Johnson, A.M.; Jacovina, M.E.; Russell, D.E.; Soto, C.M. Challenges and solutions when using technologies in the classroom. In *Adaptive Educational Technologies for Literacy Instruction*; Crossley, S.A., McNamara, D.S., Eds.; Taylor & Francis: New York, NY, USA, 2016; pp. 13–29.
39. United Nations. Digital Economy Report. In *Value Creation and Capture: Implications for Developing Countries*; United Nations Publications: New York, NY, USA, 2019; pp. 1–194.
40. Belaineh, M.S. Students' conception of the learning environment and their approach to learning and its implication on quality education. *Educ. Res. Rev.* **2017**, *12*, 695–703. [CrossRef]

sustainability

Article

Predicting Factors Influencing Preservice Teachers' Behavior Intention in the Implementation of STEM Education Using Partial Least Squares Approach

Tommy Tanu Wijaya [1], Peijie Jiang [2,*], Mailizar Mailizar [3,*] and Akhmad Habibi [4]

1 School of Mathematical Sciences, Beijing Normal University, Beijing 100875, China
2 School of Mathematics and Statistics, Hunan Normal University, Changsha 410081, China
3 Mathematics Education Department, Universitas Syiah Kuala, Banda Aceh 23111, Indonesia
4 Fakultas Keguruan dan Ilmu Pendidikan, Universitas Jambi, Jambi 36122, Indonesia
* Correspondence: peijiejiang@hunnu.edu.cn (P.J.); mailizar@unsyiah.ac.id (M.M.)

Abstract: The integration of STEM education has been promoted to improve the quality of education in the 21st century, with its usage leading to emphasis on the factors influencing the intentions of preservice teachers. Therefore, this study aims to determine the factors influencing preservice teachers' intentions, as well as the effects of gender and age on the implementation of STEM education. The Theory of Planned Behavior (TPB) was adopted to predict the relationship between knowledge, social influence, attitude, perceived usefulness, control, and behavioral intention (BI) of using STEM education among preservice secondary school teachers. A total of 30 item questionnaires on behavioral intentions were distributed to 201 respondents, with data being analyzed using the Structural Equation Model (SEM). The results showed that perceived usefulness had a positive significance, and a relationship with the attitudes of preservice teachers toward STEM education. Habit had a positive significance in influencing teachers' behavioral intentions and implementation. Subjective norms did not have a significant correlation with BI and implementation. These results are recommended for providing solutions to analytical problems, and to successfully improve future learning through an educational approach.

Keywords: preservice teacher; STEM education; structural equation model; behavioral intention; perception

Citation: Wijaya, T.T.; Jiang, P.; Mailizar, M.; Habibi, A. Predicting Factors Influencing Preservice Teachers' Behavior Intention in the Implementation of STEM Education Using Partial Least Squares Approach. *Sustainability* **2022**, *14*, 9925. https://doi.org/10.3390/su14169925

Academic Editor: Pedro Guilherme Rocha dos Reis

Received: 29 June 2022
Accepted: 9 August 2022
Published: 11 August 2022

1. Introduction

STEM education is rapidly becoming popular, globally accepted, and implemented in various countries, because of its innovative method of serving social and economic needs [1]. In this era, innovation is highly considered in various disciplines, as it provides solutions to complex realistic problems and supports sustainable development [2]. Several studies assumed that innovation breakthroughs were due to the combination of teams, ideas, and objects from different fields [1,3,4].

In 2015, STEM education was introduced in Indonesia and developed by the Southeast Asian Ministers of Education Organization (SEAMEO), although the implementation remains minimal and in the development stage [5,6]. This is considered one of the factors hindering the maximization of the educational approach, since it is based on the improvement of students' literacy [7,8]. According to Bybee [9], STEM literacy is defined as the conceptual understanding, as well as the individualistic skills and abilities to overcome daily problems related to the educational approach and global issues. Few studies explain the patterns of increasing knowledge of STEM education and its implementation intentions by preservice teachers [10]. This indicates that the approach encourages uneducated teachers to combine many subject topics that are more challenging and frightening. A preservice teacher is defined as a person in a transition from a normal university student to a school educator. They are also known as novice teachers, who still need to learn more about the various approaches and media used to achieve educational goals. The enrichment of these teachers with abundant knowledge of learning models, approaches, and effective methods are carried out through STEM education. This is considered a vital element of any preservice teacher-training program, to improve their ability to meet the demands of 21st century education. Based on these descriptions, determination of the factors related to the attitudes of these teachers to STEM education is highly necessary, accompanied by the improvement patterns of its implementation. Therefore, this study aims to determine the factors influencing the intentions and attitudes of preservice teachers concerning implementing STEM (science, technology, engineering, and mathematics) education. Using this approach, the ability to teach is often a complex learning activity including various factors, due to being insufficient to only emphasize individual attitudes, as well as the knowledge of science, technology, engineering, and mathematics. From previous studies, TPB (Theory of Planned Behavior) [11] was modified by adding several important factors influencing the intention of the preservice teachers toward implementing STEM education [12]. When teachers have positive attitudes toward this approach, their creative and convergent thinking skills are found to increase due to their being important for teaching [1].

The questions evaluated in this study are observed as follows:

- What are the significant positive factors influencing the preservice teachers' attitudes toward STEM education?
- What are the significant positive factors influencing the behavioral intention of preservice teachers to implement STEM education?
- What are the factors with the highest positive significance on the attitudes of preservice teachers toward implementing STEM education?
- What are the factors with the highest positive significance on the behavioral intention of preservice teachers toward implementing STEM education?
- Do gender and age affect preservice teachers in STEM education?

The results are recommended for providing solutions to analytical problems and to successfully improve future learning through STEM education. It is also expected to provide a new perspective on the determination of previous studies. Additionally, the proposed model helps local governments, universities, and schools in identifying and predicting the patterns of improving the implementation of STEM education.

2. Theoretical Background and Hypothesis Development

2.1. STEM Education

In the last decade, STEM education has become an important learning approach based on having relationships with and influence on sustainable development goals [13–15]. This approach is the combination of skills and knowledge in science, technology, engineering, and mathematics education [3,16]. It is also defined as a learning approach with the most important goal being encouraging students to cooperate between disciplines, improve systematic thinking and communication skills, conduct analysis and produce projects, as well as enhance creativity and problem-solving techniques [17–19].

Education exhibits the importance of the professions within the science, technology, engineering, and mathematics sectors, indicating that the improvement and development of life are related to these study fields [20–22]. Moreover, technology has affected all available sectors including education [23], with the expectation of humans becoming producers, developers, and inventors [24]. The exploitation and utilization of knowledge in these study fields are also important learning points. To meet this need, individuals need to be able to carry out the following: (1) think creatively and critically, (2) determine problems, (3) develop plans, (4) provide solutions, and (5) evaluate decisions. This shows that many countries have implemented STEM education in K-12 programs [25–27], to improve the quality of a superior workforce. Besides this, the learning approach is also carried out at the elementary level, as indicated in Figure 1. STEM education has a significant relationship with the four applicable scientific fields, i.e., science, technology, engineering, and mathematics, in terms of developing products [28,29]. This proves that the encouragement of the learning approach from an early age is important, as the attitude of preservice teachers toward utilization is observed to have a relationship with increasing the implementation of STEM education in the future.

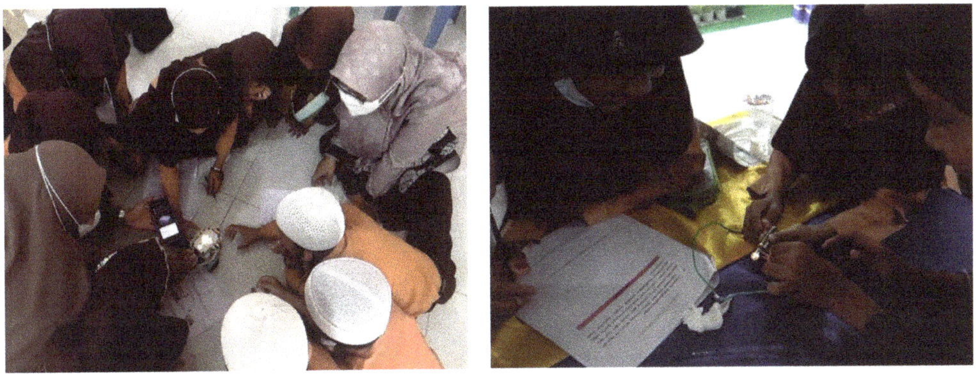

Figure 1. Preservice teachers practice STEM education in schools.

2.2. Stem Education in Indonesia

STEM education was introduced in Indonesia in 2013, subsequently becoming the focus and commitment of all stakeholders in 2014 [30]. During this period, all educational-level stakeholders agreed to improve the implementation, specifically among ordinary and preservice teachers. The government also established a partnership with the United States Agency for International Development (USAID) to jointly develop a suitable STEM-based learning model for the country's educational conditions [31]. However, the implementation of this model has not been maximized. According to Schmid et al. [10], the integration of STEM-based learning in Elementary Schools was guided by the Faculty of Education, Universiti Kebangsaan Malaysia (UKM). At the tertiary level, Syiah Kuala University emphasized STEM education by building a study center for related educational analysis. These university centers are expected to focus on developing professional or preservice teachers, to improve the implementation of the learning model in the future. An analysis of the factors related to the attitudes and behavioral intentions of these teachers concerning implementing and using STEM education is also necessary.

Based on meta-analysis data [32], only one study has been conducted on STEM education, which then improved in 2015–2018 although it still remains at a low standard. This study emphasized the development of the learning media or models; however, its application and implementation were not carried out. In 2020, the meta-analysis study data on STEM education subsequently decreased due to the COVID-19 pandemic. The implementation of STEM education in Indonesia is often carried out in schools, with the decline in 2020 being the initial start to an accelerative restoration in various cities. In addition, one of the important initial steps is to observe whether teachers already know its essentials.

2.3. Theoretical Analysis Model

To determine the factors influencing preservice teachers' behavioral intentions to implement STEM education, we extended the original theoretical Theory of Planned Behavior (TPB) model [11] (Figure 2). TPB is one of the models often used to predict behavioral intention (BI) in the educational sector [33]. In exploring the relationship between teachers' BI and beliefs, the effectiveness of this model has been examined in many studies [34,35]. This describes teachers' reaction to new learning and habits, as well as predicting their responses to a new approach or educational model. Meanwhile, this study adds several potential influential factors supported by strong literature studies to analyze the behavioral intention and attitude concerning STEM education. In addition, a moderator variable was added to subsequently analyze the differences in gender and age.

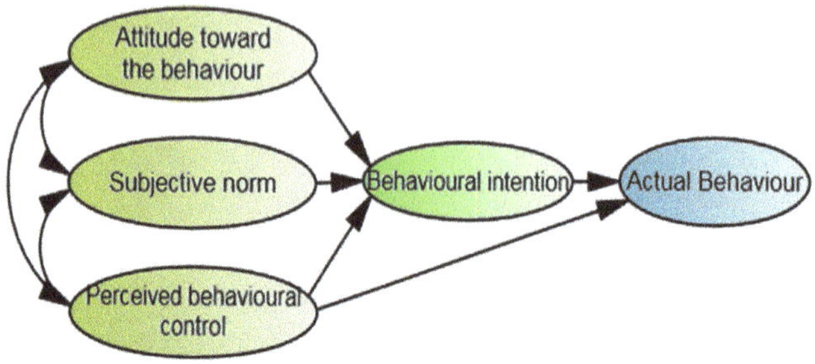

Figure 2. The original model of the Theory of Planned Behavior.

2.4. Knowledge of STEM Education

Using STEM education, learning is defined as an interdisciplinary lesson having a relationship with the fields of science, technology, engineering, and mathematics, aimed to determine or develop in-depth knowledge related to nature [36]. Technology education has the initial goal of teaching the modification patterns of natural nature to meet human needs and desires [37,38]. Engineering education also aims to teach utilization patterns concerning natural resources for the benefit of life and mankind [39]. Irrespective of these conditions, literacy in mathematics is needed to analyze the scientific model and relationship between American science, technology, and engineering [40].

Although several models have investigated this educational approach, the technique emphasizing the relationship between STEM disciplines was adopted through a specific field. This indicated that preservice teachers need to master the knowledge of their selected discipline while possessing sufficient understanding to combine their knowledge with other STEM fields.

2.5. Value of STEM Education

The attitudes and intentions of one's actions are determined by individualistic perceptions [41]. This explains that the preservice teacher's knowledge of STEM education needs to essentially affect their perceptions regarding the value of the educational approach. Furthermore, value is subjectively based on feelings, observations, and experiences. According to Williams [12], prestigious teachers' attitudes toward STEM education is generally influenced by the lessons obtained at related universities. Monroe, Day, and Grieser [42] also stated that value is related to perceived usefulness and behavioral control. This led to the inclusion of the variable in these analytical processes that involves analyzing the relationship between the approach value and the implementation intentions of preservice teachers. The questionnaire item of this variable was also adopted by Mahoney [43] and Lin and Williams [12].

2.6. Habit

This study adopts an interesting variable introduced by Venkatesh [44], where habit was the most influential factor in a person's behavioral intentions. However, a report in 2003 showed that habit was one of the important factors influencing the continuous utilization of technology [45]. This was conducted and confirmed before Venkatesh [44] included the variable in UTAUT2. Based on these conditions, the initial hypothesis stated that teachers adapted to the utilization of STEM education when attending related and familiar courses. This subsequently affected their intentions to use the educational approach.

2.7. Attitude towards STEM Education

Attitude is defined as people's expectations and predictions for an action [41,46]. This proves the positivity of their attitudes toward behavioral intentions, with several studies subsequently showing a relationship between both variables [46–48]. In this report, attitudes concerning STEM education were interpreted based on good performances, which positively affected teachers' behavioral intention regarding implementation.

2.8. Perceived Usefulness

Perceived usefulness is defined as the individual believing that an action or a specific system is capable of improving work performance [41]. In this study, it is, however, defined as the beliefs of preservice teachers regarding the assumption that the implementation of STEM education could increase their teaching effectiveness and provide good results to students. The more they believe, the higher their behavioral intention in using the educational approach to teach. According to previous studies, perceived usefulness has a good relationship with a person's intention and actual usage of technology in the educational sector [47,49,50].

2.9. Subjective Norms Related to STEM Education

Subjective norm is the pressure felt by a person from the surrounding environment to perform an action [11]. It also indicates the perception that important people often need to provide support to others when acting [33,51,52]. In this report, subjective norm is, however, defined as the degree to which preservice teachers receive demands from important people regarding the implementation of STEM education. According to Venkatesh [44], many individuals often make similar assumptions based on the perceptions of these important personalities. This was in line with the study of Wijaya [53], where subjective norm was the most influential predictor of a person's behavioral intention, concerning using technology. Another report also showed that it had a significant positive effect on behavioral intention within an educational context [54–56].

2.10. Perceived Behavioral Control of STEM Education

This is highly related to self-efficacy and is defined as the level of a person's behavioral assessment regarding the implementation patterns and the extent of utilization [57]. This shows that perceived behavioral control is generally measured as self-efficacy or people's beliefs in their abilities. Based on several observations, the variable has a positive relationship with a person's usage intention [58–61]. In this study, it is consequently a critical factor when preservice teachers decide to implement STEM education. Regarding the initial hypothesis, perceived behavioral control has a relationship with the successful implementation of STEM education. It also helps in solving the problems related to its implementation.

2.11. Innovativeness

This is often used by many study experts to predict technology use and is widely integrated with TAMs (technology-acceptance models) [62–64]. It is also defined as a person's willingness to use new designs or products [65]. In this study, innovativeness is consequently the willingness of preservice teachers to implement STEM education. This is not often influenced by the environment or other external variables, with Law [66] showing that it has a relationship with the extent to which a person is willing to experiment with new patterns or concepts.

2.12. Behavioral Intention to Implement STEM Education

Behavioral intention is defined as the personal attitude concerning carrying out a specific action [41,44,67]. However, it is the intention exhibited concerning the implementation of STEM education, according to this study. In this condition, the initial hypothesis states that behavioral intention is influenced by eight independent variables, namely knowledge, value, habit, perceived usefulness, innovativeness, PBC (perceived behavioral control), attitude, and subjective norms (Figure 3). Therefore, the hypothesis in this report includes the following,

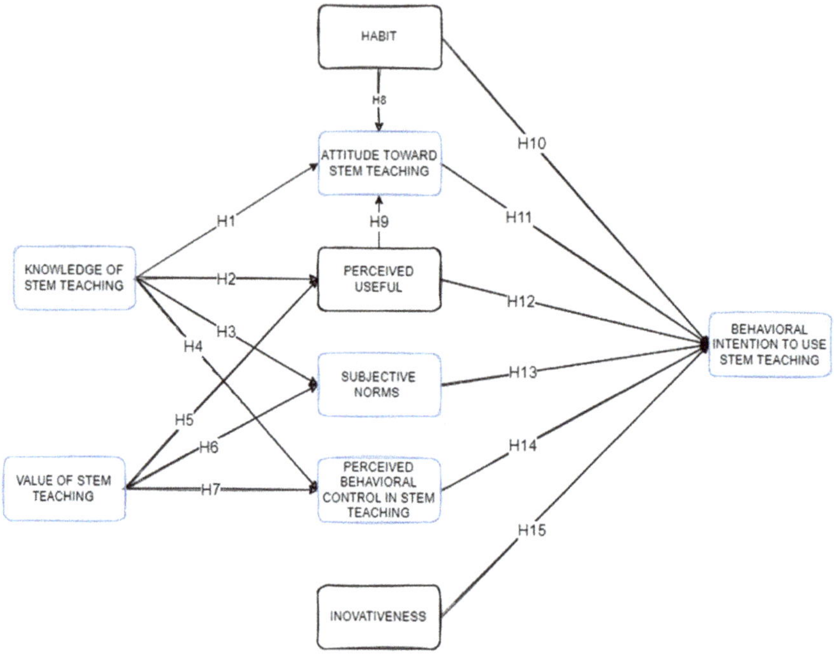

Figure 3. The proposed conceptual model developed from the original TPB model [11].

H1. *The knowledge of STEM education has a positive relationship with attitude.*

H2. *The knowledge of STEM education has a positive relationship with perceived usefulness.*

H3. *The knowledge of STEM education has a positive relationship with the subjective norm.*

H4. *The knowledge of STEM education has a positive relationship with perceived behavioral control.*

H5. *The value of STEM education has a positive relationship with perceived usefulness.*

H6. *The value of STEM education has a positive relationship with the subjective norm.*

H7. *The value of STEM education has a positive relationship with perceived behavioral control.*

H8. *Habit has a positive relationship with the attitude of preservice teachers toward STEM education.*

H9. *Perceived usefulness has a positive relationship with preservice teacher attitudes toward STEM education.*

H10. *Habit has a positive relationship with the behavioral intention to implement STEM education.*

H11. *Attitude has a positive relationship with the behavioral intention to implement STEM education.*

H12. *Perceived usefulness has a positive relationship with the behavioral intention to implement STEM education.*

H13. *Subjective norm has a positive relationship with the behavioral intention to implement STEM education.*

H14. *PBC (perceived behavioral control) has a positive relationship with the behavioral intention to implement STEM education.*

H15. *Innovativeness has a positive relationship with the behavioral intention to implement STEM education.*

2.13. Moderation Effect

The demographic information data are used to determine whether gender and age moderate preservice teachers' intention to implement STEM education. In this condition, gender is the main moderating variable in the UTAUT model [57] due to being widely used in behavioral intention studies within education contexts. Based on perceived usefulness, several previous studies proved that men and women highly emphasized the technology or solution capable of helping them quickly complete work and improve job performances, respectively [57]. Furthermore, significant differences were observed between men and women in subjective norms, as the males were less concerned with the opinions of those around them compared to the females [57,68]. Several studies also showed significant results regarding gender moderation [69–71]. During the process of age moderation, some differences were subsequently observed, as younger people easily adapted to new technologies and work patterns compared to older individuals. This was in line with some previous studies, where attitudinal differences were observed between the older and younger age groups [72,73]. Therefore, this present report classifies the age of preservice teachers into below and above 20 years.

Based on the potential moderator of these variables concerning the knowledge, value, perceived usefulness, subjective norm, habit, attitude, innovativeness, and PBC of STEM education, additional hypotheses were subsequently developed as follows:

H16. *Age moderates the relationship between knowledge, value, perceived usefulness, subjective norm, habit, attitude, innovativeness, and PBC with the preservice teachers' behavioral intentions to implement STEM education.*

H17. *Gender moderates the relationship between knowledge, value, perceived usefulness, subjective norm, habit, attitude, innovativeness, and PBC, with the preservice teachers' behavioral intentions to implement STEM education.*

3. Methods

A quantitative approach with a survey design was used in this analysis, with the Partial Least Square-Structural Equation Modeling (PLS-SEM) technique [74,75] being adopted to analyze and explain the relationship between variables, as well as the fit of the study model. This technique was suitable for reports incorporating and examining untested predictors and theoretical frameworks.

3.1. Sample and Population

Before the questionnaire distribution, consent was thoroughly obtained from the ethical committee of SK University (No. 3806/UN11.1.6/TU.01.01/2022). Consent was thoroughly obtained from the ethical committee of SK University, which has a STEM laboratory and "STEM education" courses. These possessions emphasized the improvement of professional teachers' abilities, according to the standards needed in the 21st century. Furthermore, The data were obtained using a purposive sampling technique [76,77], where the preservice teachers need to have acquired a course on STEM education for 1 year and experienced school attendance for related practice studies. These participants were invited to participate by filling out questionnaires on the factors influencing their intention to implement STEM education in the future. Participation and responses were also voluntary and anonymous. This was based on an explanation that the data were only used for study purposes. Using an online questionnaire from April to May 2022, the data collection instruments were subsequently distributed. Furthermore, the virtual filling-in of the data provided strong evidence that participation was highly voluntary. After completion, a total of 206 data were obtained; only 201 were valid responses; of the participants, 85.07% were women. The complete demographic respondent data are shown in Table 1.

Table 1. Demographic Respondent Data.

Data Demography		N	%
gender	male	30	9.14
	female	171	85.07
age	under 20	54	26.87
	20–25	147	73.13
major	mathematics	136	67.66
	physics	17	8.46
	biology	29	14.43
	chemistry	19	9.45

The instrument had an introductory section explaining the study objectives, the surveyor's procedure, and the consent form. The structure of the questionnaire was also divided into 2 parts, with the first category containing 5 questions on demographic information. Meanwhile, the second part contained 26 measurement items derived from 9 variable constructs (see Appendix A). These emphasized the results of extensions from the TPB model, using a 5-point Likert response scale from SD to SA (strongly disagree to strongly agree).

3.2. Data Analysis

The Partial Least Squares-Structural Equation Modeling (PLS-SEM) technique was used through SmartPLS and SPSS software to analyze the data and test the moderator effect. PLS-SEM was chosen as the major analysis method for this investigation for a number of reasons. In this investigation, there were 9 variables and 15 paths, demonstrating that PLS-SEM could support a model with a complex interaction involving multiple indicator variables and paths. Assumptions about data distribution were not affected by the method [75]. To establish causal information, this study used a predictive strategy for model estimation. Because one of PLS-main SEM's purposes is to deal with the dichotomy between information, prior associated ideas, and prediction as to the development foundation, PLS-SEM was chosen as a result. Comparing PLS-statistical SEM's power to that of other methods like CB-SEM and simple regression [75]. The advantages apply even when predicting common factor model data. When the outcomes are present in the population, the PLS-SEM function's statistical power also allows for the identification of correlations between constructs or variables [78]. For exploratory research that looks at less common ideas, the PLS-SEM feature within the statistical power is particularly useful. Thirdly, the software is more user-friendly than other programs like CB-SEM by SPSS and linear structural relations (Lisrel) since it has easy-access software packages available. The data were examined using SmartPLS 3, which is a user-friendly, state-of-the-art latent variable modeling tool that combines cutting-edge techniques (such as PLS-POS, IPMA, and sophisticated bootstrapping processes). Predicting a particular set of hypothesized associations that optimizes the explained variance in the dependent variables is the main goal of the user-friendly graphical user interface technique. PLS is often more suitable for testing model relationships and new path modelling, which has a more complex study structure [74,75]. In PLS-SEM, the analytical step was divided into several stages based on Hair et al. [75]. In this condition, both convergent and discriminatory validities (CV and DV) were initially evaluated, with the values of Item loadings, Cronbach alpha, composite reliability, and the AVE being used for the measurement of CV. For DV analysis, the Fornell–Larcker criterion [79] and heterotrait–monotrait (HTMT) [80] values were also evaluated. Meanwhile, the final step emphasized the evaluation of multicollinearity through the VIF value [81]. After the reliability and validity tests, the structural model and initial hypothesis analyses were then carried out through bootstrapping.

4. Results

4.1. Data Normality Analysis

Before entering the measurement model, the normality of the data needs to be tested through the kurtosis and skewness values of each item in the descriptive statistics table, as in Table 2. Based on the criteria, all the utilized item variables had a kurtosis and skewness value between −0.840 and 2.013, and −1.000 and 0.182, respectively, which were below 2.2 [82,83]. From these results, all these variables were observed to have normally distributed data.

Table 2. Descriptive statistics data for data normality testing.

	Mean	Min	Max	Standard Deviation	Excess Kurtosis	Skewness
KNW1	4.005	1.000	5.000	0.820	0.397	−0.665
KNW2	3.796	1.000	5.000	0.794	2.013	−1.000
KNW3	3.572	1.000	5.000	0.873	0.058	−0.359
KNW4	4.025	1.000	5.000	0.729	1.362	−0.737
VAL1	3.980	2.000	5.000	0.684	−0.582	−0.069
VAL2	4.020	2.000	5.000	0.654	0.049	−0.236
VAL3	3.960	2.000	5.000	0.675	−0.258	−0.147
VAL4	3.950	2.000	5.000	0.718	−0.430	−0.170
VAL5	4.164	2.000	5.000	0.660	−0.308	−0.297
VAL6	4.075	2.000	5.000	0.661	−0.343	−0.187
ATD1	4.184	3.000	5.000	0.565	−0.169	0.008
ATD2	4.159	2.000	5.000	0.610	0.158	−0.233
ATD3	3.905	2.000	5.000	0.681	−0.612	0.025
SN1	3.532	1.000	5.000	0.829	−0.297	0.108
SN2	3.920	1.000	5.000	0.722	0.135	−0.198
SN3	3.836	1.000	5.000	0.718	0.507	−0.313
PBC1	4.095	3.000	5.000	0.524	0.527	0.111
PBC2	4.065	2.000	5.000	0.607	1.233	−0.435
PBC3	4.085	2.000	5.000	0.588	1.655	−0.462
PBC4	4.090	2.000	5.000	0.592	1.589	−0.461
PU1	4.149	3.000	5.000	0.629	−0.536	−0.126
PU2	4.035	3.000	5.000	0.649	−0.617	−0.034
PU3	4.114	3.000	5.000	0.663	−0.736	−0.131
BI1	3.821	2.000	5.000	0.704	−0.806	0.182
BI2	4.090	3.000	5.000	0.671	−0.777	−0.107
HB1	3.512	1.000	5.000	0.864	0.014	−0.085
HB2	3.726	1.000	5.000	0.834	0.321	−0.276
INV1	4.114	2.000	5.000	0.616	0.111	−0.204
INV2	4.124	3.000	5.000	0.654	−0.680	−0.133
INV3	4.040	2.000	5.000	0.653	−0.280	−0.148
INV4	3.592	1.000	5.000	0.877	−0.476	−0.083
ANX1	4.060	3.000	5.000	0.681	−0.840	−0.075
ANX2	3.682	2.000	5.000	0.874	−0.721	−0.096
ANX3	3.741	2.000	5.000	0.787	−0.611	0.001

4.2. Measurement Model

For the analysis of the measurement model, Hair [75] suggested evaluation of the convergent validity, the composite reliability, and the AVE value. These estimation steps were subsequently achieved in the PLS algorithm for CFA (Confirmatory Factor Analysis); the analysis of factor loading values is shown in Figure 4. This fully explained the internal consistency analysis concerning evaluating CFA-based reliability, validity, and DV.

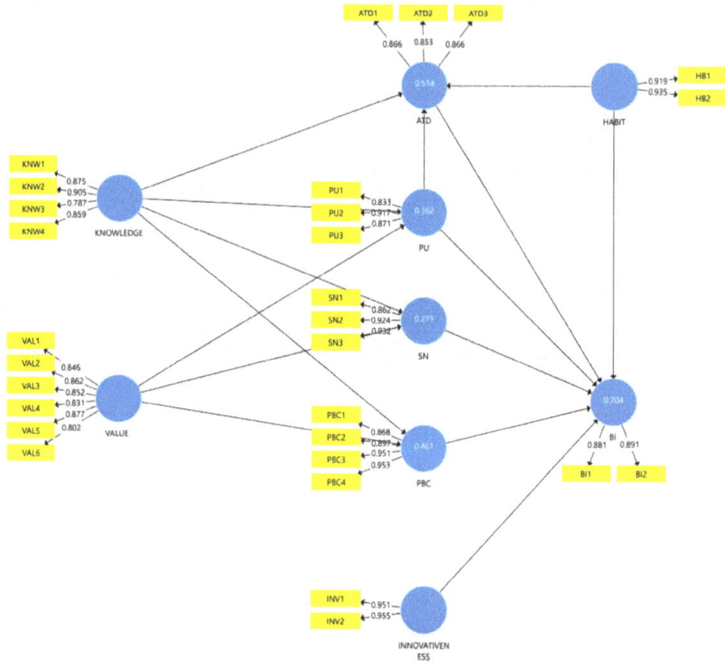

Figure 4. CFA and R-square values in PLS.

4.3. Internal Consistency, Reliability, and Validity

The CFA-based PLS algorithm was used to analyze the value of the outer loading on each measurement item, as shown in Figure 4. However, SmartPLS was used to analyze the factor loading value for all items. Based on the results, the value of outer loadings was between 0.802 and 0.955. This indicated that all constructs exceeded the recommended values from Hair [75], where the outer loading estimations need to be higher than 0.708. In the next step, the Rho-A value was used to determine the internal consistency of reliability, with Henseler [84] stating that higher estimation led to a greater model reliability level. This proved that a value above 0.7 was sufficient to define satisfaction, although a problem due to construct validity reduction was observed when the estimation was more than 0.95 [85]. A Rho-A value above this estimation also portrayed redundancy items [86]. Based on the results, the study model had the lowest and highest Rho-A values at 0.728 and 0.945, respectively. In this condition, Cronbach alpha was also used as an option to analyze internal consistency reliability, although it needs to be higher than 0.700. From the analysis, the lowest Cronbach alpha value was 0.727, as the full estimations in collaboration with Rho-A are shown in Table 3. For all constructs, both the Cronbach alpha and Rho-A values met the criteria, indicating that they had satisfactory internal consistency reliability. This was because all values were not less or more than 0.700 and 0.950, respectively.

Table 3. Cronbach alpha, CR, AVE, and Rho-A values for internal consistency reliability and validity analysis.

Construct	Cronbach's Alpha	Rho-A	Composite Reliability	Average Variance Extracted (AVE)
ATD	0.831	0.860	0.896	0.743
BI	0.727	0.728	0.880	0.786
HABIT	0.836	0.842	0.924	0.859
INNOVATIVENESS	0.899	0.900	0.952	0.908
KNOWLEDGE	0.879	0.881	0.917	0.735
PBC	0.937	0.945	0.955	0.843
PU	0.847	0.865	0.907	0.765
SN	0.891	0.899	0.933	0.822
VALUE	0.920	0.923	0.938	0.715

The next step emphasized the evaluation of CV (convergent validity), which was used to determine the extent to which the convergent construct described the variance items according to Hair, where the analysis of CV was observed from the AVE value. Using the PLS algorithm, the AVE had the lowest and highest estimations of 0.715 and 0.908 on value and innovativeness, respectively. This was in line with the Hair [75] criteria, where the AVE value needs to be greater than 0.5. For each construct, the Rho-A, CR, and AVE values proved that internal consistency was achieved for a more complete measurement model, as shown in Table 3.

The final step emphasized the analysis of the discriminant validity (DV), which is defined as the extent to which each construct is empirically different from others. This was analyzed using two methods, namely the Fornell–Larcker criterion [79] (see Table 4) and the heterotrait–monotrait (HTMT) ratio. These were used because some studies assumed that the analysis of DV was less effective when using only one method.

Table 4. Fornell–Larcker criterion value of smart PLS software.

	ATD	BI	HABIT	INV	KNW	PBC	PU	SN	VALUE
ATD	**0.862**								
BI	0.567	**0.886**							
HABIT	0.429	0.750	**0.927**						
INV	0.547	0.643	0.537	**0.953**					
KNW	0.401	0.382	0.379	0.526	**0.858**				
PBC	0.592	0.627	0.560	0.563	0.446	**0.918**			
PU	0.688	0.621	0.402	0.657	0.376	0.611	**0.874**		
SN	0.450	0.629	0.681	0.518	0.379	0.567	0.494	**0.907**	
VALUE	0.690	0.616	0.570	0.571	0.486	0.664	0.593	0.501	**0.846**

Note: The bold and diagonal values indicate the square root of AVE and the correlation between constructs unable to exceed the AVE square root, respectively.

Based on the Fornell–Larcker criterion [79], the cross-loading value (in bold) needs to be higher than the loading values of other constructs. For example, the cross-loading value for innovativeness was 0.953 higher than the other loading construct estimations, i.e., knowledge, PBC, perceived usefulness, subjective norm, and value at 0.526, 0.563, 0.657, 0.518, and 0.571, respectively. According to Hair [75], smaller HTMT values led to better discriminants. To meet the discriminant validity criteria, the HTMT value should also not be more than 0.900. Meanwhile, Henseler [84] stated that the HTMT value should not exceed the upper threshold of 1.00. In this condition, construct validity becomes less dominant when the HTMT value exceeds the threshold, due to having a similar concept. Using bootstrapping 5000 subsamples with Smart PLS software (Table 5), the obtained HTMT value was lower than 0.900, showing that the discriminant validity had met the criteria and should be continuously analyzed in the structural model process.

Table 5. Heterotrait–monotrait value (HTMT) for discriminant validity testing.

	ATD	BI	HABIT	INV	KNW	PBC	PU	SN	VALUE
ATD									
BI	0.702								
HABIT	0.477	0.863							
INV	0.619	0.792	0.616						
KNW	0.452	0.470	0.435	0.589					
PBC	0.661	0.756	0.627	0.613	0.484				
PU	0.801	0.767	0.445	0.740	0.425	0.674			
SN	0.495	0.780	0.794	0.577	0.423	0.616	0.547		
VALUE	0.773	0.748	0.642	0.626	0.535	0.710	0.661	0.542	

4.4. Structural Model

According to Hair [75], the process of assessing the structural model began by analyzing the issue of collinearity. This was accompanied by a relationship analysis, through the path coefficients of t and p values. To obtain the strength of this model in explaining the attitude of preservice teachers concerning STEM education, subsequent analyses were carried out on the coefficient of determination (R2), the effect size (F2), and the predictive relevance (Q2). In addition, the gender and age differences were analyzed using the ANOVA test [87].

4.5. Collinearity

The collinearity test between latent variables was observed in the Variance Inflation Factor (VIF) value. In this condition, the collinearity level was higher with a greater VIF value. Furthermore, a VIF value greater than 5.00 indicated a collinearity problem between the structural model variables [75,78]. Table 6 shows that all VIF values are below 5, confirming that all the model variables did not have a problem with collinearity.

Figures 5 and 6 and Table 7 shows the results of the initial hypothetical analysis, where 11 and 4 of the proposed hypotheses were supported and unsupported, respectively. Based on the results, knowledge had a relationship with the teachers' behavioral intentions, subjective norms, and PBC concerning STEM education (H1, $\beta = 0.130$, $t = 2.062$, p-value < 0.05; H3, $\beta = 0.177$, $t = 1.924$, p-value < 0.05; H4, $\beta = 0.161$, $t = 1.975$, p-value < 0.05), although it did not affect perceived usefulness (H2, $\beta = 0.114$, $t = 1.459$, p-value > 0.05). Value also affected perceived usefulness, and had a relationship with the subjective norm and PBC (H5, $\beta = 0.538$, $t = 7.070$, p-value < 0.001; H6, $\beta = 0.415$, $t = 5.637$, p-value < 0.001; H7, $\beta = 0.586$, $t = 9.159$, p-value < 0.001). Furthermore, habit had a relationship and greatly affected the attitude and behavioral intentions of the preservice teachers toward STEM education, respectively (H8, $\beta = 0.487$, $t = 7.646$, p-value < 0.01; H10, $\beta = 0.487$, $t = 7.646$, p-value < 0.001). Perceived usefulness also had the strongest positive significance and relationship with the attitude and behavioral intentions of preservice teachers toward the educational approach, respectively (H9, $\beta = 0.580$, $t = 10.257$, p-value < 0.001; H12, $\beta = 0.224$, $t = 3.335$, p-value < 0.01). Moreover, attitude, subjective norm, and PBC did not have a relationship with the teachers' behavioral intentions concerning STEM education (H11, $\beta = 0.066$, $t = 1.083$, p-value > 0.05; H13, $\beta = 0.045$, $t = 0.776$, p-value > 0.05; H14, β = 0.080, t = 1.103, *p*-value > 0.05). Irrespective of these conditions, innovativeness still had a relationship with the preservice teachers' intentions to implement the educational approach (H15, β = 0.131, t = 2.045, *p*-value < 0.05).

Table 6. Variance Inflation Factor (VIF) value.

Construct	VIF Value	Construct	VIF Value
ATD1	2.348	PBC3	6.500
ATD2	2.205	PBC4	6.605
ATD3	1.611	PU1	2.087
BI1	1.485	PU2	2.763
BI2	1.485	PU3	1.900
HB1	2.067	SN1	2.092
HB2	2.067	SN2	3.263
INV1	3.002	SN3	3.446
INV2	3.002	VAL1	2.899
KNW1	3.135	VAL2	3.720
KNW2	3.366	VAL3	3.473
KNW3	1.759	VAL4	2.357
KNW4	2.574	VAL5	3.865
PBC1	2.585	VAL6	2.776
PBC2	3.250		

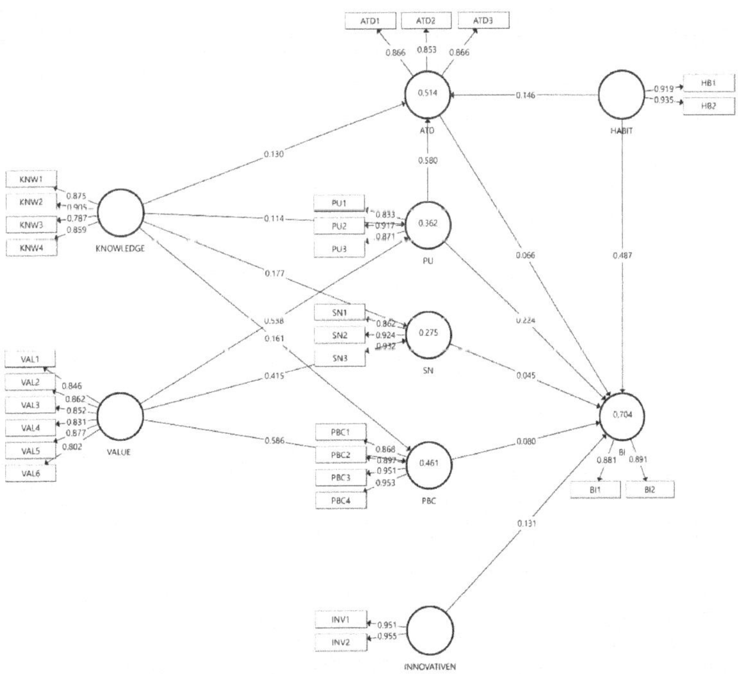

Figure 5. Final model with R square and path coefficient.

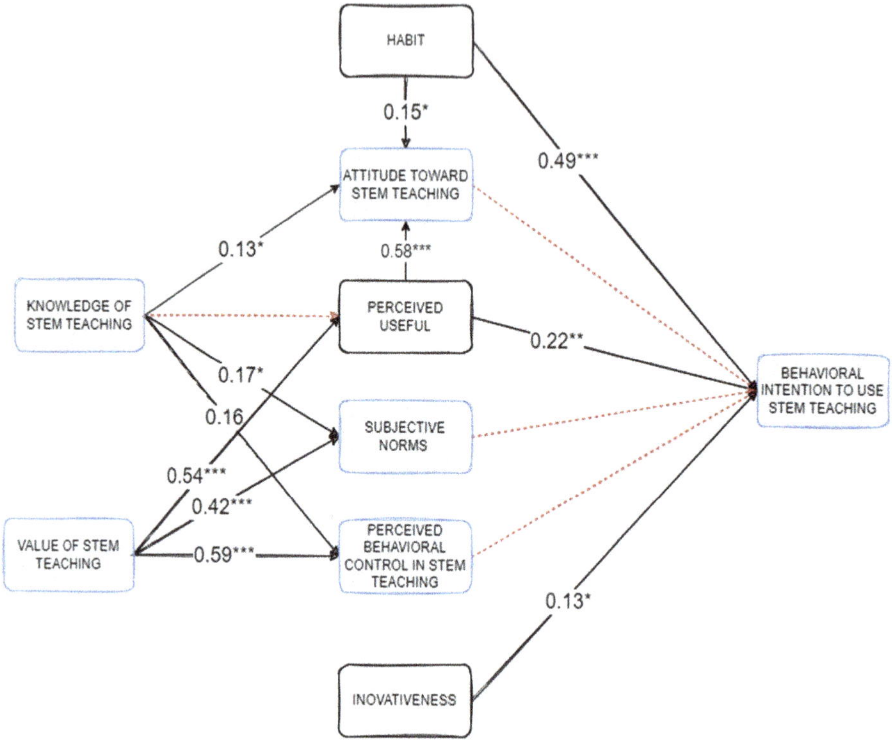

Figure 6. Hypothesis testing results. The dotted red line indicates the relationship is not significant. Note: * $p < 0.05$, ** $p < 0.01$ *** $p < 0.001$.

Table 7. Results of initial hypothesis testing and path estimation.

	Hypothesis	β	Mean	STDEV	T Statistics	*p*-Values	Hypothesis Testing Results
H1	KNOWLEDGE → ATD	0.130	0.136	0.063	2.062	0.040	supported
H2	KNOWLEDGE → PU	0.114	0.110	0.078	1.459	0.145	Not supported
H3	KNOWLEDGE → SN	0.177	0.167	0.092	1.924	0.050	supported
H4	KNOWLEDGE → PBC	0.161	0.159	0.081	1.975	0.049	supported
H5	VALUE → PU	0.538	0.540	0.076	7.070	0.000	supported
H6	VALUE → SN	0.415	0.422	0.074	5.637	0.000	supported
H7	VALUE → PBC	0.586	0.587	0.064	9.159	0.000	supported
H8	HABIT → ATD	0.146	0.145	0.057	2.557	0.011	supported
H9	PU → ATD	0.580	0.578	0.057	10.257	0.000	supported
H10	HABIT → BI	0.487	0.490	0.064	7.646	0.000	supported
H11	ATD → BI	0.066	0.065	0.061	1.083	0.280	Not supported
H12	PU → BI	0.224	0.227	0.067	3.335	0.001	supported
H13	SN → BI	0.045	0.048	0.058	0.776	0.438	Not supported
H14	PBC → BI	0.080	0.083	0.073	1.103	0.271	Not supported
H15	INNOVATIVENESS → BI	0.131	0.121	0.064	2.045	0.041	supported

4.6. Coefficient of Determination (R2)

The coefficient of determination (R2) is commonly used to evaluate the structural model due to the extent to which relationships are predicted between the dependent and independent variables [88,89]. This ranges from 0 to 1, with a higher R2 value leading to greater prediction accuracy in a study model [90]. R2 values less than 0.25, between 0.5 and 0.75, and above 0.75 are categorized as weak, moderate, and substantial, respectively. Based on the smart PLS results, all the model's dependent variables had good predictive accuracy with moderate R2 values, as shown in Table 8 and Figure 5.

Table 8. Coefficient of determination (R2).

Construct	R2	Interpretation
PU	0.362	moderate
SN	0.275	moderate
PBC	0.461	moderate
ATD	0.514	moderate
BI	0.704	moderate

4.7. Predictive Relevance (Q2)

Besides the analysis of R2 values as a significant criterion, Hair [75] also suggested the evaluation of Stone–Geisser's Q2, which is used to assess the relevant predictive stone [91]. This shows that a higher Q2 value leads to the specific predictive accuracy of item data points. The value of this construct is also obtained through the blindfolding procedure in SmartPLS software. Moreover, Q2 is used to assess cross-validated redundancy measures for each dependent construct [92]. Based on the endogenous latent variable, a relevant predictive model was achieved due to the value of Q2 being more than 0. Predictive relevance is also considered small, medium, and large when below, equal to, and above 0.02, 0.15, and 0.35, respectively. Table 9 shows the Q2 values for the five endogenous variables not less than 0.217, indicating that the independent construct (behavioral intention) had a significant predictive value for the dependent determinant.

Table 9. Q2 Value.

Construct	SSO	SSE	Q2 (=1 − SSE/SSO)	Interpretation Q2 Value
ATD	603.000	387.197	0.358	large
BI	402.000	186.412	0.536	large
PBC	804.000	495.756	0.383	large
PU	603.000	446.721	0.259	medium
SN	603.000	472.187	0.217	medium

4.8. Effect Size (F2)

According to the evaluation of the structural model, an analysis of the F2 value also needs to be performed. This construct explains the effect of exogenous on endogenous variables to determine changes in the R2 value when specific exogenous determinants are excluded from the model. The F2 values below, between, and above 0.15, 0.15 and 0.35, and 0.35 are categorized as small, medium, and large effects, respectively. Based on the results, perceived usefulness had the largest F2 for attitudes, as shown in Table 10.

Table 10. Value of effect size.

Relationship	F-Square	Effect Size
ATD → BI	0.007	small
HABIT → ATD	0.034	small
HABIT → BI	0.367	large
INNOVATIVENESS → BI	0.027	small
KNOWLEDGE → ATD	0.028	small
KNOWLEDGE → PBC	0.037	small
KNOWLEDGE → PU	0.016	small
KNOWLEDGE → SN	0.033	small
PBC → BI	0.010	small
PU → ATD	0.540	large
PU → BI	0.064	small
SN → BI	0.003	small
VALUE → PBC	0.487	large
VALUE → PU	0.346	large
VALUE → SN	0.182	medium

4.9. *Moderating Effect Analysis of Gender and Age*

This study subsequently examined the occurrence of differences between two demographic factors—gender and age—regarding preservice teachers' intentions to implement STEM education. The moderator effects were calculated and analyzed using *t*-test and ANOVA analysis through the SPSS software [93]. In this case, the results of the independent sample *t*-test concerned the differences between men and women in the implementation of the educational approach. Table 11 also indicates that the preservice teachers did not show a significant difference ($p > 0.05$) in the perceptions of men and women for the 10 variables. Therefore, gender did not affect Indonesian preservice teachers' behavioral intentions to implement STEM education.

Table 11. Results of testing gender differences in STEM education using a *t*-test.

Variable	Male		Female		t-Value	*p*-Value
	Mean	Std	Mean	Std		
KNW	3.8500	0.75886	3.8494	0.67871	0.000	0.997
VAL	4.0833	0.44150	4.0146	0.59267	0.367	0.545
ATD	3.9333	0.49052	4.1092	0.53971	2.779	0.097
SN	3.9778	0.80674	3.7251	0.65815	3.504	0.063
PBC	4.1500	0.41833	4.0716	0.55005	0.552	0.458
PU	4.1000	0.41199	4.0994	0.59177	0.000	0.996
BI	4.1000	0.63518	3.9298	0.60468	1.991	0.160
HB	3.7500	0.75144	3.5965	0.79579	0.965	0.327
INV	3.8417	0.49777	3.9898	0.60595	1.600	0.207

Based on the age moderator effect analysis, differences were observed in the subjective norm and habit variables at *p*-values < 0.05, as well as t-values of 4.239 and 5.546, respectively.

5. Discussion

The developed model aimed to predict the factors influencing the attitude and behavioral intention of preservice teachers toward implementing STEM education. Based on the PLS algorithm, 11 of the 15 hypotheses were supported, with three predictors each directly affecting the teachers' attitudes and behavioral intentions in implementing STEM education. At approximately 70%, the structural model also successfully validated and analyzed the factors having a relationship with the two constructs (attitude and behavioral intention). Additionally, the final model increased the power of TPB, contributing to the behavioral intention knowledge of the teachers. These results contributed to the closure of analytical gaps and provided valuable outputs for improving the present implementation of STEM education in Indonesia. Regarding the increase in teachers' attitudes concerning the implementation of this educational approach, knowledge, perceived usefulness, and habit were observed as the positively significant variables. Perceived usefulness had the greatest positive effect on the implementation attitudes. This was in line with previous articles, where the variable largely affected behavior in the acceptance model theory [48,50,94]. These results provided information to Indonesian governments and institutions to improve the implementation of STEM education. This ensured the need for special/continuous training and guidance on the objectives of learning, as well as the benefits of the educational approach. It also enabled the awareness of students concerning the need for the present STEM education and increased technical knowledge. The factors directly affecting behavioral intentions included habit, perceived usefulness, and innovativeness. In this case, habit had the largest positive significance on the BI of teachers concerning implementing STEM education. Innovativeness also had a significant positive effect due to the utilization of Generation Z children not older than 25 years old [95]. This is because they have high enthusiasm for new things and new concepts, with the curiosity to utilize innovative learning approaches subsequently considered [96,97]. The generation was also dominated by undergraduates, as they were required to learn many things and innovate through different learning approaches. These outcomes are in line with several previous studies [62,64].

According to these results, habit was the biggest effective factor having a relationship with the implementation of STEM education in Indonesia. Therefore, all educational institution levels need to familiarize preservice teachers with the use of STEM education. It should also be included as a mandatory or additional course while transforming the learning approach in eligible classes. Additionally, the obtained results need to adequately familiarize these teachers with STEM education-based learning to increase their behavioral intention to implement it. Meanwhile, subjective norms did not have a significant relationship with BI. This was not in line with many previous studies, where the variable significantly affected a person's behavioral intention to perform attitudinal actions [53,98]. More articles also showed that it was the first biggest positive factor influencing BI and UB (behavioral intention and usage behavior) [99]. In this condition, Indonesian preservice teachers assumed that they had the value and knowledge of STEM education, and felt that the perceived usefulness of implementation was a more important factor than environmental influences.

The results also showed no gender effect on all factors affecting preservice teachers' behavioral intentions toward implementing the educational approach. This showed that STEM education improved gender differences in the fields of science and computer programs, which were previously less favored by women in Indonesia. These results were essentially in line with Price [100], where differences were observed in the gender factor concerning intention to continue STEM education. Age was also found to moderate subjective norms and habits, with the teachers <20 years old having higher responses than other individuals (see Table 12). This explained that the preservice teachers were environmentally influenced in the first and second years of college. Therefore, lecturers and campuses need to capitalize on this golden opportunity to instill the knowledge and importance of STEM education in all educational personnel.

Table 12. The results of the age difference test on STEM education using the *t*-test.

Construct	Under 20		20–25		t-Value	*p*-Value
	Mean	Std	Mean	Std		
KNW	3.8750	0.66366	3.8401	0.70038	0.101	0.751
VAL	3.9568	0.66208	4.0499	0.53579	1.046	0.308
ATD	4.0802	0.57530	4.0839	0.52174	0.002	0.966
SN	3.9259	0.66562	3.7029	0.68594	4.239	0.041
PBC	4.1435	0.53111	4.0612	0.53282	0.944	0.333
PU	4.0494	0.60884	4.1179	0.55284	0.574	0.449
BI	4.0833	0.67816	3.9082	0.57938	3.286	0.071
HB	3.8333	0.81264	3.5408	0.76863	5.546	0.019
INV	4.0833	0.61621	3.9252	0.57963	2.842	0.093

6. Conclusions and Implication

This study predicted the factors influencing the behavioral intention of Indonesian preservice teachers to implement STEM education. It also examined the moderating role of gender and age on the behavioral intentions concerning implementation. Using Indonesian preservice teachers as samples, success was achieved in the validation, analysis, and extension of the TPB model. In this condition, three factors each had a direct relationship with attitude and behavioral intention. Based on gender and age, differences were also observed in subjective norms, perceived behavioral control, and habit. These results are expected to be an important contribution toward increasing the use of STEM education in the future. This is because the learning technique encourages interdisciplinary education with a sustainable development approach. Besides this, sustainable development goals also require STEM education to educate sensitive and professional interdisciplinary individuals. The results obtained subsequently need to close a gap in areas related to the behavioral intention of implementing STEM education in developing countries. The strength of this study model also showed that preservice teachers' attitudes and BI toward STEM education were above 50 and 70%, respectively.

7. Limitation and Suggestions

This study had several limitations, including the development of a conceptual model from the original TPB model. Another limitation was the addition of several determinants with the potential to influence behavioral intention concerning implementing STEM education. Of the many models used to analyze these implementation intentions, each of them had its advantages, leading to different results from the conceptual development of the study design.

Since the sample was only limited to the preservice teachers at SK University, Indonesia, the result generalization in other countries needs to be carefully carried out. Subsequent studies need to examine the comparative factors influencing preservice teachers' intentions to implement STEM education. Additionally, the study only emphasized the use of quantitative techniques. This indicates that a combination of qualitative and quantitative methods needs to be used to deeply explain the influential factors of STEM education implementation. Some additional moderators such as voluntary and major also need to be added for subsequent analysis. Finally, no negative statements were observed in the questionnaire, as their additions are suggested to avoid response bias.

Author Contributions: Conceptualization, T.T.W. and M.M.; methodology, A.H.; software, T.T.W. and A.H.; formal analysis, T.T.W. and M.M.; data curation, M.M.; writing—original draft preparation, T.T.W. and M.M.; writing—review and editing, T.T.W.; visualization, all author; supervision, P.J. and A.H.; project administration, P.J.; funding acquisition, P.J. All authors have read and agreed to the published version of the manuscript.

Funding: This research was funded by 2022 Hunan Provincial General Higher Education Teaching Reform Research Project "Virtual Teaching and Research Section Empowering Excellent Middle School Mathematics Teacher Training Research"; 2021 Hunan Provincial Department of Education Scientific Research Project "Academic Development Tracking Research of Top-notch Innovative Reserve Talents in Adolescent Basic Disciplines-Based on Mathematics Competitive Students as an Example" (21B0049); Ph.D. Research Start-up Fund of Hunan Normal University.

Institutional Review Board Statement: The study was conducted according to the guidelines of the Declaration of Helsinki, and approved by the Institutional Review Board (or Ethics Committee) of Universitas Syiah Kuala (protocol code 3806/UN11.1.6/TU.01.01/2022 and 1 May 2022).

Informed Consent Statement: Informed consent was obtained from all subjects involved in the study.

Data Availability Statement: Data available on request.

Acknowledgments: Thank for all preservice teacher participate in this study.

Conflicts of Interest: The authors declare no conflict of interest.

Appendix A

Construct	English Version	Indonesian Version
Knowledge of STEM education	I am familiar with science knowledge (chemistry/physics/biology) at the junior and senior high school levels	Saya farmiliar dengan pengetahuan science (kimia/fisika/biologi) di level SMP dan SMA
	I am familiar with technology-related knowledge at the junior and senior high school level	Saya farmiliar dengan pengetahuan terkait teknologi pada level SMP dan SMA
	I am familiar with the knowledge of engineering (STEM) at the junior and senior high school level	Saya farmiliar dengan pengetahuan mengenai engineering (STEM) pada level SMP dan SMA
	I am familiar with mathematics knowledge at the junior and senior high school level	Saya farmiliar dengan pengetahuan matematika pada level SMP dan SMA
Value of STEM education	I feel teaching students how to collect STEM-related data during the teaching process is very important	Saya merasa mengajarkan siswa bagaiamana cara mengumpulkan data yang berhubungan dengan STEM selama proses mengajar sangat penting
	I feel teaching students how to use STEM-related data during the teaching process is very important	Saya merasa mengajarkan siswa bagaimana cara menggunakan data yang berhubungan dengan STEM selama proses mengajar sangat penting
	I feel like teaching students how to use STEM-related data during the testing and modification process.	Saya merasa mengajarkan siswa bagaiamana cara menggunakan data yang berhubungan dengan STEM pada saat testing dan proses modifikasi.
	I like using STEM learning in class	Saya suka menggunakan pembelajaran STEM di kelas
	I think STEM activities help improve students' abilities	Saya pikir kegiatan STEM sangat membantu meningkatkan kemampuan siswa
	I think STEM activities help improve student learning scores	saya pikir STEM activites dapat membantu meningkatkan nilai belajar siswa

Construct	English Version	Indonesian Version
Attitude concerning STEM education	Using STEM learning is a good idea	Menggunakan pembelajaran STEM adalah ide yang bagus
	Using STEM learning makes teaching and learning activities interesting	Menggunakan pembelajaran STEM membuat kegaiatan belajar mengajar menjadi menarik
	I enjoy using STEM learning when teaching	Saya senang menggunakan pembelajaran STEM saat mengajar
Subjective norm	Other teachers at my school use STEM learning when teaching	Guru guru lain di sekolah saya menggunakan pembelajaran STEM saat mengajar
	The government advise teachers to implement STEM learning	Pemerintah menyarankan para guru untuk mengimplementasikan pembelajaran STEM
	The head of the curriculum section advises teachers to implement STEM education	Kepala bagian kurikulum menyarankan para guru untuk mengimplementasikan STEM education
Perceived behavior control in STEM education	I will try to collaborate with other teachers to implement STEM education	Saya akan berusaha berkolaborasi dengan guru lain untuk mengimplementasikan STEM education
	I will try to remind students to solve problems based on STEM knowledge	Saya akan berusaha mengingatkan siswa untuk memecahkan masalah berdasarkan pengetahuan STEM
	I will try to teach students how to modify products based on STEM knowledge	Saya akan berusaha mengajarkan siswa bagaimana untuk memodifikasi produk berdasarkan pengetahuan STEM
	I will try to teach students how to think based on STEM knowledge in teaching and learning activities	Saya akan berusaha untuk mengajarkan siswa bagaimana untuk berpikir berdasarkan pengetahuan STEM pada kegiatan belajar mengajar
Perceived usefulness	I think STEM learning is useful in teaching and learning activities	Saya rasa pembelajaran STEM bermanfaat pada kegiatan belajar mengajar
	STEM learning improves my teaching skills in school	Pembelajaran STEM meningkatkan kemampuan mengajar saya di sekolah
	I think using STEM learning for teaching is more effective than traditional teaching	Saya pikir menggunakan pembelajaran STEM untuk mengajar efektif dibanding pengajaran tradisional
Behavioral intention toward STEM education	I will use STEM learning at every teaching opportunity	Saya akan menggunakan pembelajaran STEM pada setiap kesempatan mengajar
	I will recommend STEM learning to my teacher friends	Saya akan merekomendasikan pembelajaran STEM kepada teman teman guru
	Using STEM learning has become my teaching habit	
Habit	I have to use the STEM approach if I teach at the junior high and high school levels	Menggunakan pembelajaran STEM sudah menjadi kebiasaan mengajar saya
	I enjoy learning about new learning approaches	Saya harus menggunakan pendekatan STEM jika mengajar di tingkat SMP dan SMA
Innovativeness	I enjoy using new teaching methods	Saya senang belajar tentang pendekatan pembelajaran baru
	When there is a new teaching approach, I try to practice it in the classroom	Saya senang menggunakan metode mengajar yang baru

Construct	English Version	Indonesian Version
	Compared to other teachers, I am usually the first to try a new learning model	Ketika ada pendekatan mengajar yang baru, saya mencoba untuk mempraktekkannya di dalam kelas
	I am curious about teaching using the STEM approach	Dibandingkan dengan guru guru lain, saya biasanya orang pertama yang mencoba model pembelajaran baru
Anxiety	I am afraid I cannot teach well if I use the STEM approach	Saya merasa penasaran mengajar menggunakan pendekatan STEM
	I am afraid that students will not like my teaching method using the STEM approach	Saya takut tidak dapat mengajar dengan baik jika menggunakan pendekatan STEM
	I will use STEM learning at every teaching opportunity	Saya takut siswa tidak menyukai cara mengajar saya menggunakan pendekatan STEM

References

1. So, H.J.; Ryoo, D.; Park, H.; Choi, H. What Constitutes Korean Pre-service Teachers' Competency in STEAM Education: Examining the Multi-functional Structure. *Asia-Pacific Educ. Res.* **2019**, *28*, 47–61. [CrossRef]
2. Grodal, S.; Thoma, G. Cross-pollination in science and technology: Concept mobility in the nanobiotechnology field. *Ann. Econ. Stat. d'E'conomie Stat.* **2014**, *115*, 57–80. [CrossRef]
3. Kier, M.W.; Johnson, L.L. Exploring How Secondary STEM Teachers and Undergraduate Mentors Adapt Digital Technologies to Promote Culturally Relevant Education during COVID-19. *Educ. Sci.* **2022**, *12*, 48. [CrossRef]
4. Williams, J.; Roth, W.-M.; Swanson, D.; Doig, B.; Groves, S.; Omuvwie, M.; Ferri, R.B.; Mousoulides, N. *Interdisciplinary Mathematics Education: A State of the Art*; ICME-13 Topical Surveys; Springer: Cham, Switzerland, 2016; pp. 1–35. [CrossRef]
5. Lee, K.T.; Nason, R.A. Reforming the preparation of future STEM teachers. In Proceedings of the 2nd International STEM in Education Conference, Beijing, China, 30 June 2012.
6. Honey, M.; Pearson, G.; Schweingruber, H. *STEM Integration in K-12 Education: Status, Prospects, and an Agenda for Research*; National Academies Press: Washington, DC, USA, 2014.
7. Ahmad, J.; Siew, N.M. Curiosity towards stem education: A questionnaire for primary school students. *J. Balt. Sci. Educ.* **2021**, *20*, 289–304. [CrossRef]
8. Makamure, C.; Tsakeni, M. COVID-19 as an agent of change in teaching and learning stem subjects. *J. Balt. Sci. Educ.* **2020**, *19*, 1078–1091. [CrossRef]
9. Bybee, R.W. Advancing STEM Education: A 2020 Vision. *Technol. Eng. Teach.* **2010**, *70*, 30–35.
10. Schmid, M.; Brianza, E.; Petko, D. Self-reported technological pedagogical content knowledge (TPACK) of pre-service teachers in relation to digital technology use in lesson plans. *Comput. Hum. Behav.* **2021**, *115*, 106586. [CrossRef]
11. Ajzen, I. *From Intentions to Actions: A Theory of Planned Behavior in Action Control*; SSSP Springer Series in Social Psychology; Springer: Berlin/Heidelberg, Germany, 1985.
12. Lin, K.Y.; Williams, P.J. Taiwanese Preservice Teachers' Science, Technology, Engineering, and Mathematics Teaching Intention. *Int. J. Sci. Math. Educ.* **2016**, *14*, 1021–1036. [CrossRef]
13. ACM International Conference Proceeding Series. 2019. Available online: https://www.scopus.com/inward/record.uri?eid=2-s2.0-85117535722&partnerID=40&md5=7c9a8692c55c8d5b5f1d39815a2549ef (accessed on 10 June 2022).
14. Syahmani, S.; Hafizah, E.; Sauqina, S.; Adnan, M.B.; Ibrahim, M.H. STEAM Approach to Improve Environmental Education Innovation and Literacy in Waste Management: Bibliometric Research. *Indones. J. Learn. Adv. Educ.* **2021**, *3*, 130–141. [CrossRef]
15. Khadri, H.O. Becoming future-proof STEM teachers for enhancing sustainable development: A proposed general framework for capacity-building programs in future studies. *Prospects* **2022**, 1–15. [CrossRef]
16. Chang, D.F.; Lee, K.Y.; Tseng, C.W. Exploring Structural Relationships in Attracting and Retaining International Students in STEM for Sustainable Development of Higher Education. *Sustainability* **2022**, *14*, 1267. [CrossRef]
17. Newhouse, C.P. STEM the Boredom: Engage Students in the Australian Curriculum Using ICT with Problem-Based Learning and Assessment. *J. Sci. Educ. Technol.* **2017**, *26*, 44–57. [CrossRef]
18. Nicholas, C.; Scribner, J.A. Enhancing PBL Authenticity by Engaging STEM-Professional Volunteers. *Interdiscip. J. Probl. Learn.* **2021**, *15*, 2. [CrossRef]
19. Kilty, T.; Burrows, A.; Welsh, K.; Kilty, K.; McBride, S.; Bergmaier, P. Transcending Disciplines: Engaging College Students in Interdisciplinary Research, Integrated Stem, and Partnerships. *J. Technol. Sci. Educ.* **2021**, *11*, 146–166. [CrossRef]
20. Thu, H.L.T.; Tran, T.; Phuong, T.T.T.; Tuyet, T.L.T.; le Huy, H.; Thi, T.V. Two decades of stem education research in middle school: A bibliometrics analysis in scopus database (2000–2020). *Educ. Sci.* **2021**, *11*, 353. [CrossRef]
21. Simoncini, K.; Lasen, M. Ideas About STEM Among Australian Early Childhood Professionals: How Important is STEM in Early Childhood Education? *Int. J. Early Child.* **2018**, *50*, 353–369. [CrossRef]

22. Petrosino, A.J., Jr.; Stroup, W.M. Group-Based Cloud Computing for Secondary STEM Education. June 2017. Available online: https://www.scopus.com/inward/record.uri?eid=2-s2.0-85030539786&partnerID=40&md5=49d7d118ea79fd3cec66892c7900a7ba (accessed on 11 June 2022).
23. Wijaya, T.T.; Zhou, Y.; Ware, A.; Hermita, N. Improving the Creative Thinking Skills of the Next Generation of Mathematics Teachers Using Dynamic Mathematics Software. *Int. J. Emerg. Technol. Learn.* **2021**, *16*, 212–226. [CrossRef]
24. Purwaningsih, E.; Sari, S.P.; Sari, A.M.; Suryadi, A. The effect of stem-pjbl and discovery learning on improving students' problem-solving skills of the impulse and momentum topic. *J. Pendidik. IPA Indones.* **2020**, *9*, 465–476. [CrossRef]
25. Smith, K.N.; Jaeger, A.J.; Thomas, D. 'Science Olympiad Is Why I'm Here': The Influence of an Early STEM Program on College and Major Choice. *Res. Sci. Educ.* **2019**, *51*, 443–459. [CrossRef]
26. Lin, P.Y.; Chai, C.S.; Jong, M.S.Y. A Study of Disposition, Engagement, Efficacy, and Vitality of Teachers in Designing Science, Technology, Engineering, and Mathematics Education. *Front. Psychol.* **2021**, *12*, 661631. [CrossRef]
27. Correia, M.S.M.; Baptista, M.L.M. The effects of a STEM approach on preservice elementary teachers' subject matter knowledge about sound. *Acta Sci.* **2021**, *23*, 179–208. [CrossRef]
28. Coles, A. Facilitating the use of video with teachers of mathematics: Learning from staying with the detail. *Int. J. STEM Educ.* **2019**, *6*, 1. [CrossRef]
29. Wong, C.Y.; Shih, Y.T. Enhance STEM Education by Integrating Product Design with Computer-Aided Design Approaches. *Comput. Aided. Des. Appl.* **2022**, *19*, 694–711. [CrossRef]
30. Arlinwibowo, J.; Retnawati, H.; Kartowagiran, B. How to Integrate STEM Education in The Indonesian Curriculum? A Systematic Review. *Chall. Sci.* **2021**, 18–25. [CrossRef]
31. Prabowo, A.; Anggoro, R.P.; Adiyanto, R.; Rahmawati, U. Interactive Multimedia-based Teaching Material for Trigonometry. *J. Phys. Conf. Ser.* **2018**, *1097*, 012138. [CrossRef]
32. Farwati, R.; Metafisika, K.; Sari, I.; Sitinjak, D.S.; Solikha, D.F.; Solfarina, S. STEM Education Implementation in Indonesia: A Scoping Review. *Int. J. STEM Educ. Sustain.* **2021**, *1*, 11–32. [CrossRef]
33. Stössel, J.; Baumann, R.; Wegner, E. Predictors of student teachers' esd implementation intention and their implications for improving teacher education. *Sustainability* **2021**, *13*, 9027. [CrossRef]
34. Zhao, W.; Mok, I.A.C.; Cao, Y. Factors influencing teachers' implementation of a reformed instructional model in China from the theory of planned behavior perspective: A multiple case study. *Sustainability* **2020**, *12*, 1. [CrossRef]
35. Yusop, F.D.; Habibi, A.; Razak, R.A. Factors Affecting Indonesian Preservice Teachers' Use of ICT During Teaching Practices Through Theory of Planned Behavior. *SAGE Open* **2021**, *11*, 21582440211027572. [CrossRef]
36. National Research Council (NRC). *National Science Education Standards*; National Academy Press: Washington, DC, USA, 1996.
37. Dias, A.; Ordaz, M.G.; López, F.M. The wisdom innovation model—Adjusting new insights and hosting new perspectives to human augmented reality. *Commun. Comput. Inf. Sci.* **2011**, *219*, 73–85. [CrossRef]
38. 15th International Conference on Human Computer Interaction, HCI International 2013. Available online: https://www.scopus.com/inward/record.uri?eid=2-s2.0-85015595703&partnerID=40&md5=9e54145641ddf1a2aeebdc27bb2b347a (accessed on 10 June 2022).
39. Nejkovic, V.; Tosic, M. Exploring factors for effective use of online information in SPOC within the engineering education. *Comput. Appl. Eng. Educ.* **2018**, *26*, 1457–1469. [CrossRef]
40. American Association for the Advanced of Science. *Benchmarks for Science Literacy*; Oxford University Press: New York, NY, USA, 1993.
41. Davis, F.D. Perceived usefulness, perceived ease of use, and user acceptance of information technology. *MIS Q.* **1989**, *13*, 319–340. [CrossRef]
42. Monroe, M.; Day, B.; Grieser, M. *Environmental Education & Communication for a Sustainable World*; Academy for Educational Development: Washington, DC, USA, 2000.
43. Smith, P.J.; Murphy, K.L.; Mahoney, S.E. Towards identifying factors underlying readiness for online learning: An exploratory study. *Distance Educ.* **2003**, *24*, 57–67. [CrossRef]
44. Venkatesh, V.Y.L.; Thong, J.; Xu, X. Consumer acceptance and use of information Technology: Extending the unified theory of acceptance and use of technology. MIS Quarterly. *Manag. Inf. Syst. Q.* **2012**, *36*, 157–178. [CrossRef]
45. Gefen, D. TAM or just plain habit: A look at experienced online shoppers. *J. End User Comput.* **2003**, *15*, 1–13. [CrossRef]
46. Gellerstedt, M.; Babaheidari, S.M.; Svensson, L. A first step towards a model for teachers' adoption of ICT pedagogy in schools. *Heliyon* **2018**, *4*, e00786. [CrossRef]
47. Drueke, B.; Mainz, V.; Lemos, M.; Wirtz, M.A.; Boecker, M. An Evaluation of Forced Distance Learning and Teaching Under Pandemic Conditions Using the Technology Acceptance Model. *Front. Psychol.* **2021**, *12*, 701347. [CrossRef]
48. Smeda, A.; Shiratuddin, M.F.; Wong, K.W. A structural equation modelling approach for adoption of e-book amongst mathematics and statistics (MAS) students at higher education institutions in Libya. *Int. J. Inf. Learn. Technol.* **2018**, *35*, 240–254. [CrossRef]
49. Osih, S.C.; Singh, U.G.; Africa, S. Students ' perception on the adoption of an e-textbook (digital) as an alternative to the printed textbook. *S. Afr. J. High. Educ.* **2020**, *34*, 201–215. [CrossRef]
50. Joo, Y.J.; Park, S.; Lim, E. Factors Influencing Preservice Teachers' Intention to Use Technology. *J. Educ. Technol. Soc.* **2018**, *21*, 48–59. Available online: http://www.jstor.org/stable/26458506 (accessed on 10 June 2022).

51. Lavidas, K.; Komis, V.; Achriani, A. Explaining faculty members' behavioral intention to use learning management systems. *J. Comput. Educ.* **2022**, 0123456789. [CrossRef]
52. Saleem, M.; Kamarudin, S.; Shoaib, H.M.; Nasar, A. Influence of augmented reality app on intention towards e-learning amidst COVID-19 pandemic. *Interact. Learn. Environ.* **2021**, 1–15. [CrossRef]
53. Wijaya, T.T.; Cao, Y.; Weinhandl, R.; Yusron, E. Applying the UTAUT Model to Understand Factors Affecting Micro-Lecture Usage by Mathematics Teachers in China. *Mathematics* **2022**, *10*, 1008. [CrossRef]
54. Mukminin, A.; Habibi, A.; Muhaimin, M.; Prasojo, L.D. Exploring the drivers predicting behavioral intention to use m-learning management system: Partial least square structural equation model. *IEEE Access* **2020**, *8*, 181356–181365. [CrossRef]
55. Taamneh, A.; Alsaad, A.; Elrehail, H.; Al-Okaily, M.; Lutfi, A.; Sergio, R.P. University lecturers acceptance of moodle platform in the context of the COVID-19 pandemic. *Glob. Knowl. Mem. Commun.* 2022; *Ahead-of-print*. [CrossRef]
56. Alenezi, A.R. Modeling the Social Factors Affecting Students' Satisfaction with Online Learning: A Structural Equation Modeling Approach. *Educ. Res. Int.* **2022**, *2022*, 2594221. [CrossRef]
57. Venkatesh, M.G.; Morris, G.B.; Davis, G.B.; Davis, F.D. User acceptance of information technology: Toward a unified view. *Manag. Inf. Syst. Q.* **2003**, *27*, 425–478. [CrossRef]
58. Azizan, S.N.; Lee, A.S.H.; Crosling, G.; Atherton, G.; Arulanandam, B.V.; Lee, C.E. Online Learning and COVID-19 in Higher Education: The Value of IT Models in Assessing Students' Satisfaction. *Int. J. Emerg. Technol. Learn.* **2022**, *17*, 245–278. [CrossRef]
59. Reyes, E.G.D.; Galura, J.C.; Pineda, J.L.S. C5-LMS design using Google Classroom: User acceptance based on extended Unified Theory of Acceptance and Use of Technology. *Interact. Learn. Environ.* **2022**, 1–10. [CrossRef]
60. El-Sofany, H.F.; El-Seoud, S.A. Implementing effective learning with ubiquitous learning technology during coronavirus pandemic. *Comput. Syst. Sci. Eng.* **2022**, *40*, 389–404. [CrossRef]
61. Mujalli, A.; Khan, T.; Almgrashi, A. University Accounting Students and Faculty Members Using the Blackboard Platform during COVID-19; Proposed Modification of the UTAUT Model and an Empirical Study. *Sustainability* **2022**, *14*, 2360. [CrossRef]
62. Salloum, S.A.; Al-emran, M.; Shaalan, K.; Tarhini, A. Factors affecting the E-learning acceptance: A case study from UAE. *Educ. Inf. Technol.* **2019**, *24*, 509–530. [CrossRef]
63. Abu-Al-Aish, A.; Love, S. Factors influencing students' acceptance of m-learning: An investigation in higher education. *Int. Rev. Res. Open Distance Learn.* **2013**, *14*, 82–107. [CrossRef]
64. Liu, Y.; Zhao, L.; Su, Y.-S. Exploring Factors of Preschool Parents' Behavioral Intention to Use Face Recognition Technology on Campus. *Front. Phys.* **2022**, *10*, 857751. [CrossRef]
65. Saprikis, V.; Avlogiaris, G.; Kataracha, A. Determinants of the intention to adopt mobile augmented reality apps in shopping malls among university students. *J. Theor. Appl. Electron. Commer. Res.* **2021**, *16*, 30. [CrossRef]
66. Law, K.M.Y.; Geng, S. Correction to: How innovativeness and handedness affect learning performance of engineering students? *Int. J. Technol. Des. Educ.* **2019**, *29*, 1207. [CrossRef]
67. Fishbein, M.; Ajzen, I. *Belief, Attitude, Intention, and Behavior: An Introduction to Theory and Research*; Addison-Wesley: Boston, MA, USA, 1975.
68. Venkatesh, V.; Davis, F.D. Theoretical extension of the technology acceptance model: Four longitudinal field studies. *Manage. Sci.* **2000**, *46*, 186–204. [CrossRef]
69. Palomares-Ruiz, A.; Cebrián-Martínez, A.; García-Toledano, E.; López-Parra, E. Digital gender gap in university education in Spain. Study of a case for paired samples. *Technol. Forecast. Soc. Chang.* **2021**, *173*, 121096. [CrossRef]
70. Garg, A. Investigating the Moderating Effects of Age and Gender on Customers' Use of Tablet Menu in Casual Dining Restaurants. *J. Qual. Assur. Hosp. Tour.* **2021**, 1–39. [CrossRef]
71. Erdogan, A.; Sahin, I. Relationship between math teacher candidates' Technological Pedagogical and Content Knowledge (TPACK) and achievement levels. *Procedia—Soc. Behav. Sci.* **2010**, *2*, 2707–2711. [CrossRef]
72. Wang, Y.S.; Wu, M.C.; Wang, H.Y. Investigating the determinants and age and gender differences in the acceptance of mobile learning. *Br. J. Educ. Technol.* **2009**, *40*, 92–118. [CrossRef]
73. Lin, T.C.; Tsai, C.C.; Chai, C.S.; Lee, M.H. Identifying Science Teachers' Perceptions of Technological Pedagogical and Content Knowledge (TPACK). *J. Sci. Educ. Technol.* **2013**, *22*, 325–336. [CrossRef]
74. Hair, J.; Black, B.; Babin, B.; Anderson, R.E.; Tatham, R.L. *Multivariate Data Analysis*, 6th ed.; Prentice Hall: Hoboken, NJ, USA, 2006.
75. Hair, J.F.; Hult, G.T.M.; Ringle, C.; Sarstedt, M. *A Primer on Partial Least Squares Structural Equation Modeling (PLS-SEM)*; SAGE Publications: Thousand Oaks, CA, USA, 2016.
76. Leow, L.P.; Phua, L.K.; Teh, S.Y. Extending the social influence factor: Behavioural intention to increase the usage of information and communication technology-enhanced student-centered teaching methods. *Educ. Technol. Res. Dev.* **2021**, *69*, 1853–1879. [CrossRef]
77. Jiang, H.; Islam, A.Y.M.A.; Gu, X.; Spector, J.M. Online learning satisfaction in higher education during the COVID-19 pandemic: A regional comparison between Eastern and Western Chinese universities. *Educ. Inf. Technol.* **2021**, *26*, 6747–6769. [CrossRef]
78. Kock, N. Common method bias in PLS-SEM: A full collinearity assessment approach. *Int. J. E-Collab.* **2015**, *11*, 1–10. [CrossRef]
79. Fornell, C.; Larcker, D.F. Evaluating structural equation models with unobservable variables and measurement error. *J. Mark. Res.* **1981**, *18*, 39–50. [CrossRef]

80. Wut, T.M.; Lee, S.W.; Xu, J.B. How do Facilitating Conditions Influence Student-to-Student Interaction within an Online Learning Platform? A New Typology of the Serial Mediation Model. *Educ. Sci.* **2022**, *12*, 337. [CrossRef]
81. Law, L.; Fong, N. Applying partial least squares structural equation modeling (PLS-SEM) in an investigation of undergraduate students' learning transfer of academic English. *J. English Acad. Purp.* **2020**, *46*, 100884. [CrossRef]
82. Morchid, N. The Determinants of Use and Acceptance of Mobile Assisted Language Learning: The Case of EFL Students in Morocco. *Arab World Eng. J.* **2019**, *5*, 76–97. [CrossRef]
83. Moorthy, K.; T'Ing, L.C.; Ming, K.S.; Ping, C.C.; Ping, L.Y.; Joe, L.Q.; Jie, W.Y. Behavioral Intention to Adopt Digital Library by the Undergraduates. *Int. Inf. Libr. Rev.* **2019**, *51*, 128–144. [CrossRef]
84. Henseler, J.; Ringle, C.M.; Sarstedt, M. A new criterion for assessing discriminant validity in variance-based structural equation modeling. *J. Acad. Mark. Sci.* **2015**, *43*, 115–135. [CrossRef]
85. Diamantopoulos, A.; Siguaw, J.A. Formative versus reflective indicators in organizational measure development: A comparison and empirical illustration. *Br. J. Manag.* **2006**, *17*, 263–282. [CrossRef]
86. Alvi, I. College students' reception of social networking tools for learning in India: An extended UTAUT model. *Smart Learn. Environ.* **2021**, *8*, 1–18. [CrossRef]
87. Mukminin, A.; Habibi, A.; Muhaimin, M.; Asrial, A.; Haryanto, E.; Setiono, P.; Sofyan, S. Vocational Technical High School teachers' beliefs towards ICT for the 21 st century education: Indonesian context. *Probl. Educ. 21st Century* **2019**, *77*, 22–38. [CrossRef]
88. Mulyani, E.A.; Alpusari, M.; Putra, E.D. The Effect of Learning Facilities and Family Environment on Motivation to Learn of Prospective Elementary Teacher Education on Online Learning. *J. Teach. Learn. Elem. Educ.* **2021**, *4*, 89. [CrossRef]
89. Shiferaw, K.B.; Mengiste, S.A.; Gullslett, M.K.; Zeleke, A.A.; Tilahun, B.; Tebeje, T.; Wondimu, R.; Desalegn, S.; Mehari, E.A. Healthcare providers' acceptance of telemedicine and preference of modalities during COVID-19 pandemics in a low-resource setting: An extended UTAUT model. *PLoS ONE* **2021**, *16*, 1–15. [CrossRef]
90. Wijaya, T.T.; Zhou, Y.; Houghton, T.; Weinhandl, R.; Lavicza, Z.; Yusop, F.D. Factors affecting the use of digital mathematics textbooks in indonesia. *Mathematics* **2022**, *10*, 1808. [CrossRef]
91. Stone, M. Cross-validatory choice and assessment of statistical predictions. *J. R. Stat. Soc.* **1974**, *36*, 111–133. [CrossRef]
92. Yusop, F.D. Exploring the factors affecting pre-service science teachers ' actual use of technology during teaching practice Exploring the factors affecting pre-service science t eachers ' actual use of technology during teaching practice. *S. Afr. J. Educ.* **2022**, *42*, 1–11. [CrossRef]
93. Almekhlafi, A.G. The effect of E-books on Preservice student teachers' achievement and perceptions in the United Arab Emirates. *Educ. Inf. Technol.* **2021**, *26*, 1001–1021. [CrossRef]
94. Akar, E.; Mardikyan, S. Analyzing factors affecting users ' behavior intention to use social media: Twitter. *Int. J. Bus. Soc. Sci.* **2014**, *5*, 85–95.
95. Turner, A. Generation Z: Technology and Social Interest. *J. Individ. Psychol.* **2015**, *71*, 103–113. [CrossRef]
96. Cilliers, E.J. The Challenge of Teaching Generation Z. *People Int. J. Soc. Sci.* **2017**, *3*, 188–198. [CrossRef]
97. Priporas, C.V.; Stylos, N.; Fotiadis, A.K. Generation Z consumers' expectations of interactions in smart retailing: A future agenda. *Comput. Hum. Behav.* **2017**, *77*, 374–381. [CrossRef]
98. Zhou, Y.; Li, X.; Wijaya, T.T. Determinants of Behavioral Intention and Use of Interactive Whiteboard by K-12 Teachers in Remote and Rural Areas. *Front. Psychol.* **2022**, *13*, 934423. [CrossRef]
99. Joa, C.Y.; Magsamen-Conrad, K. Social influence and UTAUT in predicting digital immigrants' technology use. *Behav. Inf. Technol.* **2021**, *41*, 1620–1638. [CrossRef]
100. Saxton, E.; Burns, R.; Holveck, S.; Kelley, S.; Prince, D.; Rigelman, N.; Skinner, E.A. A Common Measurement System for K-12 STEM education: Adopting an educational evaluation methodology that elevates theoretical foundations and systems thinking. *Stud. Educ. Eval.* **2014**, *40*, 18–35. [CrossRef]

sustainability

Article

Toward a Design Framework for Mathematical Modeling Activities: An Analysis of Official Exemplars in Hong Kong Mathematics Education

Chung-Kwan Lo *, Xiaowei Huang and Ka-Luen Cheung

Department of Mathematics and Information Technology, The Education University of Hong Kong, Hong Kong 999077, China
* Correspondence: chungkwanlo@eduhk.hk

Abstract: Mathematical modeling is considered a bridge to STEM education and has been incorporated into K–12 mathematics curriculums in various countries. However, it has a relatively short history in Hong Kong schools. The lack of high-quality, relevant exemplars of mathematical modeling activities is a challenge to teacher practice in this area. Hence, this study aims to establish a design framework for mathematical modeling activities suitable for teachers and students in Hong Kong. We explore the design and content of the official mathematical modeling exemplars published by the Hong Kong Education Bureau using a document analysis approach. The findings provide the basis for developing a framework to be used in the future design of mathematical modeling activities. Four exemplars were found and analyzed in terms of their structural components, level of learning experience in mathematical modeling, and design characteristics. Based on our findings, we discussed various strategies to enhance the design of a mathematical modeling activity, including setting diversified learning objectives, cross-subject collaboration when formulating the problem context and instructions, designing more activities suitable for average and underperforming students, emphasizing the evaluation of modeling outcomes, and providing relevant supporting materials. Our study thus lays the groundwork for advancing the teaching and learning of mathematical modeling in school contexts.

Keywords: mathematical modeling; modeling activity; mathematics education; STEM education; secondary education

Citation: Lo, C.-K.; Huang, X.; Cheung, K.-L. Toward a Design Framework for Mathematical Modeling Activities: An Analysis of Official Exemplars in Hong Kong Mathematics Education. *Sustainability* **2022**, *14*, 9757. https://doi.org/10.3390/su14159757

Academic Editor: Jesús C. Hernandez

Received: 2 July 2022
Accepted: 6 August 2022
Published: 8 August 2022

1. Introduction

Mathematical modeling and its applications have been receiving increasing attention worldwide. As Kertil and Gurel [1] described, mathematical modeling is a process of mathematizing, interpreting, verifying, revising, and generalizing real-life situations. It is a cyclical process (i.e., a modeling cycle) in which a real-world model is transformed into a mathematical model, and then the formulated model is validated and iteratively adjusted based on the real-life situation [2]. Mathematical modeling is widely applied in the field of sustainable development across STEM (Science, Technology, Engineering, and Technology) disciplines, such as monitoring water quality [3], waste treatment infrastructure planning [4], sustainable energy systems [5], financial risk assessment [6], and pandemic control strategy [7]. Accordingly, mathematical modeling is considered a bridge to STEM education [1].

Although training in mathematical modeling is most commonly offered at the university level, it has been the foundation for the mathematics framework of the Programme for International Student Assessment (which measures 15-year-olds' abilities) since 2015 [8]. As the interest in STEM education has increased, mathematical modeling has been incorporated into K–12 mathematics curriculums in various countries, such as Singapore [9] and Australia [10]. Sokolowski [11] conducted a meta-analysis which involved 1670 participants in mathematical modeling research across countries (e.g., USA, Germany, and Turkey). His results suggested that engaging students in mathematical modeling activities significantly promoted their understanding and application of mathematics concepts compared to traditional teaching methods. In particular, the interventions in high school settings showed an overall large effect on student achievement (g = 0.94, 95% CI [0.79, 1.08]) in favor of the mathematical modeling classrooms over the traditional classrooms [11]. This result thus provided the ground for practicing mathematical modeling activities in schools to enhance student learning of mathematics.

So, why do we need to investigate the way of designing mathematical modeling activities? In Hong Kong, there is a relatively short history of teaching and learning mathematical modeling in primary and secondary schools. The 2017 curriculum guide [12] was probably the first to explicitly introduce mathematical modeling into Hong Kong mathematics education. To promote the teaching and learning of mathematical modeling, Ang [9] pointed out that teachers need a set of good, ready-made exemplars of modeling problems and relevant resources. At the time of writing, however, we could find only four official (i.e., published by the Hong Kong Education Bureau) exemplars [12–15] of mathematical modeling activities that were accessible to teachers. Further effort is required to develop high-quality exemplars and resources relevant to the government's direction.

Therefore, we synthesized the design characteristics of the official mathematical modeling exemplars in Hong Kong using a document analysis approach, aiming at establishing a design framework for mathematical modeling activities suitable for our teachers and students. Using the established framework, researchers can have a more focused agenda for future research to examine the effect of mathematical modeling activities on students' ability and interest in mathematics in Hong Kong school contexts. Furthermore, this study can help researchers from other regions to understand how official mathematical modeling activities are designed in Hong Kong, laying the groundwork for further comparative studies involving different educational contexts.

2. Conceptual Background

The conceptual background of this study is threefold. First, we provide an overview of the mathematics education and mathematical modeling in Hong Kong schools. This can help readers understand the curriculum arrangements regarding the mathematics topics involved in this study and the government's effort in promoting mathematical modeling in recent years. Second, we explicate Ang's [9] three levels of learning experience in mathematical modeling. Third, we discuss recent design principles for mathematical modeling activities. These are the useful resources to establish the design framework in this study.

2.1. Mathematics Education and Mathematical Modeling in Hong Kong School Contexts

In Hong Kong, mathematics is a compulsory subject in both primary and secondary schools. The mathematics curriculum covers various learning units across different content areas (e.g., algebra and data handling) [12]. Nevertheless, the Education Bureau allows teachers to cater to the needs and abilities of their students through flexibility and diversification. For example, the secondary mathematics curriculum includes both foundation topics (i.e., essential concepts and knowledge) and non-foundation topics (i.e., content beyond the foundation topics). For the non-foundation topics, teachers can evaluate their suitability for and relevance to their students. In addition to the compulsory part of the curriculum, there is an optional extended part in the senior secondary school mathematics curriculum [12]. The extended part consists of two modules: Module 1 (Calculus and Statistics) and Module 2 (Algebra and Calculus). Students who wish to learn more advanced mathematics can take either Module 1 or Module 2. To illustrate how mathematics topics are arranged in the curriculum, Table 1 provides examples of several topics in three learning units (probability, exponential and logarithmic functions, and differentiation) [16,17]. These units are involved in the official exemplars retrieved in this study, and thus can help readers understand the analysis of the exemplars.

Table 1. Mathematics topics of three learning units in Hong Kong mathematics curriculum [16,17].

	Compulsory Part		
Learning Unit	**Foundation Topics**	**Non-Foundation Topics**	**Extended Part**
Probability	The concept of probability; the calculation of probabilities of events by listing the sample space and counting; etc.	The concept of expectation; the addition law and multiplication law of probability; the concept and notation of conditional probability; etc.	The concept of discrete probability distribution; the concepts of expectation $E[X]$; etc.
Exponential and logarithmic functions	None.	The definition and properties of logarithms; the properties of exponential functions and logarithmic functions; etc.	The definition of e and the exponential series; using exponential functions and logarithmic functions to solve problems; etc.
Differentiation	None.	None.	The addition rule, product rule, quotient rule, and chain rule of differentiation; solving the problems relating to rate of change, maximum, and minimum; etc.

The teaching and learning of mathematical modeling are at an initial stage in Hong Kong. Mathematical modeling was not explicitly mentioned in the government's mathematics curriculum documents until 2017. In response to the evolving changes in the contemporary world, the most recent mathematics curriculum guide for Primary 1 to Secondary 6 [12] has various updates, including an emphasis on STEM education and an introduction to mathematical modeling. It states that "STEM education could be strengthened through creating opportunities for students to apply the mathematical knowledge and skills in analysing and modelling real-life problems" [12] (p. 43). To this end, the curriculum guide recommends organizing STEM-related activities to provide students with "more opportunities to participate in mathematical modelling through identifying, formulating and solving the problem" [12] (p. 44). Therefore, topic- and project-based approaches to integrating learning elements from different key learning areas (e.g., science education and technology education) are proposed (see [12] (pp. 70–75) for a review). In other words, mathematical modeling can be (1) incorporated into the teaching and learning of specific topics inside the classroom and/or (2) conducted outside the classroom as a project-based learning activity.

In each of the last three academic years (i.e., 2019/20, 2020/21, and 2021/22), the Education Bureau implemented a seed project to promote STEM education by infusing mathematical modeling into secondary school mathematics education [18,19]. The implementation of these three consecutive seed projects reflects not only the Education Bureau's determination to promote STEM education through mathematics but also the substantial demand for professional development opportunities and school support. However, Hong Kong lacks established teacher professional development programs on mathematical modeling, unlike nearby regions. For example, the Singapore Ministry of Education and mathematics educators from the National Institute of Education in Nanyang Technological University have collaborated to provide professional development in designing and implementing mathematical modeling activities in schools since 2010 [20]. To enhance teachers' professional capacity in promoting STEM education through mathematics, The Education University of Hong Kong has launched a new professional development program on mathematical modeling (to be offered in 2023) for in-service teachers. Hence, further effort is required to investigate how to advance the teaching and learning of mathematical modeling in Hong Kong schools.

2.2. Three Levels of Learning Experience in Mathematical Modeling

To help teachers plan and set goals for a mathematical modeling activity, Ang [9] divided the learning experience in mathematical modeling into three levels of cognitive demands and expectations. Using this taxonomy, teachers have a clear trajectory to build their students' modeling capacities by moving from one level to the next. The three levels of learning experience are summarized below [9] (pp. 60–69):

- Level 1 focuses on acquiring skills in a modeling context. The skills are either purely mathematical skills or some specific skills that are used in mathematical modeling. Ang [9] observed that in Singapore, Level 1 modeling activities are trim enough to fit into a typical one- or two-period mathematics lesson. In his "Mountain climbing" activity, students were required to find an exponential function (i.e., a mathematical skill) that fitted real data on atmospheric pressures at various altitudes above sea level. In addition, the students were introduced to the use of Excel's Solver Tool (i.e., an IT skill) in finding the value of each parameter in their model.
- Level 2 focuses on developing students' modeling competencies in applying knowledge specific to mathematical modeling. The activity objectives at this level are to help students learn to make assumptions that simplify a problem, identify the factors that influence a variable, and interpret a mathematical solution in real-world terms, among other modeling skills. Ang [9] cautioned that the Level 2 modeling activities are more advanced than the Level 1 activities and thus require more instructional time. In his "Water warming" activity, a cup of ice-water was left to warm up, and its temperature was recorded every 5 s. Students developed a model that used their knowledge of differential equations to describe how the water temperature changed over time. Through this activity, teachers helped their students learn to state the factors that can affect water temperature and to make relevant assumptions about the warming process.
- Unlike Level 2 (which develops students' modeling competencies), the overarching objective of Level 3 is to tackle a mathematical modeling problem. Students are required to work in groups and apply various modeling skills, such as developing a model, solving the model, and making a presentation. The activities at this level further develop the students' modeling competencies, and they may take a few days to complete. In his "Accident at the MRT [Mass Rapid Transit] station" activity, Ang's [9] students were given the scenario of a girl who accidentally fell onto the tracks after walking in a random manner on the platform. The students were expected to communicate their ideas and construct a simulation model to study her random walk by (1) listing the variables in the model, (2) making assumptions about the situation and simplifying the problem, and (3) designing and carrying out the simulation.

2.3. Design Principles for Mathematical Modeling Activities

In addition to students' levels of learning experience in mathematical modeling, some researchers have considered the instructional dimension when designing mathematical modeling activities [21,22]. Geiger et al. [21] generated a set of task design principles based on both the seminal work of Galbraith [22] and their collaboration with frontline teachers. The researchers adopted and enriched five of Galbraith's [22] six principles, namely relevance and motivation, accessibility, feasibility of approach, feasibility of outcome, and didactical flexibility principles (Principles 2 to 5 and 7, Table 2). Based on the teachers' feedback, Geiger et al. [21] revealed the need to emphasize the nature and authenticity of the problems (i.e., open-ended real-world problems). Therefore, they formulated the nature of problem principle (Principle 1, Table 2) which emphasizes the open-endedness of a problem and the involvement of both intra-mathematical information (i.e., pure mathematics) and extra-mathematical information (i.e., information outside of mathematics). For brevity, Niss and Blum [23] noted that a situation in the extra-mathematical domain may simply be referred to as "context." In contrast, intra-mathematical problems are pure mathematical tasks that are not connected to reality [24].

Table 2. Design principles for mathematical modeling activities based on Geiger et al. [21] (p. 324) and Galbraith [22] (p. 235).

Principle	Description
Principle 1—Nature of problem	Problems must be open-ended and involve both intra- and extra-mathematical information. The degree of open-endedness depends on students' previous experience with modeling.
Principle 2—Relevance and motivation	There are some genuine links with students' everyday lives. Therefore, the problem context must be a part of their everyday experience or related to their personal circumstances.
Principle 3—Accessibility	It is possible to identify and specify mathematically tractable questions from a general statement. Suitable sub-questions can be implied by the general problem.
Principle 4—Feasibility of approach	The formulation of a solution process is feasible and involves (a) the use of mathematics available to students, (b) the making of necessary assumptions, and (c) the assembly of necessary data.
Principle 5—Feasibility of outcome	It is possible for students to solve the mathematics for a basic problem and interpret the results.
Principle 6—Feasibility of evaluation	An evaluation procedure is available that enables students to check for mathematical accuracy and the appropriateness of the solution in the contextual setting.
Principle 7—Didactical flexibility	The problem is structured into sequential questions that retain the integrity of the real-world situation. These questions are either given as occasional hints or used to provide organized assistance by scaffolding a line of investigation.

It is worth nothing that one of Galbraith's [22] principles of model evaluation was not included in Geiger et al. [21]. That principle emphasizes the feasibility of checking the mathematical accuracy and appropriateness of students' solutions in real-world settings. In line with Galbraith [22], we regard model evaluation as an essential procedure because mathematical modeling is a cyclical process in which a formulated model is validated and iteratively adjusted based on the real-world situation [2]. Therefore, we include Galbraith's [22] feasibility of evaluation principle (Principle 6, Table 2) in our analytical framework.

After synthesizing Geiger et al. [21] and Galbraith [22], Table 2 shows seven design principles which we drew as the foundation for developing a design framework relevant to Hong Kong school contexts. For example, if we wish to adopt the "Accident at the MRT station" activity [9], the relevance and motivation principle (Principle 2) suggests that we should use similar incidents in our own region to contextualize the problem for our students. Furthermore, the feasibility of approach principle (Principle 4) emphasizes the consideration of our students' mathematics backgrounds. Taking the "Mountain climbing" activity [9] as an example, if the modeling process involves only linear functions (a foundation topic in Hong Kong) instead of exponential functions (a non-foundation topic in Hong Kong), more students can manage and formulate a solution.

3. Methods

Although Hong Kong teachers lack established guidelines for designing mathematical modeling activities, the Education Bureau has published several exemplars which can inform our design. Accordingly, a document analysis approach is suitable for identifying the structural components (i.e., the parts or sections that the exemplars comprise) and design characteristics underpinning those exemplars. With the above conceptual background, the following research questions (RQ1 to RQ3) guided our study:

- RQ1: What structural components do the official exemplars comprise?
- RQ2: Which levels of learning experience in mathematical modeling do the official exemplars provide?

- RQ3: How can the design principles for mathematical modeling activities be enacted in the context of Hong Kong mathematics education?

The outcomes solicited from RQ1 can be used to formulate a template for teachers' activity plan, whereas the findings of RQ2 and RQ3 can inform the way of setting learning objectives and offering instructions of a mathematical modeling activity, respectively.

3.1. Document Analysis

Document analysis is a qualitative research method used for reviewing and evaluating documents to elicit meaning, gain understanding, and develop empirical knowledge [25,26]. This research approach has been widely applied in the field of sustainable development [27–32] and mathematics education [33]. Depending on researchers' objectives, their documents can take a variety of forms, such as policy documents [27–29], curriculum documents [30], community sustainability plans [31], and media articles [32]. For example, Yang [33] analyzed three textbook chapters on the Pythagorean theorem. She identified their common patterns to develop an understanding, compared their instructional approaches, and made appropriate recommendations. Through document analysis, researchers can further generate theories to explain phenomena and guide their actions [25].

According to Bowen [25], document analysis procedures entail first retrieving and selecting relevant documents and then appraising (i.e., making sense of) and synthesizing the data contained in those documents. These procedures are detailed in the following sub-sections.

3.2. Retrieval and Selection of Documents

To retrieve relevant documents, we searched the official website of the Hong Kong Education Bureau and examined the curriculum documents for mathematics education. As Mhlanga [32] noted, an advantage of this approach is that the documents are available in the public domain, making it easier for researchers and readers to obtain information without the need to ask permission from the original authors. The following four factors were considered when selecting exemplars of mathematical modeling activities [26] (pp. 71–72):

1. Authenticity: the extent to which a document is genuine.
2. Credibility: the extent to which a document is free from errors.
3. Representativeness: the extent to which a document is typical.
4. Meaning: whether the evidence in a document is clear and comprehensible.

As of 20 June 2022 (the date on which we finalized our search), we had found four exemplars. Table A1 in Appendix A shows the documents reviewed and the number of exemplars in each document. First, these exemplars are authentic because they are primary sources retrieved from the Education Bureau. As Table 3 shows, Exemplar 1 is published in the curriculum guide for mathematics education [12] (pp. 189–194), whereas Exemplars 2 to 4 [13–15] are published on the "Resources—STEM Examples: Examples on STEM Learning and Teaching Activities" website of the Education Bureau [34]. Notice that a slightly revised version of Exemplar 1 is also published on that website [35]. Second, these exemplars are credible because they were produced by the Education Bureau, and we retrieved the original documents. Third, these exemplars are representative because at the time of writing, they were the only available exemplars of mathematics modeling activities designed and published by the Education Bureau. Fourth, these exemplars are clearly written and understandable. Therefore, the four exemplars that we found were suitable documents for our analysis.

Table 3. The official exemplars analyzed in this study.

ID	Title	Source
Exemplar 1	"Modelling the spread of a disease"	Curriculum guide [12]
Exemplar 2	"Mathematical modelling on the accommodation demand of visitors to Hong Kong"	Official website [13]
Exemplar 3	"Mathematical modelling on decision-making: Probabilistic model"	Official website [14]
Exemplar 4	"Investigation on the relation between the maximum walking velocity and the length of legs"	Official website [15]

3.3. Data Analysis

We addressed our research questions through content analysis of the exemplars. To answer RQ1, we identified and summarized the structural components of the exemplars. To answer RQ2, we used the three levels of learning experience in mathematical modeling [9] presented in Section 2.2 as our framework for analysis. To answer RQ3, we used the design principles for the mathematical modeling activities [21,22] detailed in Section 2.3 as our lens through which to analyze the design characteristics of each exemplar. Although these frameworks provided a basis for our content analysis, we were open to refining or adding to them if new levels of learning experience or design principles emerged. To enhance the reliability of our analysis, all of the exemplars were double-coded by the first two authors and reviewed by the third author. In the event of disagreement, we re-examined the exemplars to come to a consensus. Multiple reviews were conducted until we reached perfect agreement.

4. Findings

4.1. RQ1: What Structural Components Do the Official Exemplars Comprise?

Table 4 summarizes the structural components of the official exemplars. Most of the components were found in all of the four exemplars, including their title, key stage (i.e., grade level), learning unit, objective, prerequisite knowledge, background information or scenario, description of activities, and references. However, only Exemplars 2 to 4 explain the exemplars' relationships with other key learning areas in Hong Kong secondary education. Exemplar 2 relates to a module in technology education ("Business Environments, Operations & Organisations") in junior secondary education [36], whereas Exemplars 3 and 4 relate to other subjects ("Business, Accounting and Financial Studies" [37] and "Physics" [38], respectively) in senior secondary education. In addition, only Exemplars 1 and 3 describe the generic skills involved in the activities, such as critical thinking and problem-solving skills. In terms of the number of sub-activities, Exemplar 1 consists of three different parts that discuss a simple epidemic model, a counter plague model, and a challenging problem that requires the application of calculus. By contrast, Exemplars 2 to 4 are divided into a series of three to four interrelated activities. Finally, Exemplars 2 and 3 further provide useful teaching and learning resources, including a list of resources required in the activities, an annex of the information necessary for the modeling, and student worksheets.

4.2. RQ2: Which Levels of Learning Experience in Mathematical Modeling Do the Official Exemplars Provide?

We identified the levels of learning experience in mathematical modeling that the official exemplars provide by making sense of their expectations and instructions. We distinguished between Level 1 (Exemplar 1), Level 2 (Exemplars 2 and 3), and Level 3 (Exemplar 4).

Exemplar 1 is a modeling activity related to the spread of a disease. There are three objectives: (1) "To help students relate STEM education with the real life," (2) "To let students recognise the mathematics in everyday life and apply information technology to solve problems," and (3) "To let students recognise mathematics as a powerful tool for planning." These objectives mainly focus on providing students with Level 1 learning experience in mathematical modeling. Specifically, in the three parts of this exemplar, the students are expected to learn to apply exponential functions (Model 1: a simple epidemic model), probability and expectation (Model 2: a counter plague model), and calculus (a challenging problem) in the context of modeling the spread of a disease. Although assumptions are involved in the modeling process, the methods of making those assumptions are given, along with the initial conditions. For example, "it is assumed that no one dies in 10 months" and "Initially there are 8 healthy people and 2 infected people." In other words, Exemplar 1 does not emphasize developing students' modeling competencies (e.g., making assumptions to simplify a problem).

Table 4. Structural components of the official exemplars.

Component	Description	Exemplar 1	Exemplar 2	Exemplar 3	Exemplar 4
Title	A title containing the problem context.	√	√	√	√
Key stage	The grade levels of the targeted students.	√	√	√	√
Strand	The content area (e.g., data handling).	None.	√	√	√
Learning unit	The learning unit(s) in which the exemplar is situated.	√	√	√	√
Objective	The objective(s) to be achieved.	√	√	√	√
Prerequisite knowledge	The knowledge required for the mathematical modeling activities.	√	√	√	√
Relationship with other learning areas	The related topic(s) in other subjects.	None.	√	√	√
Background information/scenario	An introduction to the problem context.	√	√	√	√
Number of sub-activities	The number of sub-activities in the exemplar.	3	4	3	3
Description of activities	The descriptions of each sub-activity, including teaching instructions, questions for discussion, and notes for teachers (teaching recommendations, suggested answers/solutions, and further information).	√	√	√	√
Generic skills involved	The generic skills (e.g., critical thinking) that the exemplar requires.	√	None.	√	None.
References	The materials (e.g., articles and websites) related to the modeling problem.	√	√	√	√
Others (appeared once)			List of resources; worksheets.	Information sheet.	

Exemplar 2 is a modeling activity related to the accommodation demand of visitors to Hong Kong. There are two objectives: (1) "To enrich students' experience in applying mathematics in handling daily life problems" and (2) "To enhance students' abilities in applying the concepts of percentage in modelling real-life situations." In a series of four sub-activities, the students are guided to formulate a basic model (i.e., involving only a growth factor in the number of visitors) and then progressively refine the model by considering other factors (e.g., whether or not the visitors would stay overnight). Although the objectives of Exemplar 2 focus on Level 1 learning experience in mathematical modeling, its activities attempt to develop students' modeling competencies, such as recognizing the assumptions involved in a model (e.g., "What assumptions are made in your model?") and interpreting a mathematical solution in real-world terms (e.g., "Which assumptions may be vulnerable to hold in real-life scenarios?"). Therefore, Exemplar 2 provides students with Level 2 learning experience in mathematical modeling.

Exemplar 3 is a modeling activity related to the decision-making process of buying a new smartphone. The following objective is stated: "To allow students to understand the applications of probability in modelling real-life scenarios such as decision-making process to make reasonable forecast and nurture students' entrepreneurial spirit." In a series of three sub-activities, the students are guided to formulate a basic model (i.e., involving two brands of smartphones) and then progressively refine the model by considering other factors (e.g., whether users are likely to buy a new smartphone) and incorporating additional information (e.g., the probability of a user replacing their smartphone in the next 12 months). Although its objective focuses on the Level 1 learning experience in mathematical modeling, its activities attempt to develop students' modeling competencies, such as recognizing the assumptions involved in a model (e.g., "the new smartphones of both Brand I, Brand S and other brands are available at roughly the same time") and identifying factors that influence a variable (e.g., "the users of Brand I smartphone using earlier generations might have different considerations from those using the latest available generation"). Therefore, Exemplar 3 provides students with Level 2 learning experience in mathematical modeling.

Exemplar 4 is a modeling activity related to the relationship between the maximum walking velocity and the length of legs. The following objective is stated: "To explain some phenomena in real-life situations through mathematical modelling." A more specific direction of investigation is provided in its background information section: "to explore a suitable simple model to express the relation between the maximum walking velocity and the length of legs." Therefore, this exemplar has some characteristics of Level 3 mathematical modeling activities in which students are expected to tackle the modeling problem. This exemplar includes three sub-activities. In Activity 1, the students are divided into groups and collect data on the lengths of their legs and their corresponding walking velocities. In Activity 2, the students formulate a model (e.g., a linear function or quadratic function) that fits their data. However, we identified several instructions (e.g., "The teacher guides students to extract the data" and the "students in each group examine whether the linear model of function … is suitable to describe the relation") that facilitate the students' investigation but might limit the level of their learning experience. Finally, Activity 3 is designed only for students who study senior secondary school physics. The students are guided to use their physics knowledge to formulate a theoretical model of the modeling problem. Their theoretical model is then compared with the modeling outcomes in Activity 2.

4.3. RQ3: How Can the Design Principles for Mathematical Modeling Activities Be Enacted in the Context of Hong Kong Mathematics Education?

We examined the design characteristics underpinning the official exemplars using the framework presented in Table 2. Tables 5 and 6 summarize how the seven design principles of mathematical modeling activities are enacted in the exemplars. Table 5 shows that their design characteristics are similar in terms of Principle 2 (relevance and motivation), Principle 3 (accessibility), Principle 5 (feasibility of outcome), and Principle 7 (didactical flexibility). Taking the didactical flexibility principle (Principle 7) as an example, all of the four exemplars provide suggested procedures for class activities, notes for teachers, and scaffolded questions, enabling teachers to assist students' learning and investigation.

Despite the similarity of certain design characteristics, Table 6 shows that the exemplars are different in their enactments of Principle 1 (nature of problem), Principle 4 (feasibility of approach), and Principle 6 (feasibility of evaluation). First, the nature of problem principle (Principle 1) emphasizes the open-endedness of the problem in a mathematical modeling activity. Although the problems in all four exemplars are open-ended, we found that Exemplars 1 and 4 are different from Exemplars 2 and 3. The former are open-ended in terms of the mathematical approach to modeling. Taking Exemplar 1 as an example, it discusses the following approaches to modeling the spread of a disease: (1) a simple epidemic model using knowledge of exponential functions; (2) a counter plague model using knowledge of probability and expectation; and (3) a theoretical solution using knowledge of calculus. In contrast, only one major mathematical approach is adopted in Exemplar 2 (growth rate) and Exemplar 3 (probability and relative frequency). Nevertheless, these two exemplars are open-ended in terms of their approaches to making assumptions in the modeling process. In a series of activities, the students are guided to consider new variables (e.g., Exemplar 3: "Consider a new smartphone?") and refine their assumptions (e.g., Exemplar 2: "Which assumptions may be vulnerable to hold in real-life scenarios? Why? How can it be refined so as to improve the accuracy of the model?") to improve their models.

Table 5. Similar enactments of the design principles in the official exemplars.

Principle (Description)	Exemplar	Representative Quotes
Principle 2—Relevance and motivation. (The problem context is related to students' everyday lives.)	1	"Bird flu, SARS and Ebola are examples of fatal epidemics that have emerged in a large scale in the past two decades."
	2	"Tourism industry is a mainstay of Hong Kong's economy."
	3	"The activities to be introduced are based on real life scenario on the decision-making process of buying a new smartphone."
	4	"When you are hurry to somewhere, you may walk very fast."
Principle 3—Accessibility. (Sub-questions can be implied by the general problem.)	1	"What is the difference if 3 persons are infected at each stage?"
	2	"Please suggest some ways to estimate the number of visitor arrivals to Hong Kong in a whole year."
	3	"the teacher may . . . ask students to represent the scenario with a tree diagram with suitably defined events."
	4	"students in each group examine whether the linear model of function . . . is suitable to describe the relation between V (their walking velocities) and L (lengths of their legs)."
Principle 5—Feasibility of outcome. (The solution can be addressed by students with relevant knowledge. Suggested questions are provided to guide their interpretation.)	1	"How many steps will it take to infect all the people in the classroom? How about the whole school?"
	2	"When will the room supply be inadequate if all growth rates are unchanged?"
	3	"by using the tree diagram, . . . Predict the future market share of Brand I and Brand S after their release of new models."
	4	"According to students' mathematical knowledge involving functions and indices, students may try the following functions for explorations: $V = aL^2 + bL + c \ldots$; $V = ab^L \ldots$ "
Principle 7—Didactical flexibility. (Procedures of class activities, notes for teachers, and scaffolded questions are provided.)	1	"Does the epidemic take off or die out in each case?"
	2	"What information is needed for constructing the model?"
	3	"Which group of smartphone users has greater brand loyalty, users of Brand I or Brand S?"
	4	"1. The class is divided into several groups . . . 2. In each group, every student walks as fast as possible . . . 3. In each group, the length of the legs of each group member is calculated . . . "

Second, the feasibility of approach principle (Principle 4) includes three aspects: (a) the use of mathematics available to students, (b) the making of necessary assumptions, and (c) the assembly of necessary data. The major difference between the four exemplars concerns the use of mathematics available to students. Accordingly, the four exemplars target different students. Only Exemplar 2 is suitable for all of the students in Hong Kong because it involves only the foundation topics (e.g., concepts of percentage changes and growth rates) in the mathematics curriculum. Exemplar 3 is suitable for more capable students because it involves non-foundation topics (e.g., the concept and notation of conditional probability). Only some specific groups of students have adequate knowledge to complete all of the sub-activities in Exemplars 1 and 4 (Exemplar 1: students taking the extended part of the mathematics curriculum; Exemplar 4: students taking senior secondary school physics).

Third, the feasibility of evaluation principle (Principle 6) emphasizes that the students should be allowed to evaluate their model in their real-world setting. In Exemplar 2, this evaluation is feasible using the data on government websites, such as the Hong Kong Tourism Commission and the Hong Kong Census and Statistics Department. These websites are provided in both the teacher notes and student worksheets. An evaluation is stated in the teacher notes: "The model [formulated in Activity 1] does not fit the data in

2015, 2016 and 2017." In Exemplar 4, the students collect data and formulate their models in Activities 1 and 2, respectively. Their modeling outcomes are then compared with the theoretical model formulated in Activity 3. However, Activity 3 requires physics knowledge (e.g., circular motion and centripetal acceleration). Therefore, not all of the students can formulate the theoretical model and make the required comparison. In Exemplars 1 and 3, no explicit statements about model evaluation were identified.

Table 6. Different enactments of the design principles in the official exemplars.

Principle	Exemplar	Description
Principle 1—Nature of problem	1 and 4	The problem is open-ended in terms of mathematical approaches to modeling.
	2 and 3	The problem is open-ended in terms of approaches to making assumptions.
Principle 4—Feasibility of approach	1	The formation of a solution requires knowledge of the concept of probability and expectation (JS; F), exponential functions (SS; NF), and calculus (SS; EP).
	2	The formation of a solution requires knowledge of the concepts of percentage changes and growth rates (JS; F). The websites with the necessary data are provided in the teacher notes and student worksheets.
	3	The formation of a solution requires the concept of probability (JS; F) and the concept and notation of conditional probability (SS; NF). The necessary data are provided in the annex of this exemplar.
	4	The formation of a solution requires knowledge of the concepts of linear (SS; F), quadratic (SS; F), and exponential and logarithmic (SS; NF) functions. Activity 3 also requires knowledge of physics (SS).
Principle 6—Feasibility of evaluation	1 and 3	None.
	2	Evaluation is feasible using the provided data on government websites, such as the websites of the Hong Kong Tourism Commission and the Hong Kong Census and Statistics Department.
	4	The students' modeling outcomes in Activity 2 are compared with the theoretical model formulated in Activity 3.

Note. JS = junior secondary; SS = senior secondary. EP = topics in the extended part; F = foundation topics; NF = non-foundation topics.

5. Discussion

The findings and implications pertaining to each research question are discussed in the following sub-sections. After that, the limitations of this study are acknowledged with recommendations for future research.

5.1. *Structural Components of a Mathematical Modeling Activity Plan (RQ1)*

In this study, we first analyzed the structural components of the official mathematical modeling exemplars (see Table 4). By and large, they can provide some basic information (e.g., content area and objective) and the details of their activities (e.g., descriptions of each sub-activity and questions for discussion). However, we found that their expected number of lessons and target students were missing. For example, students in Exemplar 4 have to collect data in an open area (where students can measure the time of walking along a 10 m straight path) and then analyze the data. The activity may require more instructional time. In addition, one of its sub-activities is only suitable for students taking senior secondary school physics, as it requires physics knowledge (e.g., circular motion and centripetal acceleration) to formulate a theoretical model. These are some practical issues that teachers may consider when adopting the mathematical modeling activity in their classrooms. Therefore, duration and target students of the activity should be mentioned in the future. To summarize, Table 7 proposes a template for a mathematical modeling activity plan. Educators can use the template to guide them in designing future activities.

Table 7. A template for a mathematical modeling activity plan.

Section	Component
Basic information	Title Key stage (or grade level) Strand (or content area) Learning unit Objective Prerequisite knowledge Relationship with other learning areas Duration (or number of lessons) Target students Background information (or scenario)
Activities (3 to 4 sub-activities)	Descriptions of each sub-activity Questions for discussion Note for teachers (including teaching recommendations, suggested answers/solutions, and further information)
Other information	Generic skills involved References
Annex	List of resources required Information sheets Sample worksheets

5.2. *Setting Diversified Learning Objectives of a Mathematical Modeling Activity (RQ2)*

Our findings suggested that the official exemplars are situated at Levels 1 to 3 of learning experience in mathematical modeling, as detailed in Section 4.2. This indicated the feasibility of using all three levels of mathematical modeling activities in Hong Kong schools. However, Ang [9] argued that teachers should progressively build students' mathematical modeling capacity by moving from one level to the next over time. Therefore, future activity plans should be flexible in terms of their objectives and learning tasks. Using the taxonomy of Ang [9], we suggested that the objectives of future mathematical modeling activities can be threefold (i.e., Levels 1 to 3), with corresponding recommended learning tasks. Teachers can conduct the activities based on their students' needs and abilities. By doing so, they can better enact the didactical flexibility principle (Principle 7) [21,22]. Using the problem context of Exemplar 1 (originally Level 1), Table 8 demonstrates a possible method of designing a set of diversified objectives and learning tasks that provide different levels of learning experience in mathematical modeling.

Table 8. Diversified objectives and learning tasks in Exemplar 1 by level of learning experience in mathematical modeling.

Level	Objective	Learning Task
1	To enhance the students' skills in applying their knowledge of differential equations in modeling the spread of a disease.	With an infection rate and recovery rate along with other necessary assumptions and initial conditions, the students apply their knowledge to formulate the model.
2	To develop the students' modeling competencies in simplifying the problem of the spread of a disease.	The students are directed to identify the variables in the model and to make assumptions to simplify the problem.
3	To explore a suitable model to express the spread of a disease.	The students formulate the model based on their knowledge.

5.3. *Enactment of Design Principles for Mathematical Modeling Activities (RQ3)*

Our findings suggested that the design principles of Geiger et al. [21] and Galbraith [22] were enacted, albeit with some variations, as shown in Tables 5 and 6. These findings can provide teachers with insights into the design of mathematical modeling activities. Several implications are discussed in the following sub-sections.

5.3.1. Principles 1 to 3: The Need for Cross-Subject Collaboration

In terms of their problem contexts, all of the exemplars involve extra-mathematical information and relate to students' everyday lives, indicating the enactment of the nature of problem principle (Principle 1) [21] and the relevance and motivation principle (Principle 2) [21,22]. Exemplar 1 is related to mathematical biology, whereas Exemplars 2 to 4 explicitly state their relationship with other learning areas in STEM, including tourism, business, and physics. Accordingly, it is unreasonable to expect that mathematics teachers alone can design the mathematical modeling activities that involve extra-mathematical knowledge. For example, they may not be able to explain the relationship between the established mathematical models and the tourism industry (Exemplar 2) and the real benefits of mathematical modeling in the global smartphone market (Exemplar 3).

As suggested in previous studies, cross-subject collaboration can facilitate the implementation of STEM education in both classroom and extracurricular activity settings [39]. Therefore, the development of mathematical modeling activities requires the involvement of teachers in different subject areas. Teachers in other subjects can help mathematics teachers to both formulate the problem context and design subject-specific instructions. However, it is worth noticing that not all problems can be addressed using the mathematics available to secondary school students. Therefore, mathematics teachers should ensure the enactment of the accessibility principle (Principle 3) [21,22] by evaluating the possibility of identifying and specifying mathematically tractable questions from the formulated problems.

5.3.2. Principles 4 and 5: The Need for More Mathematical Modeling Activities in Foundation Topics and Supporting Data in Modeling

According to the feasibility of approach principle (Principle 4) [21,22] and the feasibility of outcome principle (Principle 5) [21,22], students in a mathematical modeling activity should have adequate knowledge to formulate a solution process and address the solution. However, we found that only Exemplar 2 is suitable for all of the students in Hong Kong (see Table 6). The mathematical modeling activities in other exemplars involve some non-foundation topics (Exemplars 1, 3, and 4) and topics in the extended part of the mathematics curriculum (Exemplar 1). The average and underperforming students may not have adequate knowledge to handle the problems in these exemplars. There is thus a need for designing more mathematical modeling activities in the foundation topics (see Table 1).

Besides mathematics knowledge, the feasibility of approach principle (Principle 4) [21,22] emphasizes the importance of assembling the necessary data in the modeling process. Teachers should be aware that students' modeling processes can be hindered by the fact that not all real-world data and research reports are publicly available. For example, Exemplar 3 requires the use of marketing research data which may not be accessible to students. Therefore, an information sheet with the data on the global smartphone market is provided for students to formulate their model in the annex of this exemplar. In Exemplar 2, unfortunately, some hyperlinks to necessary data are no longer working. In future mathematical modeling activity plans, teachers can consider appending the essential data required in the modeling process to ensure the enactment of Principle 4 [21,22].

5.3.3. Principle 6: The Need for Strengthening the Evaluation of Modeling Outcomes

Our findings revealed that the enactment of the feasibility of evaluation principle (Principle 6) [22] requires reinforcement in future mathematical modeling activities. As Galbraith [22] noted, students' modeling outcomes should be tested in their real-world settings. However, we found that Exemplars 1 and 3 do not explicitly emphasize the evaluation of students' modeling outcomes (see Table 6). For Exemplar 1, such an evaluation is feasible if the students obtain real-world data on the spread of a disease, such as its infection rate and recovery rate, whereas for Exemplar 3, the evaluation is feasible if the students obtain recent data from a marketing research report. Taking the spread of the COVID-19 pandemic as an example, real-world global data are available on the Web. As of 24 June 2022, for example, the recovery rate in Hong Kong was 97.4% [40]. In addition, teachers can ask students to compare the new model to the previously formulated models (i.e., the simple epidemic model and the counter plague model) in terms of their reliability and precision. Using real-world data and relevant instructions, future mathematical modeling activities can emphasize the feasibility of testing, refining, and comparing students' modeling outcomes.

5.3.4. Principle 7: The Need for Supporting Materials

Finally, we found that some official exemplars can better support the enactment of the didactical flexibility principle (Principle 7) [21,22]. First, Exemplar 2 provides sample worksheets (see Table 4) in which teachers can understand how the modeling problem is structured into sequential activities and questions. The use of such ready-made resources not only decreases frontline teachers' workloads developing instructional materials but also facilitates their teaching practice [41]. Second, although all of the exemplars list their references, those provided in Exemplar 4 are worth mentioning. Its references [42–44] are directly related to the modeling problem in this exemplar (i.e., the relation between maximum walking velocity and the length of legs), which can help teachers prepare for the activity and support their students' independent exploration of the problem. Therefore, we recommend that these supporting materials be provided in future activity plans for mathematical modeling.

5.4. Limitations and Recommendations for Future Research

There are some limitations of this study. We acknowledge that the number of available official exemplars was limited. Therefore, the proposed set of design principles is preliminary and merits further research for improvement. We recommend that empirical studies be conducted in teacher education and professional development settings to solicit pre-service teachers' and in-service teachers' suggestions for enhancing the design framework.

Second, this study employed a document analysis approach. Therefore, our analysis focused only on the content of the official exemplars. Further authentic classroom studies are required to examine whether the mathematical modeling activities can develop students' abilities and increase their interest in mathematics. Besides the official exemplars, teachers and researchers can use our framework to design more mathematical modeling activities for teaching and research purposes. Future research can thus have a more focused agenda to examine the efficacy of mathematical modeling activities and determine whether the design framework can be applied in other regions at similar grade levels.

Finally, the COVID-19 pandemic has a profound impact on educational systems, where some instructional activities have been transitioned fully online. Future research can examine the teaching and learning of mathematical modeling in different learning environments (e.g., face-to-face, online, and hybrid) and propose updates on the curriculum for sustainable practices [45–47].

6. Conclusions

This study analyzed four official mathematical modeling exemplars published by the Hong Kong Education Bureau. Our analysis was underpinned by Ang's [9] taxonomy of learning experience in mathematical modeling and some existing principles for designing mathematical modeling activities [21,22]. Our findings indicated the feasibility of setting diversified learning objectives in future mathematical modeling activity plans, which enables teachers to conduct the activities based on their students' needs and abilities. With reference to the official exemplars, we discussed various strategies to ensure a better enactment of the design principles in future mathematical modeling activities, including cross-subject collaboration, designing mathematical modeling activities in the foundation topics, emphasizing the evaluation of modeling outcomes, and providing more supporting materials (e.g., student worksheets and references about the modeling problem in concern). Hence, the design framework proposed is useful for teachers and researchers to design mathematical modeling activities that are suitable for both teachers and students in Hong Kong. This study also serves as a reference for teachers and researchers from other regions, informing both their development of mathematical modeling activities and the methodology of building their design framework.

Author Contributions: Conceptualization, methodology, C.-K.L.; validation, formal analysis, investigation, resources, data curation, C.-K.L., X.H. and K.-L.C.; writing—original draft preparation, writing—review and editing, visualization, supervision, project administration, funding acquisition, C.-K.L. All authors have read and agreed to the published version of the manuscript.

Funding: This research was funded by the Department of Mathematics and Information Technology, The Education University of Hong Kong, reference code MIT/SGA03/21-22.

Institutional Review Board Statement: Not applicable.

Informed Consent Statement: Not applicable.

Data Availability Statement: The data used in this study can be found in the included documents (i.e., the official exemplars of mathematical modeling activities, as published by the Hong Kong Education Bureau).

Conflicts of Interest: The authors declare no conflict of interest.

Appendix A

Table A1. Documents reviewed in this study.

Documents/Sources * (Year of Publication, If Any)	Exemplars	Included Exemplars
Mathematics Education Key Learning Area Curriculum Guide (Primary 1—Secondary 3) (2002)	$N = 13$	$N = 0$
Mathematics Education Key Learning Area Curriculum Guide (Primary 1—Secondary 6) (2017)	$N = 22$ **	$N = 1$
Mathematics Key Learning Area—Pure Mathematics Curriculum and Assessment Guide (Advanced Level) (2004)	$N = 3$	$N = 0$
Target Oriented Curriculum Programme of Study for Mathematics—Key Stage 1 (primary 1–3) (1995)	$N = 5$	$N = 0$
Target Oriented Curriculum Programme of Study for Mathematics—Key Stage 2 (Primary 4–6) (1995)	$N = 7$	$N = 0$
(Website) Resources—STEM Examples: Examples on STEM Learning and Teaching Activities	$N = 29$ ***	$N = 3$

Note. * Other reviewed documents without exemplars included: Checklist of major updates on Mathematics Curriculum and Assessment Guide (S4–6) (November 2015); Comparison between Syllabuses for Secondary Schools—Mathematics (Secondary 1–5) (1999) and Syllabuses for Secondary Schools—Syllabus for Mathematics (Forms I–V) (1985); Comparison of Syllabuses for Primary Schools—Mathematics (1983) and Mathematics Education Key Learning Area—Mathematics Curriculum Guide (P1–P6) (2000); Comparison of Target Oriented Curriculum Programme of Study for Mathematics (1995) and Mathematics Education Key Learning Area—Mathematics Curriculum Guide (P1–P6) (2000); Comparison of the Content of the Revised Mathematics Curriculum and Current Mathematics Curriculum; Consultation on the Revised Mathematics Curriculum (P1–S6) (held from March to May 2017); Explanatory Notes to Junior Secondary Mathematics Curriculum (2020); Explanatory Notes to Primary Mathematics Curriculum—Key Stage 1 (2018); Explanatory Notes to Primary Mathematics Curriculum—Key Stage 2 (2020); Explanatory Notes to Senior Secondary Mathematics Curriculum—Module 1 (with updates in August 2018); Explanatory Notes to Senior Secondary Mathematics Curriculum—Module 2 (with updates in August 2018); Explanatory Notes to Senior Secondary Mathematics Curriculum—Module 1; Explanatory Notes to Senior Secondary Mathematics Curriculum—Module 2; Explanatory Notes to Senior Secondary Mathematics Curriculum—The Compulsory Part (with updates in December 2021); Explanatory Notes to Senior Secondary Mathematics Curriculum—The Compulsory Part; Guidelines on Catering for Learner Diversity and Creating Space in Senior Secondary Mathematics (2021); Information Sheet "Summary of Changes to the Contents of Syllabuses for Secondary Schools—AL Applied Mathematics (1992)"; Mathematics Curriculum and Assessment Guide (Secondary 4–6) (2007); Mathematics Curriculum and Assessment Guide (Secondary 4–6) (with updates in

December 2017); Mathematics Curriculum and Assessment Guide (Secondary 4–6) (with updates in November 2015); Mathematics Education Key Learning Area—Additional Mathematics Curriculum Guide (S4–S5) (2001); Mathematics Education Key Learning Area—Mathematics Curriculum Guide (P1–P6) (2000); Recommended implementation timeline for the revised Mathematics curriculum; Recommended implementation timeline for the revised senior secondary Mathematics curriculum; Supplement to Mathematics Education Key Learning Area Curriculum Guide: Learning Content of Primary Mathematics (2017); Supplement to Mathematics Education Key Learning Area Curriculum Guide: Learning Content of Junior Secondary Mathematics (2017); Supplement to Mathematics Education Key Learning Area Curriculum Guide: Learning Content of Senior Secondary Mathematics (2017); Supplementary Notes on Teaching of Advanced Supplementary Level—Mathematics and Statistics (1998); Supplementary Notes to Senior Secondary Mathematics Curriculum; Syllabuses for Primary Schools: Mathematics (1973); Syllabuses for Primary Schools: Mathematics (1983); Syllabuses for Secondary Schools—Additional Mathematics (Secondary 4–5) (1992); Syllabuses for Secondary Schools—Applied Mathematics (Advanced Supplementary Level) (1998); Syllabuses for Secondary Schools—Applied Mathematics (Advanced Level) (1992); Syllabuses for Secondary Schools—Mathematics & Statistics (Advanced Supplementary Level) (1991); Syllabuses for Secondary Schools—Mathematics (Secondary 1–5) (1999); Syllabuses for Secondary Schools—Mathematics (Secondary 1–5) (1999); Syllabuses for Secondary Schools—Pure Mathematics (Advanced Level) (1992); Syllabuses for Secondary Schools—Syllabus for Mathematics (Forms I–V) (1985). ** The Exemplar of "Flippable Measure Spoons" remarks that it involves "the ideas of mathematical modelling" [12] (p. 181). However, it appears that the primary focus of this exemplar is not highly related to mathematical modeling. Therefore, the exemplar was not included in our analysis. *** A slightly revised version of Exemplar 1 ("Modelling the spread of a disease," as published in the curriculum guide for mathematics education [12]) is also published on this website [34].

References

1. Kertil, M.; Gurel, C. Mathematical modeling: A bridge to STEM education. *Int. J. Educ. Math. Sci. Technol.* **2016**, *4*, 44–55. [CrossRef]
2. Stender, P. The use of heuristic strategies in modelling activities. *ZDM* **2018**, *50*, 315–326. [CrossRef]
3. Xu, R.; Pang, Y.; Hu, Z.; Hu, X. The spatiotemporal characteristics of water quality and main controlling factors of algal blooms in Tai Lake, China. *Sustainability* **2022**, *14*, 5710. [CrossRef]
4. Šomplák, R.; Kropáč, J.; Pluskal, J.; Pavlas, M.; Urbánek, B.; Vítková, P. A multi-commodity mathematical modelling approach–Hazardous waste treatment infrastructure planning in the Czech Republic. *Sustainability* **2022**, *14*, 3536. [CrossRef]
5. Nazir, M.S.; Abdalla, A.N.; Metwally, A.S.M.; Imran, M.; Bocchetta, P.; Javed, M.S. Cryogenic-energy-storage-based optimized green growth of an integrated and sustainable energy system. *Sustainability* **2022**, *14*, 5301. [CrossRef]
6. Tobisova, A.; Senova, A.; Izarikova, G.; Krutakova, I. Proposal of a methodology for assessing financial risks and investment development for sustainability of enterprises in Slovakia. *Sustainability* **2022**, *14*, 5068. [CrossRef]
7. Valenzuela-Fonseca, B.; Linfati, R.; Escobar, J.W. Strategy for locating people to reduce the transmission of COVID-19 using different interference measures. *Sustainability* **2022**, *14*, 529. [CrossRef]
8. OECD. *PISA 2015 Assessment and Analytical Framework: Science, Reading, Mathematic and Financial Literacy*, revised ed.; OECD Publishing: Paris, France, 2017.
9. Ang, K.C. Mathematical modelling in Singapore schools: A framework for instruction. In *Mathematical Modelling: From Theory to Practice*; Lee, N.H., Ng, K.E.D., Eds.; World Scientific: Singapore, 2015; pp. 57–72.
10. Geiger, V. Mathematical modelling in Australia. In *Mathematical Modelling: From Theory to Practice*; Lee, N.H., Ng, K.E.D., Eds.; World Scientific: Singapore, 2015; pp. 73–82.
11. Sokolowski, A. The effects of mathematical modelling on students' achievement—Meta-analysis of research. *IAFOR J. Educ.* **2015**, *3*, 93–114. [CrossRef]
12. Curriculum Development Council. *Mathematics Education: Key Learning Area Curriculum Guide (Primary 1—Secondary 6)*; The Education Bureau: Hong Kong, 2017.
13. Mathematical Modelling on the Accommodation Demand of Visitors to Hong Kong. Available online: https://www.edb.gov.hk/attachment/en/curriculum-development/kla/ma/res/STEM_example_JS_MM_visitors_accommodation_eng.pdf (accessed on 20 June 2022).
14. Mathematical Modelling on Decision-making: Probabilistic Model. Available online: https://www.edb.gov.hk/attachment/en/curriculum-development/kla/ma/res/STEM_example_sec_probabilistic%20model_eng.pdf (accessed on 20 June 2022).
15. Investigation on the Relation between the Maximum Walking Velocity and the Length of Legs. Available online: https://www.edb.gov.hk/attachment/en/curriculum-development/kla/ma/res/STEM_example_sec_velocity_legs_eng.pdf (accessed on 20 June 2022).
16. Curriculum Development Council. *Supplement to Mathematics Education Key Learning Area Curriculum Guide: Learning Content of Junior Secondary Mathematics*; The Education Bureau: Hong Kong, 2017.
17. Curriculum Development Council. *Supplement to Mathematics Education Key Learning Area Curriculum Guide: Learning Content of Senior Secondary Mathematics*; The Education Bureau: Hong Kong, 2017.
18. Collaborative Research and Development ("Seed") Projects/"Seed" Projects. Available online: https://www.edb.gov.hk/en/curriculum-development/kla/ma/seed.html (accessed on 25 May 2022).

19. Collaborative Research and Development ("Seed") Projects for the 2021/22 School Year Proposed by the Curriculum Development Institute. Available online: https://www.edb.gov.hk/en/curriculum-development/major-level-of-edu/seed/whatsnew-2021-22/index.html (accessed on 25 May 2022).
20. Ng, K.E.D. Towards a professional development framework for mathematical modelling: The case of Singapore teachers. *ZDM* **2018**, *50*, 287–300.
21. Geiger, V.; Galbraith, P.; Niss, M.; Delzoppo, C. Developing a task design and implementation framework for fostering mathematical modelling competencies. *Educ. Stud. Math.* **2022**, *109*, 313–336. [CrossRef]
22. Galbraith, P. Real world problems: Developing principles of design. In *Identities, Cultures, and Learning Spaces: Proceedings of the 29th Annual Conference of the Mathematics Education Research Group of Australasia*; Grootenboer, P., Zevenbergen, R., Chinnappan, M., Eds.; Mathematics Education Research Group of Australasia Inc.: Adelaide, SA, USA, 2006; Volume 1, pp. 229–236.
23. Niss, M.; Blum, W. *The Learning and Teaching of Mathematical Modelling*; Routledge: London, UK; New York, NY, USA, 2020.
24. Krawitz, J.; Schukajlow, S. Do students value modelling problems, and are they confident they can solve such problems? Value and self-efficacy for modelling, word, and intra-mathematical problems. *ZDM* **2018**, *50*, 143–157. [CrossRef]
25. Bowen, G.A. Document analysis as a qualitative research method. *Qual. Res. J.* **2009**, *9*, 27–40. [CrossRef]
26. Morgan, H. Conducting a qualitative document analysis. *Qual. Rep.* **2021**, *27*, 64–77. [CrossRef]
27. Duman, O.; Mäntysalo, R.; Granqvist, K.; Johnson, E.; Ronikonmäki, N.-M. Challenges in land use and transport planning integration in Helsinki Metropolitan Region—A historical-institutional perspective. *Sustainability* **2022**, *14*, 146. [CrossRef]
28. Lewis, C.T.; Su, M.-C. Climate change adaptation and sectoral policy coherence in the Caribbean. *Sustainability* **2021**, *13*, 8518. [CrossRef]
29. Taylor, B.; Nunes, R. Entitlement, indeterminacy and professional discretion in urban planning: Problematising a child's right to clean air for play in London. *Sustainability* **2022**, *14*, 6189. [CrossRef]
30. Bourn, D.; Soysal, N. Transformative learning and pedagogical approaches in education for sustainable development: Are initial teacher education programmes in England and Turkey ready for creating agents of change for sustainability? *Sustainability* **2021**, *13*, 8973. [CrossRef]
31. Issac, J.; Newell, R.; Dring, C.; White, C.; Ghadiri, M.; Pizzirani, S.; Newman, L. Integrated sustainability planning and local food systems: Examining areas of and gaps in food systems integration in community sustainability plans for municipalities across British Columbia. *Sustainability* **2022**, *14*, 6724. [CrossRef]
32. Mhlanga, D. Stakeholder capitalism, the fourth industrial revolution (4IR), and sustainable development: Issues to be resolved. *Sustainability* **2022**, *14*, 3902. [CrossRef]
33. Yang, K.L. Analyzing mathematics textbooks through a constructive-empirical perspective on abstraction: The case of Pythagoras' Theorem. *Eurasia J. Math. Sci. Technol. Educ.* **2016**, *12*, 913–930. [CrossRef]
34. Resources—STEM Examples: Examples on STEM Learning and Teaching Activities. Available online: https://www.edb.gov.hk/en/curriculum-development/kla/ma/res/STEMexamples.html (accessed on 20 June 2022).
35. Modelling the Spread of a Disease. Available online: https://www.edb.gov.hk/attachment/en/curriculum-development/kla/ma/res/example_SS_2_modelling_eng.pdf (accessed on 25 May 2022).
36. Curriculum Development Council. *Technology Education: Key Learning Area Curriculum Guide (Primary 1—Secondary 6)*; The Education Bureau: Hong Kong, 2017.
37. Curriculum Development Council. *Business, Accounting and Financial Studies: Curriculum and Assessment Guide (Secondary 4–6)*; The Education Bureau: Hong Kong, 2020.
38. Curriculum Development Council. *Physics: Curriculum and Assessment Guide (Secondary 4–6)*; The Education Bureau: Hong Kong, 2015.
39. Lo, C.K. Design principles for effective teacher professional development in integrated STEM education: A systematic review. *Educ. Technol. Soc.* **2021**, *24*, 136–152.
40. CoronaTracker—Hong Kong Overview. Available online: https://www.coronatracker.com/country/hong-kong/ (accessed on 24 June 2022).
41. Lo, C.K.; Cheung, K.L.; Chan, H.R.; Chau, C.L.E. Developing flipped learning resources to support secondary school mathematics teaching during the COVID-19 pandemic. *Interact. Learn. Environ.* **2021**. *advance online publication*. [CrossRef]
42. Alexander, R.M. Modelling, Step by Step. Available online: https://plus.maths.org/content/os/issue13/features/walking/index (accessed on 20 June 2022).
43. Alexander, R.M. Walking and running. *Math. Gaz.* **1996**, *80*, 262–266. [CrossRef]
44. Harrison, A.J.; Molloy, P.G.; Furlong, L.-A.M. Does the McNeill Alexander model accurately predict maximum walking speed in novice and experienced race walkers? *J. Sport Health Sci.* **2016**, *7*, 372–377. [CrossRef]
45. Khandan, R.; Shannon, L. The effect of teaching–learning environments on student's engagement with lean mindset. *Educ. Sci.* **2021**, *11*, 466. [CrossRef]
46. Klein, L.L.; Alves, A.C.; Abreu, M.F.; Feltrin, T.S. Lean management and sustainable practices in higher education institutions of Brazil and Portugal: A cross country perspective. *J. Clean. Prod.* **2022**, *342*, 130868. [CrossRef]
47. Milicevic, J.; Sremcev, N.; Cosic, I.; Lazarevic, M.; Tasic, N. The new role of teachers after the COVID19 pandemic and the application of the lean concept in education. In *Proceedings of the 32nd DAAAM International Symposium*; Katalinic, B., Ed.; DAAAM International: Vienna, Austria, 2021; pp. 402–407.

sustainability

MDPI

Article

Students' Perceptions of Online Learning during COVID-19 Pandemic: A Qualitative Approach

Mihai Curelaru [1], Versavia Curelaru [2] and Mioara Cristea [3,*]

[1] Department of Psychology, Faculty of Psychology and Educational Sciences, Alexandru Ioan Cuza University of Iași, 700506 Iași, Romania; mihai.curelaru@uaic.ro
[2] Department of Teacher Training, Faculty of Psychology and Educational Sciences, Alexandru Ioan Cuza University of Iași, 700506 Iași, Romania; versavia.curelaru@uaic.ro
[3] Department of Psychology, School of Social Sciences, Heriot Watt University, Edinburgh EH14 4AS, UK
* Correspondence: m.cristea@hw.ac.uk

Abstract: In this study, we conducted a thematic analysis of the views and perspectives of university students about online learning, specifically regarding their interpretations and experiences of the transition from traditional face-to-face courses to online teaching during the COVID-19 pandemic. The sample included 209 undergraduate and postgraduate students who were invited to complete five tasks, i.e., a free association task, answering open questions about the advantages and disadvantages of online learning, providing suggestions for improving online learning, and sharing a personal experience lived during this period. Some of the main themes extracted from the data refer to the negative aspects of online learning mentioned by participants in relation to its disadvantages, such as health and psychosocial problems (e.g., stress, anxiety, decreased motivation, isolation/loneliness, and apathy) and learning process problems (e.g., misunderstandings, a lack of feedback, additional academical requirements, a lack of challenge, and disengagement). Other recurrent themes refer to the positive aspects of online learning associated with its benefits: comfort and accessibility, economy (saving time and money), and psychological and medical safety. The personal experiences during COVID-19 shared by our respondents were organised around four main themes (positive, negative, ambivalent, and transformative experiences) related to students' adaptation to the educational context generated by the pandemic. Based on these findings, practical recommendations for universities and researchers are discussed.

Keywords: students' perceptions; online learning; educational experience; COVID-19 pandemic; qualitative study; thematic analysis

Citation: Curelaru, M.; Curelaru, V.; Cristea, M. Students' Perceptions of Online Learning during COVID-19 Pandemic: A Qualitative Approach. *Sustainability* **2022**, *14*, 8138. https://doi.org/10.3390/su14138138

Academic Editors: David González-Gómez and Jin Su Jeong

Received: 5 April 2022
Accepted: 28 June 2022
Published: 4 July 2022

Publisher's Note: MDPI stays neutral with regard to jurisdictional claims in published maps and institutional affiliations.

1. Introduction

As of March 2020, due to the ongoing COVID-19 global pandemic, most higher education institutions worldwide were forced to close their doors and start teaching online. The change from traditional learning to remote online learning was swiftly enacted, without the proper preparation or training of both academics and students. The sudden and "forced" [1] (p. 466) shift in teaching approach impacted the academic experience of both academics and students and led to a series of social, technical, educational, and psychological challenges.

There is a close relationship between education and sustainable development that requires policymakers to carefully assess the consequences resulting from its dynamics. Though sustainable development, concerning the preservation of a safe environment for future generations, is promoted through education, the educational system itself is under pressure to change due to the many challenges related to sustainability, e.g., greater awareness of natural, economic, and social resources; new perspective of social justice and universal wellbeing; and new attitudes toward consumption and lifestyle [2].

Under these circumstances, a sudden change in the educational system—introduced on a large scale due to a social crisis like the COVID-19 pandemic—needs to be discussed both globally and contextually. An analysis is needed in relation to different factors that influenced the adopted policies and strategies, e.g., the level of development of online learning networks, previous experience in integrating online and traditional learning [3], balancing synchronous and asynchronous learning [4], and training for teachers and students [5]. It is also important to discuss the effects of the change at several levels, including economic, social, psychological, teaching quality, and academic achievement. For example, some researchers highlighted the positive effects of online learning on students' academic performance, autonomous learning [6], and engagement [7]. Other researchers described the negative academic outcomes for students [8] and the psychosocial challenges of the virtual learning environment [9–11], while others considered the effects of technologies in a more nuanced manner, considering various specific factors. It is important to weigh the benefits of innovative economic, social, and technological solutions against the possible negative effects of pressure for change to not compromise some of the Sustainable Development Goals (SDGs), e.g., viable and sustainable economic solutions for the resources-constrained education systems, quality education for all, health and wellbeing, and social justice [12]. However, the experience of online education during the COVID-19 pandemic has called for innovative adaptations that can be used in the future for the digital transformation of higher education institutions by building on the empirical evidence accumulated during this period of crisis.

In this study, we conducted a psychosocial analysis of the transition from traditional face-to-face courses to online teaching during the COVID-19 pandemic. Online learning is an alternative to face-to-face courses that requires specific considerations, e.g., a higher involvement on behalf of both academics and students, a higher social online presence, and a series of personal characteristics. Most importantly, online learning requires previous training, rigorous preparation, and a conscious, deliberate acceptance of its advantages and disadvantages [13]. While many universities had the technical support necessary to adopt online learning as an alternative during the COVID-19 pandemic, most students were not prepared for it and did not freely choose it.

The general objective of this study was to explore the difficulties and challenges encountered by university students during the COVID-19 pandemic, as well as their perceptions of the advantages and disadvantages of online learning. We were also interested in the students' recommendations concerning the improvement of online learning and their shared experience of this new teaching approach. Finally, our study was aimed to provide an account of the way in which students managed online learning and to offer some recommendations for future crisis situations.

Perceptions are defined as complex mental processes by which people understand, interpret, evaluate, and form a picture of social phenomena. Furthermore, perceptions are studied by exploring individual voices that can be expressed, for example, through "narratives, storytelling, behaviour, and reactions to individuals or groups" [14] (p. 606). In this study, perceptions were operationalised through views, ways of understanding, and personal perspectives developed within the processes of social interaction and communication about online learning. In addition, we also explored perceptions of the environment, social events, and emotions, as well as self-perceptions and perceptions of others, all of which are known to be socially and culturally influenced and/or shaped [15].

Why conduct a study on students' perceptions? Perceptions represent interpretations of reality with truth value for many people, being "extremely powerful and influential in human thought and behaviour" [14] (p. 606). Negative perceptions of online learning could lead to a decrease in academic performance, while positive perceptions could have the opposite effect of increasing performance. These perceptions can also influence students' behaviour in virtual classrooms, both in relation to learning objectives and with peers and teachers. Attention, motivation, emotions, and satisfaction in response to learning can also be modulated by students' perceptions. Furthermore, by studying perceptions of the e-learning system, we can access students' views, evaluations, and interpretations, which (when corroborated with those of education professionals) can form the basis for improving the quality of learning, provide solutions to more successfully cope with pandemic-like situations, and create a basis for intervention and counselling for students who experience difficulties in adapting to such situations.

This study was based on the ontological and epistemological frameworks of social constructionist paradigm. According to the social constructionist paradigm, knowledge is not necessarily based on the objective and unbiased observation of reality, but its elaboration depends on the historical and cultural context and is achieved through experiences of social interaction and communication [16]. In this study, social interactions took place not only in a specific broader social context created by the outbreak of the COVID-19 pandemic but also within small, virtual, clearly socially bounded communities, namely those generated by the online learning environment. The participants' perceptions and experiences are the result of their conscious engagement to generate shared meanings to manage the new form of learning. Furthermore, our study was based on the weak social constructivism perspective [17] because the elaboration of shared social knowledge is not totally independent of objective societal aspects and the development of human beings; rather, it is constructed over a set of objective facts. In our case, shared meanings about online learning were marked by factors such as technological development, institutional frameworks of learning processes, and reported personal, psychological and medical issues. Finally, we adopted a qualitative approach with thematic analysis [18].

2. Online Learning vs. Face-To-Face Teaching

Previous studies underlined the numerous advantages of online learning such as lower costs, high accessibility and flexibility, rapid exchanges between teachers and students, opportunities for students to perform other activities while undertaking their studies (e.g., part-time jobs), and lower levels of stress [19–25]. Other studies also pointed out the disadvantages of online learning such as technical issues (e.g., internet connection and broadband issues), the risk of low attention levels, loss of sense of belonging, isolation, loss of motivation, and poor communication [7,20–22,24–26].

Several key factors impact the effectiveness of online learning, and some of the most important refer to: (1) technology, e.g., access, navigation and internet connection, the design of e-learning platforms, and accessibility to learning synchronous and asynchronous learning materials; (2) instructors' characteristics, e.g., teaching styles, attitudes towards students, digital skills, and encouraging interaction between participants; and (3) students' characteristics, e.g., personality traits, demographic characteristics, and digital knowledge and skills [27]. Regarding the latter, studies showed that online learning might be more beneficial for specific types of students. Motivation and self-discipline are extremely important, as students must be able to efficiently manage their deadlines and allocate time for asynchronous and synchronous materials [28]. Additionally, students must be able to learn through experience [29] and hold strong independent learning and motivation skills [20].

The objective of online learning is to maintain the same quality of education as traditional face-to-face teaching while using online methods and platforms [30]. This is harder to achieve since online learning requires a completely different learning environment, e.g., access to the learning materials, methods for online social interaction, and assessment tools. Online learning is not just a different way of delivering learning materials and contents but also a completely different social space in which individuals interact with each other, express themselves and their emotions, and seek solutions to different problems. As such, this environment needs to be as similar as possible to the traditional one to avoid any potential limitations to the communication and interaction between teachers and students.

Unlike in-person lectures, online learning is based on virtual learning environments (VLEs) accessed through a computer, smartphone, or tablet. Consequently, any act of communication and all its components (emission, reception, feedback, etc.) are mediated by a digital medium, e.g., computer, and is thus experienced quite differently than face-to-face teaching. The Media Naturalness Theory [31] was developed to explain the principles of electronic communication. According to this theory, a decrease in the degree of naturalness of a communication medium leads to the following effects in connection with a communication interaction: (a) increased cognitive effort, (b) increased communication ambiguity, and (c) decreased physiological arousal [9].

Different strategies can be adopted to increase the similarity between any communication mediated by an electronic medium and the face-to-face medium. One strategy is adopting performant digital devices, fast broadband speed, high quality learning platforms. Another one is ensuring digital training (e.g., support and training for the use of e-learning platforms) and psychological preparation (e.g., establishing appropriate expectations and developing appropriate alternative modes of communication). A final strategy is ensuring a high level of involvement from both teachers and students in actual learning, e.g., increased effort to generate social online presence, which can result in an increased senses of belonging and connectedness [32], actively interacting and working together, and supportive environments with prompt communication and common values and interests [1]. Social presence involves five integrated elements: affective association (emotional connection with learning activities), knowledge and experience (previous expertise), interaction intensity (engagement in interpersonal relations), community cohesion (sense of belonging and sharing resources), and instructor involvement (the teachers' capacity to shape students' behaviours and to engage them in critical analysis and reflection) [33,34].

3. Online Remote Learning during the COVID-19 Pandemic

Ensmann et al. [35] explored students' experiences of online learning during the COVID-19 pandemic using the Social Presence Model [33,34] as a theoretical framework. Their findings underline the importance of social presence as a literacy for learning—in any modality—and the need to provide mental health support for students. Their respondents frequently invoked a lack of face-to-face, social, and real interactions, and they linked this to increased screen time, exhaustion, and a lack of interest and motivation for learning. Similarly, Bączek et al. [36] conducted a survey to investigate the perceptions of online learning among Polish medical students eight weeks after the move to online learning. The main disadvantages of online learning evoked by their respondents included a lack of interactions with patients, technical problems, reduced interactions with the teacher, a lack of self-discipline, and social isolation. In educational areas involving significant professional practice, such as medical studies, a lack of direct contact with the professional environment is a major disadvantage. Furthermore, results from a comparative study showed that medical students are more dissatisfied with online courses than students in other degree programmes [37]. In another study, Dung [38] also identified extensive time staring at digital screens, a lack of body movements, a lack of conditions for developing social interaction skills, fear of online assessment, concentration loss, and a lack of peer interaction in virtual classrooms. Almendingen et al. [39] conducted a study examining Norwegian students' experiences of the sudden shift to online teaching during lockdown. Their results showed that 75% of students reported that their life had become more difficult and 50% felt that learning outcomes would be harder to achieve at two weeks after moving to online teaching.

Moreover, students in remote areas found online learning to be less efficient than face-to-face learning because they do not have the appropriate communication networks and infrastructure required to follow online learning [40]. Other studies also described broadband connectivity issues in rural areas as a significant challenge for students to make use of online learning initiatives [41]. These results suggest that the shift from in-person classes to online learning increased the social class achievement gap, exacerbating social class academic disparities [42], and favoured learners whose personalities were characterised by high levels of agreeableness, conscientiousness, and openness to new experiences [43]. Telyani et al. [44] investigated the perceptions of Cypriot teachers regarding the sudden shift from in-person classes to online teaching, specifically regarding the main challenges of online learning during the COVID-19 pandemic. The most frequently invoked challenges among the 20 interviewed teachers focused on students' behaviours such as reduced interaction, not engaging in solving problems, not answering questions, reduced task completion, decreased attendance, lowered engagement, and reduced performance [44]. Likewise, Biwer et al. [45] reported that during the period of online learning, students felt less able to focus their attention and invest as much time and effort in self-study as in the pre-pandemic period. Effects on mental health were also highlighted. For example, college students faced increases in anxiety, loneliness, and depression [46,47]. The increases in stress, anxiety, and depressive thoughts among students were caused by fear and worry about their own health and of their loved ones, difficulty concentrating, disruptions to sleeping patterns, decreased social interactions, and increased concerns over academic performance. Physical health problems, such as headaches, were also reported as a result of prolonged screen exposure [48].

How did the abrupt adoption of online learning during the pandemic period affect students' academic performance from various levels of education (primary school to university)? The change had both positive and negative effects on academic performance. In general, however, the effects were rather negative and were influenced by various factors such as age (or level of schooling), previous learning experience, and learner characteristics. In a systematic review, Hammerstein et al. [49] highlighted the negative effects of school closures on student achievement (or test scores) on mathematics, reading, and other subjects from primary and secondary education, showing a stronger impact on younger students and those from a lower socio-economic status. Other studies showed that online learning outcomes were influenced by factors such as learner characteristics, perceived usefulness, course content and design, ease of use, and faculty capacity. Of these, learner characteristics, e.g., proactiveness, self-study ability, and compliance, are the most important factors [50]. The academic results of students who attended at least one academic year of face-to-face learning before the outbreak of the pandemic were better than students who started their studies online [51]. In another study on K–12 students, low-performing students showed greater improvements in performance than high-performing students, suggesting that online learning had different effects and narrowed the gap between low- and high-performing students [52].

Furthermore, in a study conducted before pandemic, Broadbent and Fuller-Tyszkiewicz [53] identified five distinct profiles of self-regulated learning: minimal regulators; restrained regulators; calm, self-reliant, and capable regulators; anxious, capable collaborators; and super-regulators. The minimal regulators represent the least adaptive profile, which is characterised by the lowest motivation and self-regulated learning strategies, anxiety, lesser activity organisation, the lowest level of critical thinking, and lacking confidence in their study abilities. Super-regulators have the highest degree of adaptation. They tend to set the highest grades goals and have the highest levels of time management and organisation, effort regulation, metacognition, critical thinking, and confidence in their study abilities [53].

Similarly, Biwer et al. [45] identified four student profiles according to the reported changes in their resource-management strategies during online learning in pandemic: the overwhelmed, the surrenderers, the maintainers, and the adapters. The overwhelmed refers to the students who are less able to regulate their resources and have difficulties with attentional regulation, effort regulation, and time management. The surrenderers face similar difficulties as the overwhelmed in terms of attention, effort, and time management, but they also invest less effort and time in their self-study. The maintainers differ from the others only by a relatively small increase in effort and time investment. Finally, the adapters are characterised by the efficient management of attention, effort, and time, being more motivated in the new situation. Ishimaru et al. [54] studied the adaptation features of university students exposed to fully online education during COVID-19 pandemic, especially engagement and stress. The authors identified three groups of students: school adaptation, school maladaptation, and school over-adaptation groups. The first group presented the lowest level of mental health problems (fatigue, anxiety, and depression), and the last group experienced the highest level of mental health issues, except for fatigue. The over-adaptation group generally consisted of older female students who considered online education to be less beneficial and had shorter total sleep time on weekdays and longer total sleep time on holidays [54].

Despite numerous challenges, students identified some advantages or positive effects of online learning. The most frequently evoked were flexible schedules and convenience, comfortable environments, and enhanced technical skills [41]. Others reported advantages include the ease of sharing educational materials, effective access to study resources, updated learning materials, and flexibility in time and space [55–57]. Further studies also mentioned protecting one's health and ensuring the community's safety, saving travel time, exposing one to new forms of learning, keeping up with the original plan of the semester, having extra time for self-study, and easy access to online resources [38,58], as well as the ability to stay at home, continuous access to online materials, the opportunity to learn at one's own pace, and comfortable surroundings [36].

4. Method

4.1. Participants

In this study, the convenience sample included 209 undergraduate and postgraduate students in Psychology, Biology, and Languages from the Alexandru Ioan Cuza University of Iași, Romania (N = 188 females), aged between 19 and 52 years old (M = 22). Among the participants, 204 of them attended at least two full semesters of online learning during the COVID-19 pandemic, and 5 of them only attended one semester; 193 students did not experience online learning before, while 16 experienced this form of learning on previous occasions (e.g., independent distance learning programmes). Following the measures to prevent the spread of COVID-19 infection [59], Romanian universities, including the Alexandru Ioan Cuza University of Iași, decided to organise teaching activities predominantly online from April 2019, towards the end of the 2019–2020 academic year. At the beginning of the 2020–2021 academic year, most faculties moved their teaching activities exclusively online, with 95% of the students involved in this form of learning [3].

Prior to the pandemic period, teaching activities at the university were carried out in the traditional face-to-face format for students enrolled in full-time education and included lectures, tutorials, practical labs, and seminars. Prior to the pandemic, the university's e-learning platforms and other digital tools were mainly used for students enrolled in independent distance learning and for posting learning resources and asynchronous communication. Synchronous online activities carried out before the pandemic were limited to videoconferences, video meetings for national and international collaborations, and one-to-one mentoring/supervision activities.

During the COVID-19 pandemic, the participants involved in this study attended all their lectures, tutorials, and mentoring activities exclusively online. Only a small part of their practical labs required a hybrid approach. More specifically, Psychology and Language students attended all teaching activities exclusively online, while Biology students carried out a small part of their practical labs in face-to-face mode with very strict social distancing measures in place. Online teaching was officially carried out using the Cisco Webex, Microsoft Teams, and Moodle digital platforms. Some academics also used other applications such as Zoom, WhatsApp, and Google Suite in addition to those agreed upon by the University management. Cisco Webex allows for synchronous online teaching via video meetings in various forms, such as one-to-many (e.g., lectures), many-to-many (e.g., debates), one-to-one (mentoring), breakout room/sessions (e.g., group discussions or labs activities), and chat communication. Microsoft Teams and Moodle provide options for asynchronous learning in addition to the online video classes (e.g., course materials, forums, class assignments, and course calendar).

4.2. Procedure

The Ethics Committee of Faculty of Psychology and Educational Sciences, Alexandru Ioan Cuza University of Iași, approved the study. Participants were provided with an information sheet that contained a full description of the study and details about anonymity, data confidentiality, and their right to withdraw from the study. After participants signed the consent form, they filled in an online Google Forms questionnaire consisting of the five items described below and a few questions related to socio-demographic variables. The questionnaire was followed by a short debrief. Data were collected in June 2021 at the end of Semester 2 of the academic year.

4.3. Measures and Instruments

(1) Free association task

The free association task is a method widely used to collect data about the content of a social representation [60–62]. The free association task allows researchers to identify the latent dimensions defining the structure of the semantic universe of the social object being studied. It consists of associating words or expressions with a stimulus word or expression corresponding to the object of representation. In this study, participants were asked to write down the first five words/expressions that came to their mind when they heard the stimulus expression-word "Online learning".

(2) Advantages of online learning

Participants were asked to list five advantages of online learning (*From my perspective, the advantages of online learning are* ...).

(3) Disadvantages of online learning

Participants were asked to list five disadvantages of online learning (*From my perspective, the disadvantages of online learning are* ...).

(4) Suggestions for improvement

Participants were asked to provide three suggestions, based on their online learning experience during the pandemic, for improving the quality of the academic activities delivered online (*To improve the quality of the academic activities delivered online, I recommend* ...).

(5) Personal experience depicting online learning

Participants were invited to describe a significant event experienced during the pandemic and associated with their online learning experience (*Describe, in 10–15 lines, a personal experience—a situation, an event—related to online learning*).

(6) Socio-demographics

Participants were asked to provide socio-demographic information concerning their age, gender, years of study, field of study, and number of semesters they attended online.

4.4. Data Analysis

All data were analysed using a thematic analysis (TA) method [18,60,63–67]. TA is based on a process of generating or identifying themes, subthemes, and interconnections between themes and subthemes [63,65]. A theme represents "a specific pattern of meaning found in the data" [60] (p. 209) that "captures something important" about their relation to the research question [18] (p. 82). Themes and subthemes are obtained from codes, the smallest components of the analysis, by collapsing or clustering them [63]. Codes are "building blocks" for themes and "patterns of meaning" [66] (p. 297) shared by research participants.

We chose this approach because it corresponded to the objectives of our research and has multiple advantages. Firstly, the collected data were qualitative and TA is one of the most used methods for managing and interpreting this type of data. TA can be used to process data from various sources such as interviews, focus groups, and newspaper articles [65], diaries, discussion forums, story-based methods (vignettes or story-completion tasks) [67], open-ended responses to questionnaire items, video, images, essays, and free associations [60]. As stated in the previous section, we used two different techniques of data collection, i.e., free associations and open-ended questions, to ensure a greater data diversity and in-depth analyses. Secondly, we chose TA because it is a tool "unbounded by theoretical commitments" [66] (p. 297) and a methodological perspective usually used in exploratory research. TA is an excellent tool for identifying, describing, and interpreting people's experiences in relation to an issue, their views and perspectives, current practices/behaviours, and shared representations of an object of social interest [68]. Thirdly, TA is a flexible method of qualitative data analysis that can be conducted in several ways [18,63,67], depending on the research objectives. This feature of the analysis allowed us to treat the data in two different ways, which we describe below, based on some methodological decisions [18,67].

4.4.1. Inductive and Deductive Analysis

In our research, the first objective was to explore the views and the perspectives of students about online learning (items 2 to 4). To process this type of data, we opted for a thematic analysis characterised by alternating the deductive and inductive approaches. Therefore, we based our work and questions on previous experiences and findings in this area. Similarly, to identify and name our codes, themes, and subthemes, we used findings from previous studies. In this sense, the analysis can be regarded as deductive. At the same time, the inductive approach was also used because most of the components of the analysis emerged from the data through a bottom–up approach.

4.4.2. Inductive Analysis

The second objective of our research was to identify, describe, and interpret the structure of students' representations of online learning using the data obtained from the free association task (item 1). The third objective was to explore the emotional experiences described by participants in relation to an event associated with online learning (item 5). All data was analysed using an inductive approach—conducted exclusively "bottom–up"—without theoretical inferences [63].

5. Results

Results were organised in tables, each table containing all the themes and subthemes resulting from the analysis of data obtained from each single item/task.

5.1. Free Association Task

By analysing data from the free association task (item 1), we were able to extract the structure of the online learning representation they shared. We identified eight main themes at an early stage of the analysis. After reviewing and refining these themes, we retained three of them: *Negative*, *Positive*, and *Neutral aspects of online learning*. Themes such as Mental health and Psychological wellbeing, Physical health and Medical issues, and Technical were included in a more comprehensive theme called *Negative aspects* concerning online learning. Table 1 presents the themes and subthemes extracted from the free association task (Item 1).

The first theme refers to *negative aspects* perceived by students and is the richest in content. Given that online learning was not an option freely adopted by the students, the elements associated with this theme must be judged in the light of the social and medical context of this period. Not surprisingly, psychological wellbeing was affected, with students experiencing increased levels of stress and anxiety, a lack of concentration, and exhaustion. This may be due to the novelty of the situation, lack of preparation, lack of an adaptation period, and the adoption of new teaching/learning strategies, all of which increased the degree of difficulty in completing academic tasks. Additionally, the prolonged exposure to screens (sometimes up to 10 h a day) led to fatigue, reduced interest, and medical problems such as backpain, headaches, and eye pain. Furthermore, since all teaching materials and activities had been prepared for face-to-face learning, academics transferred all contents to online learning without any adjustment since they lacked the time to do so. This can also explain some of the students' negative experiences with online learning.

Table 1. Themes and subthemes defining participants' free associations with online learning.

Themes and Subthemes
1. Negative aspects concerning online learning Stress, low levels of motivation, isolation/loneliness, low levels of focus and attention, experience of negative emotions, tiredness, exhaustion, frequent backpain or head and eye aches, lower quality of academic experience, additional academic tasks, monotony/boredom, lack of organisation, difficulties in understanding academic tasks, lack of preparation, difficulties with connectivity, loss of internet connection, and power cuts
2. Positive aspects concerning online learning Working from the comfort of one's home, no traffic, saving time and money, spending more time with the family, multitasking, personal development, personal change, opportunity, challenge, novelty, progress, and creativity
3. Neutral aspects concerning online learning Online lectures, virtual learning environment, online teaching platforms, oral online presentations, online tests and exams, internet, technologies, and online devices

The *positive aspects* are represented in the students' free associations by two major areas of interest. The first refers to savings (e.g., time, money, and resources), which would not be possible when conducting face-to-face learning. The other area refers to personal opportunities (e.g., challenges, openness to novelty, change, and development). At the borderline between these areas, we noticed a subtheme, multitasking. Being home allowed students to multitask while attending online lectures, e.g., listen to their lectures while also surfing the internet, collecting information about the lecture topics, or attending to other personal issues. Some participants mentioned being able to take care of their children or other family members.

The third theme comprises data that we considered *neutral.* It includes discourse about new teaching activities and contents, the adoption of new technologies and concepts. These mentions were not accompanied by positive or negative evaluations.

5.2. Advantages of Online Learning

The analysis of the benefits of learning from the students' perspective revealed the following main themes: *Comfort and accessibility, Economy (time and money),* and *Safety* (psychological and medical) (see Table 2). Initially, we had a larger number of themes; however, we reduced them to three. For example, Saving time, Family time, Saving money, Avoiding traffic, and Multitasking were all integrated into one theme, Economy (time and money).

As can be seen in Table 2, some of the advantages mentioned by students were also found in the previous analysis of their free associations (see Table 1). This repetition—which was anticipated during the elaboration of the questionnaire—is not redundant or unnecessary for our thematic analysis. On the contrary, it reinforces the results, as we discovered identical or similar data obtained through two different tasks that participants completed. For example, economic advantages were mentioned both here and in the free association task. While these advantages were included as subthemes within the Positive aspects of the previous analysis of the free associations, in this analysis, they constituted one of the main themes, *Economy*.

As a main theme for advantages, *Comfort and accessibility* includes subthemes related to the possibility of staying home within a familiar environment while accessing learning materials and activities that were difficult to use pre-pandemic without going to the university. Familiarity with one's own living space and a lack of dependence related to location and distance seem to be the defining characteristics of this theme.

Table 2. Themes and subthemes defining the advantages of online learning as perceived by students.

Themes and Subthemes
1. Comfort and accessibility Commodity, physical comfort, familiar environment, online participation, mobility, high accessibility, online resources, more opportunities for synchronous and asynchronous materials, more teaching resources, timetable flexibility, freedom to organise one's personal time, faster and more efficient communication, feedback, problem solving, and higher attendance to classes
2. Economy (time and money) More free time, saving time, saving money, having more time to engage in other activities, more time to spend within one's family, avoiding public transportation, no delays due to the use of public transportation, having a job, and multitasking
3. Safety (psychological and medical) Trust, peacefulness, reduced anxiety, and stress due to face-to-face interactions; less discrimination; less bullying; reduced competitiveness; openness; tolerance; empathy; relaxation; low risk of COVID-19 infection; and sanitary safety

The third theme, *Safety*, refers to students being able to control their levels of anxiety, stress, and mistrust that were triggered by face-to-face learning. One could argue that online learning has helped some of the students to overcome psychological issues such as the ones previously mentioned. Online learning is also linked to lower levels of competitiveness, discrimination, and bullying. We also observed some mentions about reductions in the risk of COVID-19 contamination; however, these mentions were quite scarce. An explanation may be related to the medical discourse in the media, in which younger people were frequently associated with a lower risk of infection compared to older people.

5.3. Disadvantages of Online Learning

The main themes regarding the disadvantages of online learning refer to *Health and psychosocial problems*, *Learning process problems*, *Technical problems and low digital abilities*, and *Discrimination* (see Table 3).

Table 3. Themes and subthemes defining the disadvantages of online learning as perceived by students.

Themes and Subthemes
1. Health and psychosocial problems Stress, anxiety, lack of focus and attention, lower motivation levels, apathy, boredom, lower work efficiency, tiredness, exhaustion, high levels of screen exposure, back pain, head and eye aches, lack of social contact and physical interaction, lack of face-to-face communication, and loneliness
2. Learning process problems Lack of challenge, low levels of accountability, disengagement, unfriendly learning environment, lower efficiency and quality of teaching, monotony, misunderstandings, disorganisation, lack of feedback, additional academical requirements, higher expectations from lecturers, improper evaluations, academics are unfamiliar with online assessments, high level of suspicion regarding plagiarism intentions/behaviours, lack of appropriate study spaces (home), lack of intimacy, and noises
3. Technical problems and low digital abilities Unstable internet connection, power cuts, lack of adequate technology, and low knowledge and skills to use virtual learning environments and technologies associated with online learning among academics and students
4. Discrimination Disadvantage for those without digital skills, disadvantage for those without technological equipment, and disadvantage for those who are less assertive.

Firstly, we should note some overlap between the *Health and psychosocial problems* and *Technical problems and low digital abilities* themes identified here and the *Negative aspects* theme drawn from the free association data (see Table 1). In the context of the disadvantages of online learning, the *Technical problems and low digital abilities* theme draws attention to the users' digital skills. Although the university organised various courses for the use of online learning platforms, students struggled with unstable internet connections, power cuts, a lack of adequate technology, screen blocking, etc., which required more extensive knowledge, experience, and ability to improvise (in addition to initial technical training) to find innovative solutions.

The content of the second theme, *Learning process problems*, mostly speaks for itself, though some clarifications are necessary. One of the main disadvantages of online learning mentioned by students refers to assessment methods. To avoid plagiarism and compensate for the lack of face-to-face interactions in continuous assessments during the semester, academics introduced additional or new types of assessments (homework, projects, etc.). These new types of assessments were associated with increased levels of stress, anxiety, and fatigue among students. Furthermore, some of the subthemes found here were linked to the ones included in the *Comfort and accessibility* theme drawn from the analysis of the advantages of online learning (see Table 2). While students enjoyed the comfort of their homes, the presence of other family members or colleagues (such as roommates and flatmates) was noted as significant distractions from learning. Thus, students invoked a lack of appropriate study spaces (home), a lack of intimacy, and being exposed to noises as having negative consequences of their learning experiences.

Finally, the last theme about the disadvantages of online learning refers to *Discrimination*. Online learning seems to have disadvantaged those with fewer financial resources (who cannot purchase high-performance electronic devices or pay for a high-speed internet subscriptions) and those with specific personal characteristics (high anxiety, low assertiveness and initiative, older, and low digital skills).

5.4. *Suggestions to Improve Online Learning*

Four themes were identified here: *Suggestions for the learning process*, *Suggestions for lecturers*, *Suggestions for learners*, and *Suggestions for improving communication and interpersonal relations* (Table 4). Some initial themes, e.g., Lectures and Tutorials, Lecturers' engagement, and Change of teaching approach, were reorganised into a broader theme, *Suggestions for the learning process*.

Table 4. Themes and subthemes reflecting student' suggestions to improve online learning.

Themes and Subthemes
1. Suggestions for the learning process Reducing number of students per teaching activity, reducing numbers of academic activities and assessments, better organisation, more course handbooks available online, detailed handbooks and supporting materials for each course, reducing risks of plagiarism, unique online teaching platform for all faculties in the university, more complex online teaching platform, returning to the traditional teaching approach, and introducing a blended learning approach
2. Suggestions for lecturers Change of teaching approach, interactive/attractive lectures, more professional development for academics, higher involvement, and more empathy towards students going through difficult situations
3. Suggestions for learners Counselling and wellbeing, academic skills support, additional resources for disadvantaged students, and higher involvement on behalf of students
4. Suggestions for improving communication and interpersonal relations Patience/calm, seriousness/responsibility, and understanding others and openness to communication

Most suggestions were directed towards improving the learning process. Some of these suggestions concerned administrative or organisational aspects of online teaching such as reducing the number of students per group and/or per learning activity, reducing the number of teaching hours, and reducing the number of activities and assessments. Another problem reported by students was the simultaneous use of several online platforms. For example, a student enrolled in courses offered by different faculties had to work with four or five different virtual learning platforms. To increase learning efficiency, participants suggested the use of a single online platform for the whole university.

The other two themes include suggestions focussed on academics and students. Academic staff are expected to improve the quality of the lectures (in particular, the attractiveness of the presentation and professionalism in online conditions), and students are expected to be more pro-active in order to increase the effectiveness of learning. Finally, the last theme focussed on improving the quality of the interactions and communication between academics and students by showing mutual understanding and empathy.

5.5. Online Learning Experiences

The data obtained from exploration of students' emotional experiences of online learning (item 5) formed the basis for identifying a typology of experiences. To better capture the subjective and integral nature of the analysed experiences, we chose to present the results by describing the main themes and accompanying them by extracts from the participants' responses. The most significant personal events experienced during COVID-19 were organised around four main themes: *Positive*, *Negative*, *Ambivalent*, and *Transformative experiences*.

5.5.1. Theme 1: Positive Experiences

The first theme generated from the data, *Positive Experiences*, included a variety of experiences that were emotionally intense or associated with complex emotions, such as satisfaction, interest, inspiration, joy, elevation, enthusiasm, and optimism. For example, one of the participants reported experiencing gratitude: *"I taught myself how to use new technologies (...). This represented a huge personal success. Thanks to the skills I learned online, I was able to do better assignments for my courses as well as better entrepreneurial projects. I feel quite grateful for having had this opportunity."* (F, 21) (The letter in brackets indicates the participant's gender and the number represents the participant's age). Another participant reported joy in being able to more easily express their opinions and satisfaction with using the platforms: *"As an introvert, I really enjoyed online teaching, it was easier to express my opinion. (...) For me, online learning was much enjoyable than face-to-face."* (F, 22).

Positive experiences are about not only emotions but also the possibility of personal development. For example, one subtheme of positive experiences is opportunities that were available to them due to online teaching (e.g., attending extra-courses, participating at online conferences, and participating at book launches)—opportunities that would otherwise have been impossible/inaccessible for them: *"(...) I was able to attend conferences, projects, seminars, to which I do not normal have access. The move to online teaching made attending specific events more accessible."* (F, 21). *"I had the opportunity to attend an online conference (...). But for online learning and using online platforms of interaction, some events and meetings with external experts wouldn't have been possible, or might have been more difficult to organise. As such, online learning facilitated the interactions with professional experts from different places."* (F, 20). Online learning facilitated students' development of their digital skills: *"My online presentations were much more efficient as opposed to the ones I previously did face-to-face because I had more scientific resources which were available directly on my computer."* (F, 22).

Reducing anxiety and shyness was another subtheme mentioned here. Students who identified as being introverted or socially shy said that online learning helped them to overcome these issues. Being in the comfort of their own homes created a feeling of safety and boosted their self-esteem. Two participants reported the following: *"Being a shy individual, in-person exams always made me anxious, no matter how much I revised for the exams. I would feel sick to my stomach, nauseous, and very anxious. Take-home exams made me feel comfortable and calm, I was more focussed, my anxiety levels decreased."* (F, 21). *"Face-to-face learning made me feel shy and less involved, when we started online classes, I started being more confident and feeling safer."* (F, 20).

5.5.2. Theme 2: Negative Experiences

The second theme, *Negative experiences*, refers to a lack of social interactions (e.g., a lack of physical interactions, a lack of face-to-face communication, a lack of non-verbal communication, and a lack of engagement from both students and academics). Cameras being off generated a lack of nonverbal feedback that caused communication difficulties: *"(...) face-to-face communication provides information about the student, their nonverbal behaviour (posture, gestures). Unfortunately, most of the students have their cameras off during online learning, therefore it is difficult to get any non-behavioural feedback."* (F, 22). *"(...) I was the only student to have my camera on. I did my presentation and there was no feedback, I could only see the professor's camera on (...). It was a very difficult presentation; given that I couldn't see my colleagues' faces, I didn't know on which aspects of my presentation I should focus or not."* (F, 20).

Some participants regretted the lack of face-to-face communication and felt that this affected the quality of their learning despite the academics' efforts to overcome this: *"I miss face-to-face interaction."* (F, 22). *"Online learning is not as efficient as face-to-face learning because it lacks interaction which is extremely important."* (F, 21). *"(...) I believe online teaching will never successfully replace face-to-face teaching... It is extremely painful to see professors trying their best to make online teaching interactive facing the lack of feedback and interaction. I believe professors are currently experiencing a state of confusion in their teaching approach."* (M, 20).

Other participants talked about the high levels of stress, exhaustion, and anxiety experienced during this time: *"My levels of anxiety and stress have increased. Sometimes, I'm struggling to communicate with people around me. Being a student is not only about learning but also about interacting with others."* (F, 21). *"(. . .) anxiety was one constant issue that I experienced during online learning."* (M, 21). *"I believe one of the things that marked my entire experience during this time was the lack of peer interaction. I used to be quite shy when doing oral presentations but since we moved to online learning, this has become even worse, and I believe it is because of the lack of interaction with my peers. During my oral presentation I felt my heart pounding, my mouth was dry, and I had difficulties speaking. I think the lack of physical contact with my peers and professors (...) accentuated my social anxiety."* (F, 20).

Among negative experiences, students also mentioned their confrontation with technical problems, e.g., poor performance of personal computers, broadband issues, power cuts, and difficulties in accessing learning platforms. These difficulties generated strong negative emotions such as a lack of empathy, frustration, fear, aversion, and anxiety, especially when they were associated with exams or assessments. The following two stories are illustrative of this point: *"One of the most stressful experiences (...) was an exam during which I lost my internet connection and (...) it took quite a while to get back online and be able to send my exam answers. It feels that academics are not very empathetic with these kinds of situations, particularly when we can't do anything about it. (...). It seems unfair to be penalised for circumstances that do not reflect our knowledge but rather the technological equipment that we possess. At the same time, I also try to understand my professors' position who can't confirm whether we have or not the necessary equipment (...). I suppose it is difficult to manage these kinds of situations and make sure everyone is happy"* (F, 20). *"I was unable to intervene during some classes because of the high number of students that were connected, the online platformed crashed. Even more so, my laptop and smartphone are not very performant which meant that I couldn't use my camera and I was told that I couldn't attend the exam unless I put the camera on (...)."* (F, 22).

Some students mentioned experiencing physical issues such as sedentarism, headaches, back pains, and eye problems: "(...) *after sitting in front of the computer for hours, I have headaches which sometimes can last for hours. I feel that my eyesight has worsen."* (F, 20).

5.5.3. Theme 3: Ambivalent Experiences

This theme refers to some experiences described as ambivalent in terms of both the emotions and meanings attributed to them. While participants recognised that online learning is necessary and offers numerous opportunities for personal development (e.g., the pleasure of acquiring new knowledge or the satisfaction of overcoming personal limits), they also associated online learning with negative emotions, such as a fear of the unknown, anxiety about their own academic performance, and feeling a lack of social support: *"I was fine (...), I had more time for myself, to learn, to grow (...) There were some negative effects among which stress and anxiety, (...) I miss physical interaction."* (F, 22). *"I know face-to-face interaction between students and academics is missing, but I feel that everyone adapted to the new circumstances quite well."* (F, 25).

5.5.4. Theme 4: Progressive experiences

This theme included experiences related to a progress described by the participants, from initially seeing online learning as a negative experience to seeing it as a more positive experience later. As students became acquainted with online learning, their experience with it changed. Three excerpts from the students' stories are relevant here: *"My first experiences with online learning were a bit bizarre. It felt weird not going to university, not talking to my colleagues, seeing my professors only on my laptop. In time, I got used to it, it now feels weird going back to campus. I can't say I had any issues, on the contrary. I feel more confident doing oral presentations now. Even so, I still miss face-to-face lectures."* (F, 21). *"I remember, at the beginning, don't know why, I found it difficult to talk in front of my laptop. I had a constant feeling of talking to myself, as if nobody else was there. I was surprised by this because I am a very chatty person. Online learning made it difficult for me but in time I learned to adapt. I can stay that while it started as something negative, it turned into something positive, I started to develop new skills (...) online learning helped me develop new digital skills which I didn't think it was possible to learn in such a short time."* (F, 27). *"In the beginning, it was difficult to do my courses online, but I soon adapted to all the things associated with it: technology, online platforms,* etc. *I finally got used to it and I now see it as an advantage as it saved me quite a lot of time."* (F, 41).

6. Discussion

The aim of this study was to provide a psychosocial analysis of the transition from face-to-face learning to online teaching during the COVID-19 pandemic as experienced by a sample of Romanian university students. Our findings suggested that moving from face-to face teaching to online learning during the COVID-19 pandemic was associated with a wide range of beliefs and perspectives, behaviours, and affective experiences. To begin with, we can argue that students' representation of online learning, as seen from their free association responses, was polarised; the semantic field of this representation was organised around two major themes. Some elements have positive connotations and generally refer to comfort, savings (time and money), challenges, and personal development, which is consistent with previous findings [38,55,69]. Other elements have negative connotations; students associated online learning with high levels of stress, low motivation, attention, and focus, as well as with negative emotions such as feelings of isolation and loneliness. These results are in line with previous findings [35,36,38,39,41,44,70]. Furthermore, emotional and physical exhaustion, headaches, backpains, and eye problems were among the most frequently invoked psychological and physical problems. The frequency of these issues among our participants was rather low but comparable to similar studies [33,34,71–73].

Further analyses considered the advantages and disadvantages of online learning and suggestions for improving the quality of academic activities. Firstly, students mentioned negative aspects such as a lower quality of higher education, monotony, boredom, difficulties in understanding the courses (e.g., a lack of clarity, a lack of feedback, and a lack of non-verbal behaviour), and improper space to connect from home. Other studies reported similar findings, e.g., difficulties in hearing the voice of the instructors and in acquiring the contents of the lessons [38], houses unfit for home-office purposes [39], poor learning conditions at home [36], and work–home interference [70]. Perceptions of a lower quality of higher education can be explained by the reduced engagement of both teachers and students, a lack of knowledge, and difficulties in using the appropriate technology. In a similar study, lecturers expressed difficulties such as a lack of energy and reduced performance while students invoked not engaging in solving problems, not answering questions, reduced performance, and engagement, among others [44]. Furthermore, in a study on students' perceptions about online learning during the COVID-19 pandemic, Almendingen et al. [39] reported an overall sense of reduced motivation and effort. Secondly, students reported a series of positive aspects such as comfort, commodity, avoiding traffic, higher accessibility, saving time and money, psychological safety, and opportunities for personal training and development. These results are in line with previous studies that reported positive aspects of flexibility, remote learning, accessibility [69], comfort and flexibility of space and time [55], saving travel time, and exposure to new forms of learning [38].

Some of the themes invoked by our participants can be seen as both advantages and disadvantages for online learning. For example, some students reported higher levels of anxiety and stress during online learning, while others mentioned that their levels of anxiety and stress became lower during online learning. For some, online learning brought psychological safety, while others felts less safe. Some of the students reacted more positively to the new situation; they adapted quite easily, focusing on its advantages and overcoming its challenges. Others experienced more difficulties, associating online learning with negative emotions, e.g., anxiety, shyness, and isolation. The shift from face-to-face teaching to online learning did not agree with everyone. Previous studies showed that online learning fits more with students exhibiting specific personality traits and socio-demographic backgrounds [20,27–29]. Our study results suggest that online learning provided an advantage for students already exhibiting digital knowledge, thus confirming the hypothesis that the pandemic has deepened digital inequalities [74].

When asked about suggestions to improve the quality of higher education, the most significant invoked suggestions referred to institutional management (e.g., reducing teaching hours, reducing the number of assessments, and using better equipment and online platforms). Other suggestions were focused on lecturers and referred to making online teaching more interactive, providing higher engagement, and providing more detailed course materials. Further suggestions included providing counselling and more support for disadvantaged students.

While many of these aspects are positive, the dominant image of online learning is a negative one. However, these negative perceptions are strongly linked with the development of a global health pandemic and the restrictions that accompanied it such as multiple lockdown periods. Firstly, our participants, as well as many students worldwide, were forced into online learning because of national lockdowns regardless of their preferences. The results might have been different had the students been able to choose between face-to-face and online learning. Secondly, negative perceptions were probably also caused by the way online lectures and tutorials were delivered during this period, namely through *synchronous video communication* in all its forms—e.g., one-to-one, one-to-many, and many-to-many [10]. This way of delivery has many advantages such as instructional tools, e.g., screen sharing, polling, chat, and breakout rooms. While some students appreciated these features and showed higher engagement [75], most students expressed discontent, particularly related to being forced to spend several hours per day in front of their screens, as seen in other studies [38,76]. The desire to be engaged in academic activities and the interest to explore new learning modalities were significantly diminished by psychological and physical difficulties. A blended approach including both synchronous and asynchronous learning activities might be the solution. This is in line with some of the suggestions made by our respondents: reducing live online sessions and increasing the number of online materials, e.g., asynchronous activities and recordings of live sessions. This would allow students to revisit recorded live sessions at their own pace or consult recorded materials at convenient times. Finally, another potential cause for these predominantly negative perceptions was the sudden shift—without any previous training—from traditional face-to-face teaching to online learning. Both students and lecturers were unfamiliar with virtual learning environments and associated learning platforms, and they were not ready for the academic demands of online synchronous learning, e.g., increased engagement, creating online communities, and a sense of belonging, as well as other social activities delivered online. The emergence of the pandemic and the need to teach almost exclusively online found the academic community insufficiently prepared for such an experience. Although lecturers had access to training opportunities, e.g., learning how to use virtual learning environments and platforms delivering online lectures and tutorials, they were not prepared for the human component of online learning, e.g., teaching presence, cognitive presence, and social presence, which are essential aspects of remote blended learning [33,34,71,72]. Online learning includes a series of skills regarding the development of teaching presence, e.g., direct instruction, instructional management, and building understanding [72]. The development of teaching presence is essential because it contributes to the development of other dimensions such as cognitive and social presence [77] and significantly corelates with student learning outcomes [78]. The lack of previous knowledge about remote blended learning led to a low level of teaching presence among teachers and academics that, in turn, led to perceptions of low social presence and a poor quality of the overall academic experience. These were followed by technical issues (e.g., poor internet connection, power cuts, and poor technical resources) that impacted media naturalness [31]. This might partially explain the reported physical and psychological problems (e.g., tiredness, exhaustion, and stress) caused by the high cognitive load and the extra mental effort required to assimilate the often-monotonous contents delivered online [9].

The themes drawn from our data support the identification of three categories of students in terms of their complex reactions to online learning. When analysing the data provided by each participant regarding all five tasks, we observed consistency between participants' representation of online learning, their views, perspectives, and affective polarity of the experience. These results allowed us to describe the three distinct groups of participants according to their adaptation to online learning: *the most adapted, the least adapted*, and *the uncertain* groups. The first group included the *most adapted,* as their representation of online learning was dominated by positive elements and their associated experiences had a rather positive affective tone. These students reported feeling comfortable in the new situation, having favourable opinions and attitudes towards online learning, effectively reacting to learning tasks, having better digital skills, and finding new opportunities and challenges. A second group of students were characterised as *least adapted*. Their views and perspectives were generally unfavourable to online learning, and their associated experiences had a rather negative affective tone. They reported higher levels of stress, anxiety, and loneliness and lower levels of academic effectiveness. Several students in this group mentioned exhaustion, headaches, and back or eye pain. Students in the *uncertain* group did not yet have clear views about online learning, their emotional experiences being characterised by ambivalence. While they recognised that online learning has numerous academic opportunities and possibilities for personal development, they expressed high levels of stress, anxiety, and exhaustion. They reported being engaged in a struggle to adapt, where the stakes were academic performance and wellbeing and the path required them to overcome their own limitations. These limitations seem to be linked to certain personality traits, a lack of digital skills, and the use of ineffective self-regulation strategies during a period when time and other resource management was essential. This possible typology is, to some extent, consistent with typologies originated from previous studies [45,53,54].

7. Conclusions, Limitations, and Recommendations

The global COVID-19 health pandemic forced universities worldwide to swiftly change their ways of teaching and adopt online learning. Academics increased their efforts to make sure that the quality of their teaching was not impacted by this change while creating a feeling of normality within the new online environment. Similarly, students made significant efforts to adapt to the new changes and continue their education. Despite their increased efforts, academics and students faced numerous challenges. A low online social presence, in all its forms (e.g., sense of belonging and emotional connectedness), was one of the most significant aspects invoked by our respondents. This was followed by decreases in the learning quality and efficiency caused by the sudden move to remote blended learning and the technical issues experienced by both teachers and students.

Some of our findings are worth discussing in relation to sustainable development, as digitalisation and the introduction of online education on a large scale can be judiciously used to solve economic, environmental, and social problems. Although the experience of online learning during a crisis differs from planned online education [79], knowing the perceived negative and positive effects, as well as relating personal lived experiences in a local educational context with certain economic and social characteristics, can provide valuable information for the design of virtual learning environments. Outlining the differences between adaptive and non-adaptive students—based on their perceptions of online learning—draws attention to the need to narrow the inter-individual gap for several dimensions (e.g., developing flexibility, socio-emotional, self-regulatory, digital, and problem-solving skills) while considering the particularities of students' own development and the individual pace of adaptation to change.

This study highlights the possible decreases in the quality of teaching through predominant and prolonged online learning, as psychological and social wellbeing may be affected (e.g., boredom and fatigue, decreased engagement and motivation, lower social presence, and naturalness of the social climate). The economic benefits of online learning (e.g., saving time and money) mentioned in this study support previous findings regarding the contribution of online learning to increasing access to education and narrowing the gap between the rich and poor. However, in countries with low minimum income and significant regional economic differences, online learning could be negatively impacted by factors such as inadequate space to connect from home, poor internet connection and electronic devices, unsuitable houses for home offices, and poor learning conditions at home.

Online learning was investigated in a wide variety of quantitative studies during the COVID-19 pandemic. Our study is unique as it looked at students' perceptions using a qualitative approach, thus providing the opportunity to deepen and broaden our knowledge about online learning in crisis situations. For example, while previous quantitative studies identified the characteristics of the more adapted versus less adapted learners (e.g., [45,53]) our findings provide additional information on learners' motivations, their personal affective experiences, and the psychological and medical issues that they encountered. These findings will inform future quantitative studies looking into, e.g., the development of new scales assessing students' perceptions, and might represent the foundation for the development of psychological or social interventions.

The current study had several methodological limitations. Firstly, the qualitative approach makes it difficult to formulate scientific causal conclusions based on our results. Secondly, the predominance of female participants primarily from the same university restricts the generalisation of our results to a larger student population. Future studies should investigate larger student samples equally distributed according to gender from different universities across different regions. Thirdly, due to the qualitative nature of the tasks presented to the students in this study, numerous respondents provided similar answers for the different tasks.

Nonetheless, these findings allowed us to formulate some recommendations to ensure that the quality of the academic experience will not be impacted in the future when online learning will be more frequently and systematically adopted than it was before the pandemic. Among these, we recommend (1) using the same online platform—one that offers a dynamic, interactive, and multifunctional learning environment (intuitive user interface, streaming video, efficient online assessment tools, integrated collaborative tools, private and secure sessions, etc.) across departments and faculties; (2) modifying the teaching timetable to make it more flexible (shorter live online sessions, longer breaks between online live sessions, smaller groups for tutorials/practical activities, etc.); (3) improving online materials (detailed course handbooks, additional asynchronous materials, pre-recorded materials, recordings of online live sessions, etc.) and gradually migrating towards a digitalised curriculum (see [80]); (4) providing technical and pedagogical training for teachers and students before introducing new systems or teaching approaches; (5) providing pedagogical and psychological counselling to teachers and students during critical periods; and (6) motivating teachers and students in order to develop a higher online social presence. Some of these recommendations are similar to those proposed in previous studies [38,42].

Author Contributions: Conceptualisation, M.C. (Mihai Curelaru), V.C. and M.C. (Mioara Cristea); data curation, V.C.; formal analysis, M.C. (Mihai Curelaru) and V.C.; investigation, V.C. and M.C. (Mioara Cristea); methodology, M.C. (Mihai Curelaru) and V.C.; project administration, M.C. (Mihai Curelaru) and M.C. (Mioara Cristea); resources, V.C. and M.C. (Mioara Cristea); supervision, M.C. (Mihai Curelaru) and M.C. (Mioara Cristea); validation, M.C. (Mihai Curelaru) and M.C. (Mioara Cristea); visualisation, V.C. and M.C. (Mioara Cristea); writing—original draft preparation, V.C. and M.C. (Mihai Curelaru); writing—review and editing, M.C. (Mihai Curelaru), V.C. and M.C. (Mioara Cristea). All authors have read and agreed to the published version of the manuscript. All three authors contributed equally to the realisation of this article.

Funding: This research did not receive any specific grant from funding agencies in the public, commercial, or not-for-profit sectors.

Institutional Review Board Statement: All procedures performed in studies involving human participants were in accordance with the ethical standards of the institutional and/or national research committee and with the 1964 Helsinki declaration and its later amendments or comparable ethical standards.

Informed Consent Statement: Informed consent was obtained from all subjects involved in the study, thus guaranteeing the confidentiality and anonymity of the data and other fundamental ethical aspects.

Conflicts of Interest: The authors declare no potential conflict of interest with respect to the research, authorship, and publication of this article.

References

1. Carrillo, C.; Flores, M.A. COVID-19 and Teacher Education: A Literature Review of Online Teaching and Learning Practices. *Eur. J. Teach. Educ.* **2020**, *43*, 466–487. [CrossRef]
2. Michelsen, G.; Fischer, D. Sustainability and Education 1. In *Sustainable Development Policy: A European Perspective*; Von Hauff, M., Kuhnke, C., Eds.; Routledge: New York, NY, USA, 2017; pp. 135–158.
3. Toader, T.; Safta, M.; Titirișcă, C.; Firtescu, B. Effects of Digitalisation on Higher Education in a Sustainable Development Framework—Online Learning Challenges during the COVID-19 Pandemic. *Sustainability* **2021**, *13*, 6444. [CrossRef]
4. Watts, L. Synchronous and Asynchronous Communication in Distance Learning: A Review of the Literature. *Q. Rev. Distance Educ.* **2016**, *17*, 23.
5. Kebritchi, M.; Lipschuetz, A.; Santiague, L. Issues and Challenges for Teaching Successful Online Courses in Higher Education: A Literature Review. *J. Educ. Technol. Syst.* **2017**, *46*, 4–29. [CrossRef]
6. Gonzalez, T.; De La Rubia, M.; Hincz, K.P.; Comas-Lopez, M.; Subirats, L.; Fort, S.; Sacha, G.M. Influence of COVID-19 Confinement on Students' Performance in Higher Education. *PLoS ONE* **2020**, *15*, e0239490. [CrossRef]
7. Dumford, A.D.; Miller, A.L. Online Learning in Higher Education: Exploring Advantages and Disadvantages for Engagement. *J. Comput. High. Educ.* **2018**, *30*, 452–465. [CrossRef]
8. Francis, M.K.; Wormington, S.V.; Hulleman, C. The Costs of Online Learning: Examining Differences in Motivation and Academic Outcomes in Online and Face-to-Face Community College Developmental Mathematics Courses. *Front. Psychol.* **2019**, *10*, 2054. [CrossRef]
9. Blau, I.; Weiser, O.; Eshet-Alkalai, Y. How Do Medium Naturalness and Personality Traits Shape Academic Achievement and Perceived Learning? An Experimental Study of Face-to-Face and Synchronous e-Learning. *Res. Learn. Technol.* **2017**, *25*, 1–23. [CrossRef]

10. Lowenthal, P.; Mulder, D. Social Presence and Communication Technology: Tales of Trial and Error. In *Social Presence in Online Learning: Multiple Perspectives on Practice and Research*; Whiteside, A.L., Dikkers, A.G., Swan, K., Eds.; Stylus Publishing, LLC: Sterling, VA, USA, 2017; pp. 32–44.
11. Bedenlier, S.; Wunder, I.; Gläser-Zikuda, M.; Kammerl, R.; Kopp, B.; Ziegler, A.; Händel, M. Generation Invisible? Higher Education Students'(Non) Use of Webcams in Synchronous Online Learning. *Int. J. Educ. Res. Open* **2021**, *2*, 100068. [CrossRef]
12. United Nations. *Transforming Our World: The 2030 Agenda for Sustainable Development*; United Nations: New York, NY, USA, 2015.
13. Curelaru, V. Aspecte Psihologice Ale Tranziției de La Învățământul Tradițional La Cel Online În Perioada Pandemiei de COVID-19. *Psihol. Socială* **2021**, *48*, 97–112.
14. Munhall, P.L. Perception. In *The Sage Encyclopedia of Qualitative Research Methods*; Given, L.M., Ed.; Sage: Thousand Oaks, CA, USA, 2008; pp. 606–607.
15. Kastanakis, M.N.; Voyer, B.G. The Effect of Culture on Perception and Cognition: A Conceptual Framework. *J. Bus. Res.* **2014**, *67*, 425–433. [CrossRef]
16. Burr, V. *Social Constructionism*, 3rd ed.; Routledge: New York, NY, USA, 2015.
17. Smith, C. *What Is a Person? Rethinking Humanity, Social Life, and the Moral Good from the Person Up*; University of Chicago Press: Chicago, IL, USA, 2011.
18. Braun, V.; Clarke, V. Using Thematic Analysis in Psychology. *Qual. Res. Psychol.* **2006**, *3*, 77–101. [CrossRef]
19. Davis, J.H. Traditional vs. On-line Learning: It's Not an Either/or Proposition. *Employ. Relat. Today* **2000**, *27*, 47–60. [CrossRef]
20. Klein, D.; Ware, M. E-learning: New Opportunities in Continuing Professional Development. *Learn. Publ.* **2003**, *16*, 34–46. [CrossRef]
21. Song, L.; Singleton, E.S.; Hill, J.R.; Koh, M.H. Improving Online Learning: Student Perceptions of Useful and Challenging Characteristics. *Internet High. Educ.* **2004**, *7*, 59–70. [CrossRef]
22. Zhang, D.; Zhou, L.; Briggs, R.O.; Nunamaker, J.F., Jr. Instructional Video in E-Learning: Assessing the Impact of Interactive Video on Learning Effectiveness. *Inf. Manag.* **2006**, *43*, 15–27. [CrossRef]
23. Holmes, B.; Gardner, J. *E-Learning: Concepts and Practice*; Sage: Thousand Oaks, CA, USA, 2006.
24. Alexander, M.W.; Truell, A.D.; Zhao, J.J. Expected Advantages and Disadvantages of Online Learning: Perceptions from College Students Who Have Not Taken Online Courses. *Issues Inf. Syst.* **2012**, *13*, 193–200.
25. Dhawan, S. Online Learning: A Panacea in the Time of COVID-19 Crisis. *J. Educ. Technol. Syst.* **2020**, *49*, 5–22. [CrossRef]
26. O'Donoghue, J.; Singh, G.; Green, C. A Comparison of the Advantages and Disadvantages of IT Based Education and the Implication upon Students. *Digit. Educ. Rev.* **2004**, *9*, 63–76.
27. Volery, T.; Lord, D. Critical Success Factors in Online Education. *Int. J. Educ. Manag.* **2000**, *14*, 216–223. [CrossRef]
28. Golladay, R.M.; Prybutok, V.R.; Huff, R.A. Critical Success Factors for the Online Learner. *J. Comput. Inf. Syst.* **2000**, *40*, 69–71. [CrossRef]
29. Arghode, V.; Brieger, E.W.; McLean, G.N. Adult Learning Theories: Implications for Online Instruction. *Eur. J. Train. Dev.* **2017**, *41*, 593–609. [CrossRef]
30. Serwatka, J.A. Assessment in On-Line CIS Courses. *J. Comput. Inf. Syst.* **2003**, *44*, 16–20. [CrossRef]
31. Kock, N. Media Richness or Media Naturalness? The Evolution of Our Biological Communication Apparatus and Its Influence on Our Behavior toward e-Communication Tools. *IEEE Trans. Prof. Commun.* **2005**, *48*, 117–130. [CrossRef]
32. Swan, K.; Shih, L.F. On the Nature and Development of Social Presence in Online Course Discussions. *J. Asynchronous Learn. Netw.* **2005**, *9*, 115–136. [CrossRef]
33. Whiteside, A.L. Introducing the Social Presence Model to Explore Online and Blended Learning Experiences. *Online Learn.* **2015**, *19*, n2. [CrossRef]
34. Whiteside, A. Understanding Social Presence as a Critical Literacy. In *Social Presence in Online Learning: Multiple Perspectives on Practice and Research*; Whiteside, A.L., Garrett Dikkers, A., Swan, K., Eds.; Stylus Publishing, LLC: Sterling, VA, USA, 2017; pp. 133–142.
35. Ensmann, S.; Whiteside, A.; Gomez-Vasquez, L.; Sturgill, R. Connections Before Curriculum: The Role of Social Presence during COVID-19 Emergency Remote Learning for Students. *Online Learn.* **2021**, *25*, 36–56. [CrossRef]
36. Bączek, M.; Zagańczyk-Bączek, M.; Szpringer, M.; Jaroszyński, A.; Wożakowska-Kapłon, B. Students' Perception of Online Learning during the COVID-19 Pandemic: A Survey Study of Polish Medical Students. *Medicine* **2021**, *100*, e24821. [CrossRef]
37. Aivaz, K.-A.; Teodorescu, D. The Impact of the Coronavirus Pandemic on Medical Education: A Case Study at a Public University in Romania. *Sustainability* **2022**, *14*, 542. [CrossRef]
38. Dung, D.T.H. The Advantages and Disadvantages of Virtual Learning. *IOSR J. Res. Method Educ.* **2020**, *10*, 45–48. [CrossRef]
39. Almendingen, K.; Morseth, M.S.; Gjølstad, E.; Brevik, A.; Tørris, C. Student's Experiences with Online Teaching Following COVID-19 Lockdown: A Mixed Methods Explorative Study. *PLoS ONE* **2021**, *16*, e0250378. [CrossRef] [PubMed]
40. Harefa, S.; Sihombing, G.L.A. Students' Perception of Online Learning amidst the COVID-19 Pandemic: A Study of Junior, Senior High School and College Students in a Remote Area. *F1000Research* **2021**, *10*, 867. [CrossRef] [PubMed]
41. Muthuprasad, T.; Aiswarya, S.; Aditya, K.; Jha, G.K. Students' Perception and Preference for Online Education in India during COVID-19 Pandemic. *Soc. Sci. Humanit. Open* **2021**, *3*, 100101. [CrossRef]
42. Goudeau, S.; Sanrey, C.; Stanczak, A.; Manstead, A.; Darnon, C. Why Lockdown and Distance Learning during the COVID-19 Pandemic Are Likely to Increase the Social Class Achievement Gap. *Nat. Hum. Behav.* **2021**, *5*, 1273–1281. [CrossRef]

43. Yu, Z. The Effects of Gender, Educational Level, and Personality on Online Learning Outcomes during the COVID-19 Pandemic. *Int. J. Educ. Technol. High. Educ.* **2021**, *18*, 14. [CrossRef] [PubMed]
44. Telyani, A.E.; Farmanesh, P.; Zargar, P. The Impact of COVID-19 Instigated Changes on Loneliness of Teachers and Motivation–Engagement of Students: A Psychological Analysis of Education Sector. *Front. Psychol.* **2021**, *12*, 4353. [CrossRef]
45. Biwer, F.; Wiradhany, W.; Oude Egbrink, M.; Hospers, H.; Wasenitz, S.; Jansen, W.; De Bruin, A. Changes and Adaptations: How University Students Self-Regulate Their Online Learning during the COVID-19 Pandemic. *Front. Psychol.* **2021**, *12*, 642593. [CrossRef] [PubMed]
46. Lee, J.; Solomon, M.; Stead, T.; Kwon, B.; Ganti, L. Impact of COVID-19 on the Mental Health of US College Students. *BMC Psychol.* **2021**, *9*, 95. [CrossRef]
47. Alsoud, A.R.; Harasis, A.A. The Impact of COVID-19 Pandemic on Student's e-Learning Experience in Jordan. *J. Theor. Appl. Electron. Commer. Res.* **2021**, *16*, 1404–1414. [CrossRef]
48. Hashish, A.; Aly, E.; Baatiah, N.Y.; Bashaweeh, A.H.; Kattan, A.M. The Online Learning Experience and Reported Headaches Associated with Screen Exposure Time among Saudi Health Sciences Students during the COVID-19 Pandemic. *BMC Med. Educ.* **2022**, *22*, 226. [CrossRef]
49. Hammerstein, S.; König, C.; Dreisörner, T.; Frey, A. Effects of COVID-19-Related School Closures on Student Achievement-A Systematic Review. *Front. Psychol.* **2021**, *12*, 4020. [CrossRef] [PubMed]
50. Pham, T.T.T.; Le, H.A.; Do, D.T. The Factors Affecting Students' Online Learning Outcomes during the COVID-19 Pandemic: A Bayesian Exploratory Factor Analysis. *Educ. Res. Int.* **2021**, *2021*, 2669098. [CrossRef]
51. Mostafa, S.; Cousins-Cooper, K.; Tankersley, B.; Burns, S.; Tang, G. The Impact of COVID-19 Induced Emergency Remote Instruction on Students' Academic Performance at an HBCU. *PLoS ONE* **2022**, *17*, e0264947. [CrossRef]
52. Spitzer, M.W.H.; Musslick, S. Academic Performance of K-12 Students in an Online-Learning Environment for Mathematics Increased during the Shutdown of Schools in Wake of the COVID-19 Pandemic. *PLoS ONE* **2021**, *16*, e0255629. [CrossRef] [PubMed]
53. Broadbent, J.; Fuller-Tyszkiewicz, M. Profiles in Self-Regulated Learning and Their Correlates for Online and Blended Learning Students. *Educ. Technol. Res. Dev.* **2018**, *66*, 1435–1455. [CrossRef]
54. Ishimaru, D.; Adachi, H.; Nagahara, H.; Shirai, S.; Takemura, H.; Takemura, N.; Mehrasa, A.; Higashino, T.; Yagi, Y.; Ikeda, M. Characteristics of Adaptation in Undergraduate University Students Suddenly Exposed to Fully Online Education during the COVID-19 Pandemic. *Front. Psychiatry* **2021**, *1571*, 731137. [CrossRef]
55. Khan, M.A.; Nabi, M.K.; Khojah, M.; Tahir, M. Students' Perception towards e-Learning during COVID-19 Pandemic in India: An Empirical Study. *Sustainability* **2020**, *13*, 57. [CrossRef]
56. Almahasees, Z.; Mohsen, K.; Amin, M.O. Faculty's and Students' Perceptions of Online Learning during COVID-19. *Front. Educ.* **2021**, *6*, 638470. [CrossRef]
57. Al-Alami, Z.M.; Adwan, S.W.; Alsous, M. Remote Learning during COVID-19 Lockdown: A Study on Anatomy and Histology Education for Pharmacy Students in Jordan. *Anat. Sci. Educ.* **2022**, *15*, 249–260. [CrossRef]
58. Al Zahrani, E.M.; Al Naam, Y.A.; AlRabeeah, S.M.; Aldossary, D.N.; Al-Jamea, L.H.; Woodman, A.; Shawaheen, M.; Altiti, O.; Quiambao, J.V.; Arulanantham, Z.J. E-Learning Experience of the Medical Profession's College Students during COVID-19 Pandemic in Saudi Arabia. *BMC Med. Educ.* **2021**, *21*, 443. [CrossRef]
59. Măsurile Stabilite de Ministerul Educatiei Si Cercetarii Cu Privire La Reluarea Cursurilor. Available online: https://www.edu.ro/m%C4%83surile-stabilite-de-ministerul-educa%C8%9Biei-%C8%99i-cercet%C4%83rii-cu-privire-la-reluarea-cursurilor (accessed on 4 April 2022)
60. Joffe, H. Thematic Analysis. In *Qualitative Research Methods in Mental Health and Psychotherapy: A Guide for Students and Practitioners*; Harper, D., Thompson, A.R., Eds.; Wiley-Blackwell: Chichester, UK, 2012; Volume 1, pp. 209–223.
61. Dany, L.; Urdapilleta, I.; Lo Monaco, G. Free Associations and Social Representations: Some Reflections on Rank-Frequency and Importance-Frequency Methods. *Qual. Quant.* **2015**, *49*, 489–507. [CrossRef]
62. Lo Monaco, G.; Piermattéo, A.; Rateau, P.; Tavani, J.L. Methods for Studying the Structure of Social Representations: A Critical Review and Agenda for Future Research. *J. Theory Soc. Behav.* **2017**, *47*, 306–331. [CrossRef]
63. Braun, V.; Clarke, V. Thematic Analysis. In *APA Handbook of Research Methods in Psychology*; Research designs: Quantitative, qualitative, neuropsychological, and biological; Cooper, H., Camic, P.M., Long, D.L., Panter, A.T., Rindskopf, D., Sher, K.J., Eds.; American Psychological Association: Washington, DC, USA, 2012; Volume 2, pp. 57–71.
64. Willig, C. *EBOOK: Introducing Qualitative Research in Psychology*; Open University Press: Berkshire, UK, 2013.
65. Howitt, D. *Introduction to Qualitative Research Methods in Psychology*, 3rd ed.; Pearson: London, UK, 2016.
66. Clarke, V.; Brown, V. Thematic Analysis. *J. Posit. Psychol.* **2017**, *12*, 297–298. [CrossRef]
67. Terry, G.; Hayfield, N.; Clarke, V.; Braun, V. Thematic Analysis. *SAGE Handb. Qual. Res. Psychol.* **2017**, *2*, 17–37.
68. Braun, V.; Clarke, V.; Weate, P. Using Thematic Analysis in Sport and Exercise Research. In *Routledge Handbook of Qualitative Research in Sport and Exercise*; Smith, B., Sparkes, A.C., Eds.; Routledge: London, UK, 2016; pp. 191–205.
69. Mukhtar, K.; Javed, K.; Arooj, M.; Sethi, A. Advantages, Limitations and Recommendations for Online Learning during COVID-19 Pandemic Era. *Pak. J. Med. Sci.* **2020**, *36*, S27. [CrossRef] [PubMed]
70. Wang, B.; Liu, Y.; Qian, J.; Parker, S.K. Achieving Effective Remote Working during the COVID-19 Pandemic: A Work Design Perspective. *Appl. Psychol.* **2021**, *70*, 16–59. [CrossRef]

71. Gunawardena, C.N.; Zittle, F.J. Social Presence as a Predictor of Satisfaction within a Computer-mediated Conferencing Environment. *Am. J. Distance Educ.* **1997**, *11*, 8–26. [CrossRef]
72. Garrison, D.R.; Anderson, T.; Archer, W. Critical Inquiry in a Text-Based Environment: Computer Conferencing in Higher Education. *Internet High. Educ.* **1999**, *2*, 87–105. [CrossRef]
73. Yaseen, Q.B.; Salah, H. The Impact of E-Learning during COVID-19 Pandemic on Students' Body Aches in Palestine. *Sci. Rep.* **2021**, *11*, 22379. [CrossRef]
74. Beaunoyer, E.; Dupéré, S.; Guitton, M.J. COVID-19 and Digital Inequalities: Reciprocal Impacts and Mitigation Strategies. *Comput. Hum. Behav.* **2020**, *111*, 106424. [CrossRef]
75. Fadde, P.J.; Vu, P. Blended Online Learning: Benefits, Challenges, and Misconceptions. In *Online Learning: Common Misconceptions, Benefits, and Challenges*; Lowenthal, P.R., York, C.S., Richardson, J.C., Eds.; Nova Publishers: Hauppauge, NY, USA, 2014; Volume 2014, pp. 33–48.
76. Elfirdoussi, S.; Lachgar, M.; Kabaili, H.; Rochdi, A.; Goujdami, D.; El Firdoussi, L. Assessing Distance Learning in Higher Education during the COVID-19 Pandemic. *Educ. Res. Int.* **2020**, *2020*, 13. [CrossRef]
77. Shea, P.; Bidjerano, T. Cognitive Presence and Online Learner Engagement: A Cluster Analysis of the Community of Inquiry Framework. *J. Comput. High. Educ.* **2009**, *21*, 199–217. [CrossRef]
78. Shea, P.; Hayes, S.; Vickers, J. Online Instructional Effort Measured through the Lens of Teaching Presence in the Community of Inquiry Framework: A Re-Examination of Measures and Approach. *Int. Rev. Res. Open Distrib. Learn.* **2010**, *11*, 127–154. [CrossRef]
79. Hodges, C.B.; Moore, S.; Lockee, B.B.; Trust, T.; Bond, M.A. The Difference between Emergency Remote Teaching and Online Learning. Available online: https://er.educause.edu/articles/2020/3/the-difference-between-emergency-remote-teaching-and-online-learning (accessed on 28 April 2022).
80. Khoza, S.B. Exploring the Migration to a Digitalised Curriculum at UKZN. *Educ. Sci.* **2021**, *11*, 682. [CrossRef]

sustainability
MDPI

Article

Developing and Evaluating Educational Innovations for STEAM Education in Rapidly Changing Digital Technology Environments

Zsolt Lavicza [1], Robert Weinhandl [1,*], Theodosia Prodromou [2], Branko Anđić [1], Diego Lieban [1], Markus Hohenwarter [1], Kristof Fenyvesi [3], Christopher Brownell [4] and Jose Manuel Diego-Mantecón [5]

1 Linz School of Education, STEM Education, Johannes Kepler University, Altenberger St. 54, 4040 Linz, Austria; zsolt.lavicza@jku.at (Z.L.); brankoan01@gmail.com (B.A.); diegolieban@yahoo.es (D.L.); markus.hohenwarter@jku.at (M.H.)
2 Faculty of Humanities, Arts, Social Sciences and Education, University of New England, Armidale 2351, Australia; theodosia.prodromou@une.edu.au
3 Finnish Institute for Educational Research, University of Jyväskylä, Seminaarinkatu 15, 40014 Jyväskylä, Finland; kristof.fenyvesi@jyu.fi
4 School of Education, Fresno Pacific University, 1717 S Chestnut Ave, Fresno, CA 93702, USA; chris.brownell@fresno.edu
5 Department of Mathematics, Statistics and Computing, University of Cantabria, Avda. de los Castros, s/n, 39005 Santander, Spain; josemanuel.diego@unican.es
* Correspondence: robert.weinhandl@jku.at

Citation: Lavicza, Z.; Weinhandl, R.; Prodromou, T.; Anđić, B.; Lieban, D.; Hohenwarter, M.; Fenyvesi, K.; Brownell, C.; Diego-Mantecón, J.M. Developing and Evaluating Educational Innovations for STEAM Education in Rapidly Changing Digital Technology Environments. *Sustainability* **2022**, *14*, 7237. https://doi.org/10.3390/su14127237

Academic Editors: David González-Gómez and Jin Su Jeong

Received: 19 April 2022
Accepted: 10 June 2022
Published: 13 June 2022

Abstract: In this paper, we offer three examples from our research projects on both technological and pedagogical innovations to illustrate the impact of rapid technology changes on research. Members of our research team both developed and used technology applications in their research projects, utilizing design-based research (DBR). During the experiments, we encountered new challenges by the end of the research cycle due to updates in technologies. Although we had an idea of how to redesign the project for the next cycle based on the analyses of data, we noticed that we needed to not only redesign our approaches based on the research results but the changes in technologies were so rapid that materials and pedagogies needed to be altered as well. In our article, we propose an additional aspect to be considered in DBR while researching technology integration or innovative technologies. Moreover, the rapid change in technology raises further challenges to teachers' professional development and the integration of those innovative technologies in classrooms. We anticipate our work to contribute to the development of technology resources and related pedagogies as well as the refinement of research methodologies in technology environments. Our contributions for the development of technology resources and refinement of research methods in technology-supported learning environments should, among other things, contribute to a less complex and at the same time more sustainable integration of pedagogical innovations into scientific and school practices.

Keywords: technological and pedagogical innovations; 3D printing; flipped classrooms; augmented reality; steam education

1. Introduction

Numerous studies have shown that, with increasing regularity, digital technologies are being integrated into STE(A)M (Science, Technology, Engineering, Arts and Mathematics) education (see definition below), but these innovations do meet with a variety of challenges [1,2]. These challenges include developing appropriate technologies, resources, pedagogies, and, importantly, preparing teachers to be able to utilize technologies with new teaching approaches [3]. Several studies focus on the various integrations of technologies into current educational settings and some on the future potentials of new emerging technologies [4]. These experiences have deepened students' understanding and boosted their confidence and enjoyment of engaging with mathematics and other science subjects [5–9]. Increasing students' confidence and enjoyment in engaging with mathematics and other science subjects should contribute to sustainable learning, according to [10]. In this context, sustainable learning can be considered as, among other things, approaches that may focus on elements related to learning processes rather than on accumulating knowledge. Based on current research and the experience of developing STE(A)M education technology, the Research Centre at Johannes Kepler University, Austria and the GeoGebra Development Centre, together with an international cadre of colleagues, are experimenting and evaluating the potentials and innovations of educational technologies to address issues with these technologies' current and future uses in STEAM-related education. These issues include addressing the growing emphasis on inter and trans-disciplinary learning environments, connecting subjects more closely to the other disciplines within the STEM framework, and, more recently, the inclusion of the Arts, (broadly, through a sense of design and creativity) to develop teaching from STEM to STE-A-M transitions [11]. Developing pedagogies to nurture skills, in particular creativity and critical thinking, that are increasingly identified as necessary inclusions within the future of education [12]. Critical thinking by students and teachers as well as pedagogies may transcend boundaries of subjects and are also central elements of sustainable education [10]. Adopting, developing, and integrating new innovative technologies, including the technologies of augmented reality, 3D-printing, gamification, and adaptive learning, each integrated into the dynamic geometry environment known as "GeoGebra" [13]. Developing both off- and online teacher training courses and resources that enable teachers to use technologies immediately and to consider how digital innovations may be integrated into teaching in the future [14]. While a large number of studies examine the acceptance of a particular digital technology by teachers, there are only few studies aiming at investigating how teachers adapt to constant changes in educational technologies.

Developing theoretical approaches to be able to better examine these issues. Within this paper, we will focus on the challenges that are created when the technological environment is developing and evolving at a pace that outstrips educational institutions' abilities to adopt and adapt to the innovations available to them. The main goal of the current study was to identify approaches through which teachers and researchers adapt to the changes brought along by fast-changing technologies. Therefore, we hope to contribute to the development of both research methodologies and questions raised by technology integration to further education in our age.

We offer three examples from our research projects (e.g., Da Vinci Machines and 3D printing, flipped learning approaches, and augmented reality applications), which demonstrate the situation wherein a teaching practice employing a new technology was undergoing implementation research only to need significant revision due to the advancement in the technology prior to the research being published. We will argue that research methodologies, especially in our case, and design-based research (DBR) needs to be adapted to suit the fast-paced technology changes. Further, we argue that teachers' professional development also needs to emphasize assisting teachers to keep up with the pace of changes in technology-related practices and pedagogies. In this context, our research projects as well as the need for adaptation of research approaches that we aimed to propose are in line with sustainable education. According to [15], it is a specific characteristic of sustainable education that it should not be viewed only as a simple addition to traditional learning and teaching practices but also that sustainable education should represent a cultural change. For sustainable education it is vital that it utilizes a more ecological or relational view of the world. Contemporary learning approaches, for instance, flipped classrooms, could facilitate large-scale issues, such as ecological questions or problems to be addressed in classrooms, and modern technologies, such as augmented reality applications, may enable concrete elements of complex systems, e.g., ecology, to be explored three-dimensionally and actively by students.

2. Context

Despite the initial slow integration of technologies into education around the turn of the 21st century, currently, mostly in developed countries technologies are being implemented more rapidly. Thanks to substantial investments by both government and industry, combined with the widespread use of cellular technology and educational application development, fewer barriers to accessibility exist than ever before and this is on a global scale [16–18]. Nevertheless, the use of technology is still rather marginal in most countries, but trends show that they are becoming more accepted and utilized [19]. Certainly, there are numerous issues hindering the use of technologies in schools, for instance, the uniformity and continuous reliability of machines and software [20], its demand needing to keep up with its place in curricula and assessment [21], and the novelty of pedagogical approaches needed in their uses [22]. However, according to research, the two areas mainly hindering the spread of technologies in teaching practices are the insufficient preparation and support of teachers [22,23] and the role of technologies in assessment and curricula [24]. According to [25], it is also teachers who are up to date with technological trends that are essential to achieving better and sustainable education for all by 2030. According to [26,27], teachers who keep up with technological trends also have an impact on society in general and a modern educational environment, consisting of highly trained teachers and technological trends, which, among other things, can help students develop skills needed to achieve sustainable development goals.

With respect to the latter, assessment and curricula, in many countries, teachers need to prepare students for tests and standardized assessment. National curricula often include the promotion of technologies in teaching but because assessment is not yet technologically supported, teachers do not have the time and motivation to use technology in their teaching to afford students these exam preparations. There are countries, such as Austria, Denmark, and Finland, that are changing their assessment practices to allow the utilization of advanced technologies in state-wide assessment; however, there are still difficulties in integration because of the persuasion and preparation of teachers for such new demands [28].

The professional development and support of teachers is extremely important because we need to show teachers how to use technologies in their practices and continuous support is needed to strengthen its initial integration into these [29,30]. There are numerous studies offering ideas and knowledge on teachers' professional developments with technologies [31–34] as well as programs offering continuous support in schools [35]. Additionally, there are numerous initiatives developing pedagogical innovations for technology integration [14]. Many of these ideas are powerful and innovative, but a new difficulty started to arise, in that educational technology and the opportunities offered by these technologies are changing so rapidly that it is difficult to keep up with the preparation of teachers.

Interestingly, the development of technologies and educational technology is racing ahead rapidly to support these demands and assist difficulties, but changes in the pace of development pose new challenges. Currently, technologies are developing quickly to offer advanced opportunities to be securely used in large-scale assessment and are aligned with classroom uses and offer similar interfaces. For instance, in Austria and Finland, assessment developments are being explored, which involve utilizing locked mobile phones to use only graphing software or developing sticks that restrict computer use beyond using mathematical software, respectively. Importantly, such technology involvement in assessments requires the important preparation of teachers.

Another issue in education is that software and emerging technologies, such as augmented and virtual reality, 3D printing, adaptive learning solutions, etc., offer entirely new opportunities to be utilized in teaching, besides the continuous upgrades of basic software applications. This trend needs to be followed and adopted in teacher training, not only to prepare teachers to be able to use technologies but also to teach them how to adapt their practices to these new technological opportunities.

Our research team and related groups are developing technology applications and, at the same time, carrying out research on both technological and pedagogical innovations. There are numerous projects in this area, but most of our research projects utilize design-based research (DBR) because it offers suitable frameworks for developing and testing innovations in STEAM education.

3. Design-Based Research for Examining Innovations in Education

DBR is one of the emerging methodologies utilized by numerous researchers in education. According to [36] "Design-based Research is a methodology designed by and for educators that seeks to increase the impact, transfer, and translation of education research into improved practice. In addition, it stresses the need for theory building and the development of design principles that guide, inform, and improve both practice and research in educational contexts".

Cobb et al. [37] characterized design experiments as having the potential to examine the complexity of educational settings and the numerous variables that may be observed with the implementation of such methodology. Similarly, ref. [38,39] claim that DBR and associated research results proved to be promising in overcoming the problems of educational research in complex and multi-layered educational settings. Furthermore, a variety of stakeholders may be involved in research processes (e.g., teachers, researchers, educational developers, program designers, and more). Elements of design-based research may involve tasks given to students; problems that they are asked to solve; tools and related material provided, including instructional materials; and practical means through which teachers can orchestrate classroom activities. The emphasis on the articulation of all these elements leads design experiments to be applied in a variety of configurations that often vary in type and scope. Implementations of the DBR methodology could support our understanding of how students and teachers develop their practices by collecting multiple forms of data through DBR in order to explore the variety of learning processes and practices. DBR necessitates close cooperation and collaboration among researchers and practitioners. In DBR, the roles and the tasks of researchers and practitioners are divided more clearly than in other research approaches which also involve multiple stakeholders [40,41] In DBR the main tasks of researchers and practitioners are to design innovations in education, further develop them, implement them, support students during the implementation of innovations, assess the impact of the innovation, and often begin the process again [42–44]. In this process, special attention is paid to design principles and reflecting on the possible reasons for the success or failure of a design in a specific educational setting [45–48]. Through a cycle of iterations, DBR should not only provide answers to what works and what does not but should also generate practical and theoretical knowledge [49,50].

In our projects, we followed an interpretation offered by [51], who views DBR as "a systematic but flexible methodology aimed to improve educational practices through iterative analysis, design, development, and implementation, based on collaboration among researchers and practitioners in real-world settings and leading to contextually- sensitive design principles and theories" (pp. 6–7). We encountered new and heretofore unaccounted-for challenges within DBR during our experiments. By the end of our research cycle, we would have gathered information that would lead to a redesign of the project for the next cycle based on the analyses of data. However, we also were confronting changes in the technologies being implemented. Often these changes were so rapid that they would precipitate alterations in the teaching materials and pedagogies independent of the feedback data acquired during the implementation. This may have serious implications for the validity of DBR projects that are centered on the use of leading-edge technologies. What follows are three examples from our projects based upon a rapidly changing technology. We propose an additional aspect to be considered in DBR while researching technology integration or innovative technologies.

4. Examples of Influences of Rapid Technology Changes upon Educational Innovation Research

4.1. Modelling da Vinci Machines

Lieban and Lavicza researched students' experiences while they used a geometric modeling approach with dynamic geometry software, complemented with physical modeling. Aiming to enhance students' understanding of interconnections between the current trends in Science, Technology, Engineering, Art, and Mathematics education (STEAM), they assisted students in better understanding the functioning of certain physical mechanisms. Emerging from concepts from the history of Mathematics and a book by Leonardo Da Vinci [52], the authors started their study by encouraging students to engage with a double reconstruction (both physical and with GeoGebra) of certain Da Vinci machines prototypes. The tool that was available for physical reconstruction was wood.

The conjecture of the study was that the use of historical models could offer assistance to mathematical concepts and promote students' creative thinking and problem-solving strategies while they immerse themselves in the investigative process of interesting ideas. The study concentrated on Brazilian students, combining the use of physical and digital tools as well as investigating how utilizing physical and digital tools could support students' creative thinking and problem-solving in STEAM subjects.

Inspired by recent studies, ref. [53,54], which supported the benefits of design and implementation of multi-representational approaches to exploring 3D objects using crafts, computer technology, and paper-and-pencil methods, the study attempted to integrate geometry with algebra and trigonometry reaching beyond technical instrumentation. Lieban and Lavicza particularly emphasized the example of joints with circular movements. Principles providing the background for the modeling process shown in Figure 1 were adapted from [55]. The solution can be achieved through both directions.

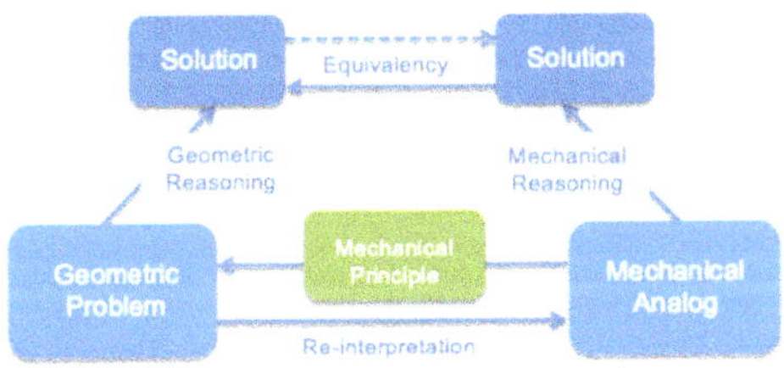

Figure 1. Modeling process from a mechanical analog to geometric problem through a mechanical principle.

This approach considered the importance of applying an methodology to problem-solving at the preliminary stage through constructing mechanical analogies for geometric problems.

The authors added the arrow in the opposite direction, since the reverse case (geometric reasoning supports mechanical reasoning) is equally possible according to their experience. The authors found that although, on the one hand, mechanical reasoning was essential to discussing the ratio for a pulley system in one case; on the other hand, in another case, with the help of rotational simulations (i.e., geometric reasoning) by means of digital modeling, the students discovered how to build a functioning physical prototype.

The researchers conducted their research with 16-year-old students who participated in a vocational (informatics) course and were supported by a teacher of mathematics and a teacher of physics. The project utilized design-based research and included cycles of both physical and digital designs. Students were asked to select Da Vinci machines to investigate. One of the most successful modeling approaches was the construction of the Da Vinci Rotary Bridge (Figure 2) developed by four students who were asked to develop both physical and digital models in order to improve the joints of the existing mechanisms. Students could select any software for their digital modeling, but most groups chose GeoGebra as it was available and suitable for their work.

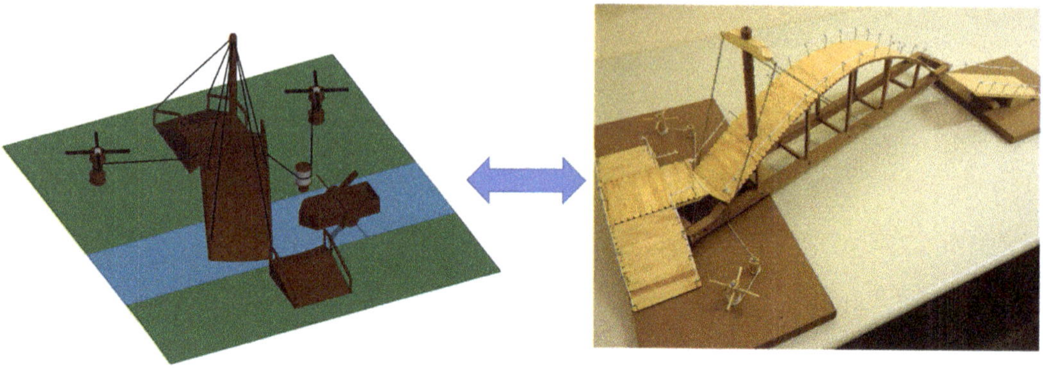

Figure 2. The digital prototype developed using the GeoGebra 3D feature (**left**) and the physical model made of wood (**right**) were developed in parallel (this and other Da Vinci models can be found at https://www.geogebra.org/m/AnHK7nCX, accessed on 12 May 2022).

Students were encouraged to develop the two models in parallel and GeoGebra materials and GeoGebra 3D features were integrated. Students were able to follow the digital modeling process and they concentrated on principles of rotation, translation, and spatial geometry.

Another construction was of DaVinci's catapult, which had a 4D frame with a structure made of flexible material (similar to plastic straws) that is easy for students to manipulate in classrooms. They simplified the physical model using simple elements that made the construction of the GeoGebra model easier (Figure 3).

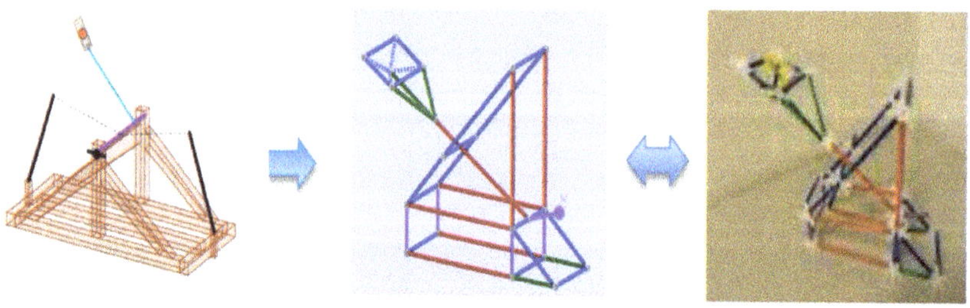

Figure 3. Catapult evolution and becoming easy to represent (by Diego Lieban).

However, the principles of joint motions remained in the models. Additionally, the coloring of specific moving elements contributed to the visualization and spatial understanding of students.

For the purpose of this paper, it is important how the evolution of technology altered the end results of the DBR approach. It can be seen in both modeling approaches that the initial modeling was created by GeoGebra without the features of GeoGebra 3D software available, then by the end of the modeling cycle GeoGebra 3D was released and new modeling opportunities arose. The experiment lasted for an entire semester and the design of the machines was improved continuously in both physical and digital forms. However, the release of new software features of GeoGebra 3D made the initial digital modeling completely obsolete and allowed students to improve their models with more appropriate tools, which also made learning through modeling more interesting. Thus, when writing up DBR results we had to consider the changes in the technological environment and re-calibrate the upcoming cycles with an updated software tool and reconsider challenges emerging from these new features. Furthermore, after the initial release of GeoGebra 3D the software continued improving considerably, allowing further ease of modeling but at the same time adding complexity for solutions because of the sophistication of the tool. In sum, we not only needed to consider results from DBR for the next cycles but also consider designing the next steps with an improved tool. Thus, was added the need to anticipate additional challenges that may appear in both practice and research. We experienced similar changes when working with 3D printing modeling for STEAM-based teaching.

4.2. Rapid Developments of 3D Printing for STEAM-Based Teaching

In line with the previously outlined research [56], the focus on developing resources to connect concrete and abstract ideas through physical and digital modelling and the development of mathematical and technological competencies through physical and digital manipulatives continued. Tasks involved possibilities for students developing mathematical models digitally who converted these constructions, via 3D printing, into physical models to explore the properties of both physical and digital representations. Similar to the Da Vinci project, we also utilized elements of DBR and action research to explore the evolution of these learning environments and offer improved designs for such modeling and learning.

We developed a simple task: dissect a cube into equal volume and surface area parts. Throughout the modeling process, students were immersed in a dynamic, exploratory process, involving constant questioning and reshaping of problems and solutions. The design process started with a brainstorming session as a starting point, followed by a brainstorming session during which students discussed alternatives and restrictions to develop their personal ideas or puzzles.

The next example, for instance, shows how students could combine pieces of a standard pyramid (Figure 4a) in different ways to obtain new solutions. While the initial model represented 1/6 of the cube, when students joined two or three pieces together it became 1/2 and 1/3 of the cube, as shown in Figure 4b,c. When continuing the dissection process as illustrated by Figure 4d–f, the solutions appeared to be a bit less intuitive. At this stage of developing different solutions from the same basic shape, it was important to discuss and show why the volume and the surface area were still the same for all the parts. Observing the symmetry and the fractions involved we realized that opportunities for learning Mathematics go beyond metric geometry.

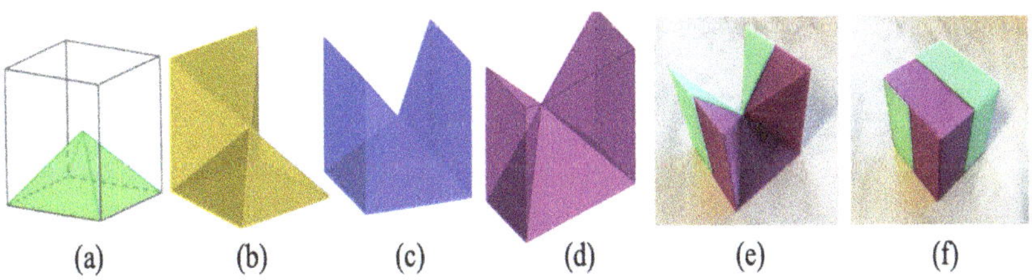

Figure 4. Developing different solutions from the same basic shape: (**a**) shows how a standard pyramid can be combined in different ways with the aim to obtain new solutions; (**b**–**d**) when two or three of them are connected together, they present 1/3 and 1/2 of the cube; (**e**,**f**) continuing the dissection process will provide less intuitive solutions (by Diego Lieban).

This following task provided a fruitful opportunity to extend ideas from a 2D plane to a 3D space. In particular, we used a solution obtained initially in 2D to split a square into four pieces with equal perimeters and surface areas. When students divided a square in such a way, they found that similar conditions were applied in both 2D and 3D spaces, which meant what they had done was equivalent to splitting a cube into four pieces which had the same area and volume, as illustrated by Figure 5.

(a) (b)

Figure 5. From a 2D to a 3D space, analogies are transferred by extrusion. Example of a digital and physical solution obtained initially in 2D to split a square (**a**) and cube (**b**) into four pieces with equal perimeter and surface area; b (by Diego Lieban).

Concerning the physical and digital explorations, we observed that while the physical models allowed more freedom in the sense of testing and assembling the cube, the preset model assembled in the digital version was able to highlight certain regularities or behaviors when dragging all parts together to arrange them as a single final piece.

Modeling was performed in GeoGebra 3D, the features of which improved slightly during the DBR process. However, opportunities to print models from GeoGebra 3D (it needs to be emphasized that only a few mathematical software have the capability to export models to 3D printing, and we know of no software that allows for dynamic manipulations in connection to 3D printing) improved considerably during the DBR experiments. We are fortunate in our research group because we can not only utilize the software but also make recommendations for the direction of its improvements. GeoGebra is not a CAD software and does not include Boolean operations, and, as such, may not yet be the best for 3D printing when involving the addition or subtraction of shapes, but it is a mathematical tool that can contribute to students' understanding, and its 3D printing features offer new insights for students to understand Mathematics. However, while defining mathematical objects in GeoGebra may be easier, in CAD software it is possibly more complicated, depending on the desired models. These experiments (and others) contributed to the inspiration for developing features in GeoGebra that utilize the advantages of both mathematical and CAD software to offer new opportunities for teachers to explore mathematics with new depth. Nevertheless, we aim to prepare teachers to be able to decide which software to use for different modeling purposes.

At the beginning of our DBR, the software did not allow the controlling of the thickness of 3D printed models and models had to be exported to CAD software for secondary processing, but through our requests to the developer team, this was resolved after some months and contributed to our experiments. In addition, when modeling digitally before printing, the transparency of models was not appropriate for exploring certain mathematical concepts inside the dissected cubes, but later assisted our explorations. Additionally, the dragging mode of the dynamic software allowed users to customize their solutions and improve their design. The evolution of 3D printing features considerably helped us and there are still continuous improvements to be made in the software, but it also resulted in changes in resources and results at the end of the DBR cycles. Furthermore, during our investigation and experiments with 3D printing the augmented reality (AR) application of GeoGebra became available, which further influenced our DBR results. In AR, besides the original software features utilized for modeling, shapes could be placed onto physical models making the process more fluid and interesting for students. The dynamic design and constructions of mathematical objects emerged in AR, and this again offered new opportunities for digital modelling and complemented our 3D printed models by projecting/merging AR models onto 3D printed models, further altering our DBR cycles. Further details of GeoGebra AR can be found in [57,58], Figure 6 shows an example of AR placement in students' surroundings, and Figure 7 depicts a learning scenario in which a physical 3D printed model is combined with an AR virtual model to further enhance mathematical ideas.

Figure 6. Example of GeoGebra AR (by Diego Lieban).

Figure 7. Desktop, 3D prints, and AR (by Diego Lieban).

The description of our project showed that it is becoming increasingly important to count on changes in technologies and complement DBR with attention to technology changes. The next example looks at how pedagogical innovations, such as flipped classroom environments, need to adapt to technology changes as well.

4.3. Adapting Flipped Learning Approaches for Technology Changes

Another example illustrating our argument that rapid changes in technology development impact both research and teaching practices come from our work on experimenting with flipped learning approaches and technology resources. We carried out several projects involving technology uses and flipped learning and experienced the impact of rapid technology changes. We chose one particular topic where students were encouraged to build bridges with physical resources and utilize dynamic mathematics software to model and analyze the mathematical content of their work. This study also employed design research approaches, and changes in technology appeared when the augmented reality application of GeoGebra became powerful enough to be used for modeling, which offered new opportunities beyond utilizing the desktop or mobile versions of GeoGebra. Before outlining the project, we offer a brief introduction to flipped learning approaches (FLA) and the technology-related considerations of the project.

As the pedagogical approach of our study, we chose to develop experiments with flipped learning approaches and utilized DBR as a methodology as the combination was able to offer interesting results for technology integration. Learning with FLA enabled students to be assigned tasks and tools to investigate before coming to classes and class time was mainly devoted to discussion and deepening students' knowledge. Originally, flipped classroom environments were used in the literature, but they were mostly associated with video preparations before classes [59,60] and stricter prescriptions on how flipped classroom methods should be applied. However, for our study, we wanted to extend the pool of technologies and develop wider opportunities for experimentation, and thus we utilized FLA as a further development of flipped classroom methods in the literature [61]. In addition, FLA enables students to decide by themselves whether to learn individually or in groups, which was an important consideration for our study design. Using FLA, ref. [62–64] were able to demonstrate the positive effects of education and we utilized numerous results from these studies for our design. Furthermore, as highlighted earlier, DBR was utilized, which was also supported by [44], who utilized DBR for complex activities in blended learning environments, which could form frameworks for flipped education. Additionally, technology integration into STEAM teaching and learning in such environments could be beneficial and is often valued by students.

Our study aimed to investigate how the modeling of real phenomena and a technology-enhanced mathematization of real-life models could be carried out in higher secondary level classrooms. Thus, we worked with 9th grade students who were asked to examine the properties and constructions of DaVinci or mathematical bridges, build such bridges in groups of three to five students, and then mathematize the self-made bridges using GeoGebra. When planning the study, we considered the desktop version of GeoGebra (Version 6) because this version of GeoGebra allowed us to simplify inserting and modifying images. Students built their bridges in teams and modeled their constructions with the GeoGebra desktop version, and the mobile app was also available for them to check their solutions. Examples of students' modeling can be seen in Figure 8.

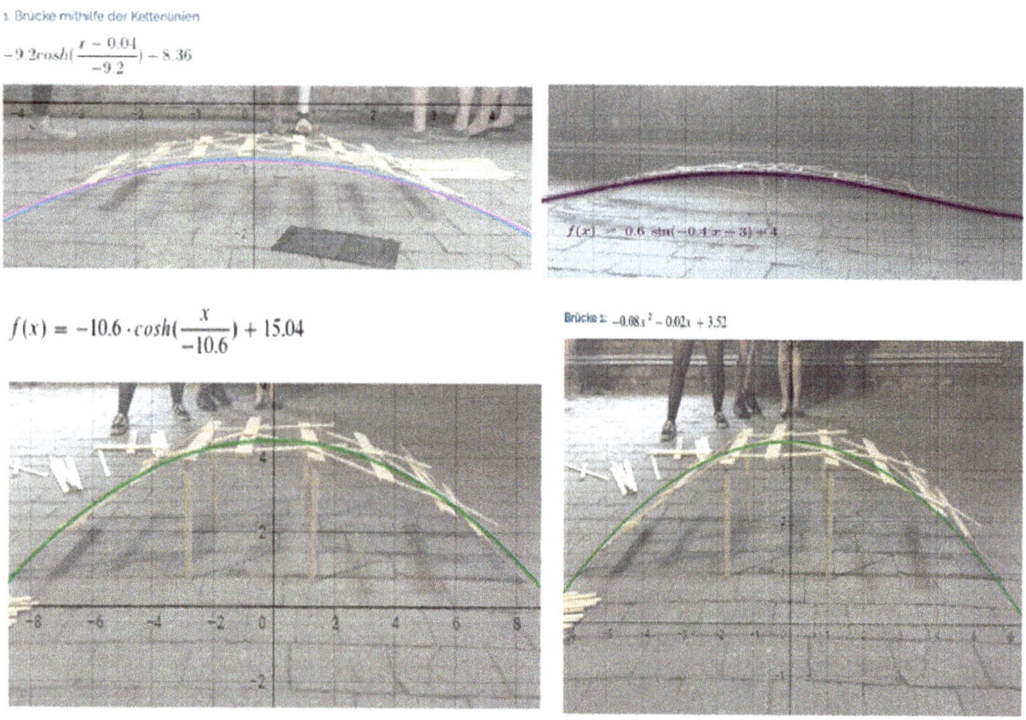

Figure 8. Solutions of Da Vinci bridges (by Robert Weinhandl).

According to DBR, we collected written feedback from students and carried out numerous interviews, besides the observations of their projects. The data were analyzed (we are currently working on further analyses by constant comparative methods) and findings were integrated into subsequent design cycles. Extended results of this study will be published in [65]. In the development of the design cycles, we had to take into account new opportunities that software development offered. Students discovered new features and immediately started to experiment with AR features, but participating teachers were slightly concerned about such quick changes in technology tools.

5. Discussion

With previous examples, we offered an overview of how some of our projects experienced the impact of fast-changing technologies. We suggested with these examples that developing technology-enhanced resources, pedagogies, and the preparation of teachers as well as research approaches need to be updated to keep up with technology changes. This would contribute to more appropriate applications of technologies in education and assist teachers in remaining updated with technological trends that are necessary for leveraging better and sustainable education for all by 2030 [25]. According to McKnight et al. [26], teachers trained to keep up with technological trends are more successful in developing creative, collaborative, personalized, and supportive learning environments. Similarly, Sarker [27] points out that providing resources and support to teachers to follow technological trends not only contributes to the education of students but also has an impact on society in general. The same authors also explained that the educational environment which is consistent with technological trends develops students' skills that are necessary for attaining to sustainable development goals (SDGs). Aligning the development of certain areas of human activity with the development of technologies is one of the key areas for the implementation of SDGs, and education has been identified as one of the key components in this process [66]. Education is critical to achieving SDGs: there is a sustainable goal dedicated solely to education (SDG4), and education and educational technologies are linked to all SDGs in numerous ways [67]. This indicates the great social importance of harmonizing education with constant changes in technologies and adopting research methods related to its rapid changes. In addition, enhancing technological infrastructure and improving digital competencies of teachers through professional development based on contemporary scientific research is crucial for the digitalization of the educational process [68].

In our work, and more generally for researchers working on innovative educational approaches, it is important to develop teaching and learning environments with more interconnectedness of subjects and topics; our examples showed topics connecting subjects and highlighted some aspects of creativity and creation integration into STEAM classrooms.

This is consistent with earlier research suggesting that digitalization opportunities should be maximized in order to make transdisciplinary materials as well as technological resources available to all [69]. Thus, we work with the framework of STEAM education [11] that involves the incorporation of creativity both in innovative classroom resources and related pedagogies [12]. In our case, the connection of physical and digital resources are key components in our projects, and by developing innovations within these frameworks we were able to show positive results in students' learning, motivations, and attitudes as well as influencing teachers' thinking on how technology could be integrated into their practices. The preparation of teachers became increasingly important for us when we observed that teachers were highly concerned about the technological changes happening even over the short period of time we worked together with technologies. We believe, and will further test in our projects, that the skills of adapting technologies and nurturing acceptance for such situations could be a way forward, but we also realized that individual teachers could hardly cope with such challenges; therefore, working within a community of like-minded teachers, these concerns could be reduced considerably. Thus, nurturing teacher communities and sharing resources and approaches could be the key to preparing teachers for the rapidly changing educational technology environment, and pooling various skills of teachers and splitting their knowledge could become increasingly important. In our projects, and future projects, in particular, we will focus even more on fostering such communities and making recommendations and specific actions for designing the training and environments to create a space for such communities. We are fortunate in our team that we could closely work together with the software developer team and offer immediate feedback from our research to improve the development of the software as well as design online training environments for developing teachers and communities. Our software development group has already begun constructing online spaces for virtual and in-class collaboration environments as well as research methodologies related to various research approaches in schools. In particular, to reflect the rapid development of technologies, we started to combine education and user experience (UX) research methods, particularly persona development and A–B testing, beyond the examples presented in this paper. Importantly, results of our study are in line with previous studies [70], in which team collaboration is vital for improving teaching and research capacities with the implementation of fast-changing digital technologies. Such studies also proposed a graduated team-building method, taking into account teams' professional and seniority integration and development. This entails accounts of nationally recognized and integrated innovative educational professionals as a team of curriculum research and development advisors to oversee research and development from macro and professional standpoints.

In connection with developing innovative technology environments and connected educational approaches, we also highlighted the necessity for updating and further developing research methodologies, especially in our case of design-based research. While developing cycles of DBR, we needed to take into account how much technology developed and how we could integrate these new feathers into the next cycles of DBR. It would be important to work on this issue as the development of technology is unlikely to slow down and more research will be carried out in this area. We are already making some recommendations, but in our future projects, we will pay particular attention to understanding the impact of technology changes. Additionally, we see great potential in combining educational research and software development research. Thus, we have begun working with some UX researchers and have made attempts to combine aspects of these two research methodologies to better understand the impact of technology changes as well as to contribute to the development of educational technologies with research insights from various methodologies. As [25] points out, it is of great importance that scientific research support the education system in the process of adapting to changes in technology and science. Thus, well-designed, state-of-the-art, and up-to-date research approaches can greatly contribute to the development of technology integration in education.

6. Conclusions

The present study was aimed at pointing out the impacts of fast-changing technologies on education and related research methodologies. We highlighted that changes in research approaches and teacher training are necessary to enhance innovations and integration in our technological education environments. Therefore, research methodologies need to be continuously updated along with transdisciplinary and technological resources for teacher development. We also highlighted the importance of the involvement of teachers, researchers, and developers as well as teacher training professionals in the process of designing learning environments for the successful applications of technologies in education. Our examples and research suggest that design-based research has an important role in such developments, but DBR needs to be continuously updated to keep up with fast-changing technologies. In our current work, we have already started new methodological experiments and have begun combining DBR with UX research approaches. Nevertheless, further studies are needed to better understand the impact of teachers' skills to adapt to constant, fast changes in technology and their abilities to follow changes in technology through pedagogical approaches as well as meaningfully updating research methodologies.

Author Contributions: Conceptualization, Z.L., R.W., D.L. and M.H.; methodology, Z.L., R.W. and D.L.; software, M.H. and Z.L.; validation, T.P., B.A. and D.L.; formal analysis, K.F., C.B. and J.M.D.-M.; investigation, Z.L., R.W., D.L. and M.H.; resources, R.W. and D.L.; data curation, K.F., C.B. and J.M.D.-M.; writing—original draft preparation, Z.L., R.W., D.L., M.H. and B.A.; writing—review and editing, all authors; visualization, R.W. and D.L.; supervision, Z.L.; project administration, Z.L.; funding acquisition, Z.L. All authors have read and agreed to the published version of the manuscript.

Funding: This research was funded by Open Access Funding by the University of Linz.

Institutional Review Board Statement: Not applicable.

Informed Consent Statement: Not applicable.

Data Availability Statement: The data that support the findings of this study are available on request from the corresponding author.

Acknowledgments: We thanked Open Access Funding by the University of Linz.

Conflicts of Interest: The authors declare no conflict of interest.

References

1. OEC. *Students, Computers and Learning: Making the Connection, PISA*; OECD Publishing: Paris, France, 2015.
2. Drijvers, P. Evidence for benefit? Reviewing empirical research on the use of digital tools in mathematics education. In Proceedings of the 13th International Congress on Mathematical Education, Hamburg, Germany, 24–31 July 2016.
3. Lavicza, Z.; Prodromou, T.; Juhos, I.; Koren, B.; Fenyvesi, K.; Hohenwarter, M.; Diego-Mantecon, J.M. The Need for Educational Research on Technology: Trends and Examples. *Int. J. Technol. Math. Educ.* **2022**, *in press*.
4. Ball, L.; Drijvers, P.; Ladel, S.; Siller, H.S.; Tabach, M.; Vale, C. *Uses of Technology in Primary and Secondary Mathematics Education: Tools, Topics and Trends*; Springer: Berlin/Heidelberg, Germany, 2018.
5. Chance, B.; Garfield, J.; delMas, R. Developing simulation activities to improve students' statistical reasoning. In Proceedings of the TIME 2000, Auckland, New Zealand, 11–14 December 2000.
6. Clements, D.H.; Sarama, J.; Yelland, N.J.; Glass, B. Learning and teaching geometry with computers in the elementary and middle school. In *Research on Technology and the Teaching and Learning of Mathematics: Vol 1. Research Synthesis*; Heid, M.K., Blume, G.W., Eds.; N.C.T.M. Publications: Carolina, NC, USA, 2008; pp. 109–154.
7. Roschelle, J.; Kaput, J.; Stroup, W. SimCalc: Accelerating students' engagement with the mathematics of change. In *Learning the Sciences of the 21st Century: Research, Design, and Implementing Advanced Technology Learning Environments*; Jacobsen, M.J., Kozma, R.B., Eds.; Erlbaum: Hillsdale, NJ, USA, 2000; pp. 47–75.
8. Sacristan, A.I.; Noss, R. Computational construction as a means to coordinate representations of infinity. *Int. J. Comput. Math. Learn.* **2008**, *13*, 47–70. [CrossRef]
9. Villarreal, M. Mathematical thinking and intellectual technologies: The visual and the algebraic. *Learn. Math.* **2000**, *20*, 2–7.
10. Thomas, I. Critical thinking, transformative learning, sustainable education, and problem-based learning in universities. *J. Transform. Educ.* **2009**, *7*, 245–264. [CrossRef]
11. Fenyvesi, K. *Bridges: A World Community for Mathematical Art*; Mathematical Intelligencer: Berlin, Germany, 2016.
12. Burnard, P.; Ross, V.; Dragovic, T.; Minors, H.; Powell, K.; Mackinlay, L. *Building Interdisciplinary and Intercultural Bridges: Where Practice Meets Research and Theory*; BIBACC Publishing: Cambridge, UK, 2017. [CrossRef]
13. Lavicza, Z.; Hohenwarter, M.; Fenyvesi, C.; Prodromou, T.; Diego-Mantecon, J.M.; Lieban, D. Mathematics learning through arts, technology and robotics: Multi-and transdisciplinary STEAM approaches. In Proceedings of the 8th ICMI-East Asia Regional Conference on Mathematics Education, Taipei, Taiwan, 7–11 May 2018; pp. 110–122.
14. Diego-Mantecón, J.M.; García-Piqueras, M.; Blanco, T.F.; Ortiz-Laso, Z. Problemas en contextos reales para trabajar las matemáticas—Plataforma STEMforYouth. *Soc. Inf.* **2018**, *58*, 29–38.
15. Sterling, S. *Sustainable Education. Science, Society and Sustainability*; Routledge: London, UK, 2010; pp. 127–140.
16. Bonilla-del-Río, M.; García-Ruiz, M.R.; Pérez-Escoda, A. Los dispositivos móviles en el aula. Oportunidades y retos para el desarrollo de la competencia mediática. In *Dispositivos Móviles en el aula. Docentes y Estudiantes Prosumidores en la era Digital*; García-Ruiz, M.R., Pérez-Escoda, A., Guzmán-Franco, M.D., Eds.; Egregius Ediciones: Sevilla, Spain, 2018; pp. 11–30.
17. Engen, B.K. Comprendiendo los aspectos culturales y sociales de las competencias digitales docentes. *Comun. Rev. Científica Iberoam. De Comun. Y Educ.* **2019**, *27*, 9–19.
18. Santana-Vega, L.E.; Gómez-Muñoz, A.M.; Feliciano-García, L. Uso problemático del móvil, fobia a sentirse excluido y comunicación familiar de los adolescentes. *Comun. Rev. Científica Iberoam. De Comun. Y Educ.* **2019**, *27*, 39–47.
19. Lavicza, Z. Integrating technology into mathematics teaching: A review. *ZDM Int. J. Math. Educ.* **2010**, *42*, 105–119. [CrossRef]
20. Wachira, P.; Keengwe, J. Technology integration barriers: Urban school mathematics teachers perspectives. *J. Sci. Educ. Technol.* **2011**, *20*, 17–25. [CrossRef]
21. Ulbrich, E.; Lieban, D.; Lavicza, Z.; Vagova, R.; Handl, J.; Andjic, B. Come to STEAM. We have cookies! In Proceedings of the Bridges 2020: Mathematics, Art, Music, Architecture, Education, Culture, Helsinki, Finland, 1–5 August 2020; pp. 297–304.
22. Francom, G.M. Barriers to technology integration: A time-series survey study. *J. Res. Technol. Educ.* **2019**, *52*, 1–16. [CrossRef]
23. Diego-Mantecón, J.M.; Arcera, Ó.; Blanco, T.F.; Lavicza, Z. An Engineering Technology Problem-Solving Approach for Modifying Student Mathematics-Related Beliefs: Building a Robot to Solve a Rubik's Cube. *Int. J. Technol. Math. Educ.* **2019**, *26*, 55–64.
24. Instefjord, E.; Munthe, E. Preparing pre-service teachers to integrate technology: An analysis of the emphasis on digital competence in teacher education curricula. *Eur. J. Teach. Educ.* **2016**, *39*, 77–93. [CrossRef]
25. Nicolai, S.; Wales, J.; Aiazzi, E. *Education, Migration and the 2030 Agenda for Sustainable Development*; ODI: London, UK, 2016.

26. McKnight, K.; O'Malley, K.; Ruzic, R.; Horsley, M.K.; Franey, J.; Bassett, K. Teaching in a digital age: How educators use technology to improve student learning. *J. Res. Technol. Educ.* **2016**, *48*, 194–211. [CrossRef]
27. Sarker, M.N.; Wu, M.; Cao, Q.; Alam, G.M.; Li, D. Leveraging digital technology for better learning and education: A systematic literature review. *Int. J. Inf. Educ. Technol.* **2019**, *9*, 453–461. [CrossRef]
28. Weinhandl, R.; Lavicza, Z. Exploring essential aspects when technology-enhanced flipped classroom approaches are at the heart of professional mathematics teacher development courses. *Int. J. Technol. Math. Educ.* **2019**, *26*, 139–144. [CrossRef]
29. Ramatlapana, K.A. Provision of in-service training of mathematics and science teachers in Botswana: Teachers' perspectives. *J. Math. Teach. Educ.* **2009**, *12*, 153–159. [CrossRef]
30. Yan, C.; He, C. Enhancing part-time in-service teacher training programmes to facilitate rural teacher development in China. *Prospects* **2011**, *41*, 553–566. [CrossRef]
31. Stormowski, V. Formação de Professores de Matemática para o uso de Tecnologia: Uma Experiência com o GeoGebra na Modalidade EAD. Ph.D. Thesis, PPGIE UFRGS, Porto Alegre, Brazil, 2015.
32. Artigue, M. Learning Mathematics in a CAS Environment: The Genesis of a Reflection about Instrumentation and the Dialectics between Technical and Conceptual Work. *Int. J. Comput. Math. Learn.* **2002**, *7*, 245–274. [CrossRef]
33. Greefrath, G.; Hertleif, C.; Siller, H.S. Mathematical modelling with digital tools—a quantitative study on mathematising with dynamic geometry software. *ZDM Math. Educ.* **2018**, *50*, 233–244. [CrossRef]
34. Prodromou, T. *Augmented Reality in Educational Settings*; Brill | Sense: Leiden, The Netherlands, 2020.
35. Blatchford, P.; Bassett, P.; Brown, P.; Koutsoubou, M.; Martin, C.; Russell, A.; Rubie-Davies, C. *Deployment and Impact of Support Staff in Schools*; Institute of Education, University of London: London, UK, 2009.
36. Bell, P.; Hoadley, C.M.; Linn, M.C. Design-based research in education. In *Internet Environments for Science Education*; Routledge: London, UK, 2013; pp. 101–114.
37. Barab, S.; Squire, K. Design-based research: Putting a stake in the ground. *J. Learn. Sci.* **2004**, *13*, 1–14. [CrossRef]
38. Reinmann, G. Innovation ohne Forschung? *Ein Plädoyer für den Design-Based Research-Ansatz in der Lehr-Lernforschung. Unterrichtswissenschaft* **2005**, *33*, 52–69.
39. Scanlon, E.; O'Shea, T. (Eds.) *New Directions in Educational Technology*; Springer Science & Business Media: Berlin/Heidelberg, Germany, 2012.
40. Anderson, T.; Shattuck, J. Design-Based Research. *Educ. Res.* **2012**, *41*, 16–25. [CrossRef]
41. O'Dwyer, L.M.; Bernauer, J.A. *Quantitative Research for the Qualitative Researcher*; SAGE Publications: Thousand Oaks, CA, USA, 2013.
42. Collins, A. Toward a Design Science of Education. In *New Directions in Educational Technology*; Scanlon, E., O'Shea, T., Eds.; Springer: Berlin/Heidelberg, Germany, 1992; pp. 15–22.
43. Kim, P.; Suh, E.; Song, D. Development of a design-based learning curriculum through design-based research for a technology-enabled science classroom. *Educ. Technol. Res. Dev.* **2015**, *63*, 575–602. [CrossRef]
44. Zheng, L. A systematic literature review of design-based research from 2004 to 2013. *J. Comput. Educ.* **2015**, *2*, 399–420. [CrossRef]
45. Baumgartner, E.; Bell, P.; Brophy, S.; Hoadley, C.; Hsi, S.; Joseph, D.; Tabak, I. Design-Based Research: An Emerging Paradigm for Educational Inquiry. *Educ. Res.* **2003**, *32*, 5–8. [CrossRef]
46. Hakkarainen, P. Designing and implementing a PBL course on educational digital video production: Lessons learned from a design-based research. *Educ. Technol. Res. Dev.* **2009**, *57*, 211–228. [CrossRef]
47. McKenney, S.; Reeves, T.C. Systematic review of design-based research progress: Is a little knowledge a dangerous thing? *Educ. Res.* **2013**, *42*, 97–100. [CrossRef]
48. Richey, R.C.; Klein, J.D. *Design and Development Research: Methods, Strategies, and Issues*; Routledge: London, UK, 2014.
49. Cobb, P.; Confrey, J.; diSessa, A.; Lehrer, R.; Schauble, L. Desig, experiments in educational research. *Educ. Res.* **2003**, *32*, 9–13. [CrossRef]
50. Annetta, L.A.; Frazier, W.M.; Folta, E.; Holmes, S.; Lamb, R.; Cheng, M.-T. Science Teacher Efficacy and Extrinsic Factors Toward Professional Development Using Video Games in a Design-Based Research Model: The Next Generation of STEM Learning. *J. Sci. Educ. Technol.* **2013**, *22*, 47–61. [CrossRef]
51. Wang, F.; Hannafin, M.J. Design-based research and technology-enhanced learning environments. *Educ. Technol. Res. Dev.* **2005**, *4*, 5–23. [CrossRef]
52. Starnazzi, C. *Leonardo Codices & Machines*, 1st ed.; Cartei & Bianchi: Florence, Italy, 2005.
53. Sinclair, N.; Bartolini Bussi, M.; de Villiers, M.; Jones, K.; Kortenkamp, U.; Leung, A.; Owens, K. Recent research on geometry education: An ICME-13 survey team report. *ZDM Math. Educ.* **2016**, *48*, 691–719. [CrossRef]
54. Lesh, R.A.; Shiraman, B. Re-conceptualizing mathematics education as a design science. In *Theories of Mathematics Education, Advances in Mathematics Education*; Shiraman, B., English, L., Eds.; Springer: New York, NY, USA, 2010; pp. 123–149. [CrossRef]
55. De Sapio, V.; De Sapio, R. Mechanical advantage: The archimedean tradition of acquiring geometric insight from mechanical metaphor. In *The Genius of Archimedes–23 Centuries of Influence on Mathematics, Science and Engineering*; Paipetis, S.A., Ceccarelli, M., Eds.; Springer: New York, NY, USA, 2010; pp. 493–506.
56. Lieban, D.; Lavicza, Z. Dissecting a Cube as a Teaching Strategy for Enhancing Students' Spatial Reasoning: Combining Physical and Digital Resources. In Proceedings of the BRIDGES Conference, Linz, Austria, 16–20 July 2019.

57. Prodromou, T.; Robutti, O.; Panero, M. Making sense out of the emerging complexity inherent in professional development. *Math. Educ. Res. J.* **2018**, *30*, 445–473. [CrossRef]
58. Tomaschko, M.; Hohenwarter, M. Augmented reality in mathematics education: The case of GeoGebra AR. In *Augmented Reality in Educational Settings*; Prodromou, T., Ed.; Brill | Sense: Leiden, The Netherlands, 2020; pp. 325–346.
59. Fischer, M.; Spannagel, C. *Lernen mit Vorlesungsvideos in der Umgedrehten Mathematikvorlesung*; Gesellschaft für Informatik: Bonn, Germany, 2012; Available online: http://dl.mensch-und-computer.de/handle/123456789/3242 (accessed on 11 April 2022).
60. Muir, T.; Geiger, V. The affordances of using a flipped classroom approach in the teaching of mathematics: A case study of a grade 10 mathematics class. *Math. Educ. Res. J.* **2016**, *28*, 149–171. [CrossRef]
61. Flipped Learning Network (FLN). *The Four Pillars of F-L-I-P*; Flipped Learning Network: Cairo, Egypt, 2014; Available online: https://flippedlearning.org/definition-of-flipped-learning/ (accessed on 27 May 2022).
62. Esperanza, P.; Fabian, K.; Toto, C. Flipped Classroom Model: Effects on Performance, Attitudes and Perceptions in High School Algebra. In *Adaptive and Adaptable Learning*; Verbert, K., Sharples, M., Klobučar, T., Eds.; Springer International Publishing: Cham, Switzerland, 2016; pp. 85–97.
63. Cronhjort, M.; Filipsson, L.; Weurlander, M. Improved engagement and learning in flipped-classroom calculus. *Teach. Math. Its Appl. Int. J. IMA* **2017**, *37*, 113–121. [CrossRef]
64. Maciejewski, W. Flipping the calculus classroom: An evaluative study. *Teach. Math. Its Applications. Int. J. IMA* **2015**, *35*, 187–201. [CrossRef]
65. Weinhandl, R.; Schallert, S.; Lavicza, Z. Towards Flipped Learning in Upper Secondary Mathematics Education. *JME (J. Math. Educ.)* **2020**, *5*, 1–15. [CrossRef]
66. Schina, D.; Esteve-González, V.; Usart, M.; Lázaro-Cantabrana, J.L.; Gisbert, M. The integration of sustainable development goals in educational robotics: A teacher education experience. *Sustainability* **2020**, *12*, 10085.
67. United Nations Organization. *Transforming Our World: The 2030 Agenda for Sustainable Development*; Transforming Our World: The 2030 Agenda for Sustainable Development; UN (United Nations Organization): New York, NY, USA, 2015; Available online: https://sustainabledevelopment.un.org/content/documents/21252030%20Agenda%20for%20Sustainable%20Development%20web.pdf (accessed on 23 May 2022).
68. Kalaš, I.; Laval, E.; Laurillard, D. *ICT in Primary Education: Analytical Survey, Volume 1: Exploring the Origins, Settings and Initiatives*; UNESCO Institute for Information Technology in Education: Moscow, Russia, 2012.
69. Fundación, T. *A Manifesto for a New Digital Deal; Fundación Telefónica*; Telefónica: Madrid, Spain; Barcelona, Spain, 2020; Available online: https://www.telefonica.com/digital-manifesto/ (accessed on 20 May 2022).
70. Wang, X.; Xu, W.; Guo, L. The status quo and ways of STEAM education promoting China's future social sustainable development. *Sustainability* **2018**, *10*, 4417. [CrossRef]

sustainability

MDPI

Article

Service Learning as an Education for Sustainable Development (ESD) Teaching Strategy: Design, Implementation, and Evaluation in a STEM University Course

Alberto Martín-Sánchez, David González-Gómez and Jin Su Jeong *

Departamento de Didáctica de la Ciencias Experimentales y Matemáticas, Facultad de Formación del Profesorado, Universidad de Extremadura, 10003 Caceres, Spain; amartinif@alumnos.unex.es (A.M.-S.); dggomez@unex.es (D.G.-G.)
* Correspondence: jin@unex.es

Abstract: The continuous deterioration of the environment is one of the major concerns that societies are facing nowadays. As a response to this challenging situation, the general assembly of the United Nations (UN) created the 2030 Agenda, proposing 17 Sustainable Development Goals (SDGs) to foster sustainable development. Thus, the significance of educating in values related to sustainability and care for the environment must acquire a relevant importance in the education system to provide an Education for Sustainable Development (ESD) in Science, Technology, Engineering and Mathematics (STEM). Therefore, it is necessary to incorporate teaching methodologies that are able to connect with students and to generate enriching experiences. In this regard, it can promote knowledge of the environment and provide a service to the community to ensure sustainable development. This research presents the design, implementation and evaluation of a Service Learning (SL) methodology as an ESD strategy in a university course. Precisely, it describes the development and evaluation of an SL project implemented in a general science subject during a 2020/2021 course taught at the Teacher Training College of the University of Extremadura (Spain). A total of 46 students participated in the study on a voluntary basis. A pre- and post-test methodology was used to assess the suitability of SL as an ESD strategy, resulting in a significant increase in the students' knowledge about the innovative teaching strategies to work with suitable contents after the project, as well as in their knowledge about SDGs. Moreover, the students' participation in the SL project made them aware of the community implications in maintaining the environment and generating benefit for the whole community. In addition, this research shows how the SL teaching methodology is an important tool for the achievement of both curricular competences and environmental awareness, since theoretical knowledge is applied to tangible work to perform a real community service, and therefore is a very suitable teaching strategy to be applied in EDS.

Keywords: service learning; sustainability; STEM; heritage; initial teacher training

Citation: Martín-Sánchez, A.; González-Gómez, D.; Jeong, J.S. Service Learning as an Education for Sustainable Development (ESD) Teaching Strategy: Design, Implementation, and Evaluation in a STEM University Course. *Sustainability* **2022**, *14*, 6965. https://doi.org/10.3390/su14126965

Academic Editors: Marc A. Rosen and Jordi Colomer Feliu

Received: 30 March 2022
Accepted: 6 June 2022
Published: 7 June 2022

1. Introduction

Climate change constitutes an important issue among the problems affecting humanity nowadays. The acceleration of this process in recent decades has become a social reality that severely threatens different ecosystems [1], and generates a series of effects that affects the environmental quality [2]. In fact, the high consumption of energy and the depletion of resources that the current economic system demands is generating a problematic situation between economy and environment [3–5]. These practices, far from progressing towards a sustainable model, continued their course, causing irreversible problems in our environment. Therefore, the concept of sustainable development must be considered to change a devastating dynamic between economy development and the destruction of our ecosystems.

In this context, in September 2015, the General Assembly of the United Nations (UN) created the 2030 Agenda, indicating 17 Sustainable Development Goals (SDGs) and 169 targets with an ambitious vision in the economic, social, and environmental spheres. Governments are expected to assume their responsibility by establishing policies and measures for the implementation of the 2030 Agenda [6].

Among the SDGs, the following are the most related to the project described in this research:

SDG 4. Ensure inclusive and equitable quality education and promote lifelong learning opportunities for all.

SDG 11. Make inclusive, safe, resilient, and sustainable cities and human settlements.

SDG 16. Promote peaceful and inclusive societies for sustainable development, provide access to justice for all and build effective, accountable, and inclusive institutions at all levels.

A fundamental tool for the development of a common awareness to sustainable development is working on sustainability concepts from the early educational levels. In fact, Jickling and Wals [7] point out that the challenge that teachers must face is to find a different way of educating our students, searching for strategies and methodologies that connect with the interests and motivation of the student, allowing them to achieve a meaningful learning. In this sense, the Service Learning (SL) methodology can be an important tool to change this situation. The SL methodology combines experiential learning processes that provide a service to the community [8–10]. These direct and/or indirect services involve students in their community, and take them from their own educational environment to their neighbourhood and/or other nearby municipalities [11]. Performing social actions without the search for material incentives increases the probability of generating positive reciprocity in interpersonal and social relationships [12]. Sigmon [13], cited in Sotelino [14] (p. 60), states that there have been many definitions and approaches used in the general framework of joining service and learning, undoubtedly, as Furco and Billig [15] argue, due to the fact that its multidimensional and multidisciplinary nature adds further complexity to its conceptualisation. Thus, SL could be understood as a pedagogy of teaching whereby students acquire a better understanding of academic content by applying skills and knowledge for the benefit of society [15]. Furthermore, it could be defined as an educational proposal that promotes social commitment, combining learning and community service processes in a well-functioning programme where participants are trained by working on actual needs of the environment in order to improve and transform it [16–18].

1.1. SL and Other Experiential Practices

There are many environmental education initiatives that focus on raising the awareness of pupils about environmental issues, but they are limited to the level of commitment, as awareness-raising actions that do not lead to real participation. To clearly recognise the relationship between SL and other experiential practices, Stanford University created SL programmes [19]. An adaptation of this initiative is shown in Figure 1, and is a widely used tool in the academic bibliography to differentiate SL from other experiential practices (volunteering, field work, etc.), with which it is often confused [20], and to identify which aspects need to be introduced to consider these activities as SL actions.

As shown in Figure 1, four types of educational practices can be found. Their recognition makes it possible to clarify their conceptual clarification and steps to follow, which turn each of these experiential practices into an SL proposal. To analyse the differences and similarities between them, the work of Mayor [21], and Montes, Tapia and Yaber [22] is applied.

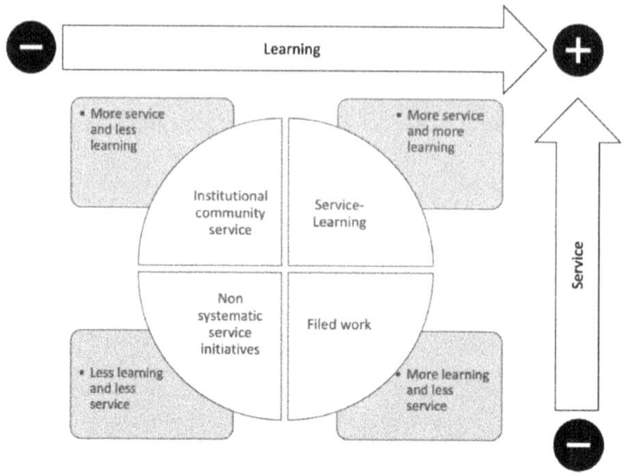

Figure 1. The quadrants of Service Learning (SL) displaying the four types of experiences based on the degree of learning and service reached.

The number of SL projects that have been designed and implemented in recent years has increased in an important way [9]. This increase has allowed us to distinguish large fields of experiential educational practices, and the necessary steps to favour the transition from these educational practices to the formulation of an SL project [21–23]. Thus, to ensure a balance between the curricular or formative learnings and the activities that constitute the service to the community, it is necessary not only to properly design the educational practices, but also the learning activities [9].

The need for a new type of citizen emerging from the universities is defined by a philanthropic profile based on personal responsibilities. For this university-based figure to emerge, learning must balance the academic with the practical [24]. As a result, SL methodology is currently used to connect students' knowledge and skills to provide the service to others, orienting the theoretical aspects, talent, and creativity towards social commitment. Thus, students not only acquire knowledge and exercise their skills, but also contribute to improving aspects of the reality in which they live [16]. From this perspective, SL is a valuable strategy for the inclusion of sustainability in the curriculum at any level of education [24–26].

1.2. SL Methodology in the Study of Science

Over the last years, the number of students taking university studies related to science and the interest of young people in this discipline has decreased significantly. The lack of interest for this discipline is a complex phenomenon and cannot be attributed to a single factor, but rather to multiple factors, such as the social consideration of science and the teaching of science at school [27]. This is causing difficulties in providing a solid scientific knowledge and, therefore, in ensuring a proper scientific literacy in citizens that allows them to participate in the scientific world. Consequently, it is the responsibility of the education system to reverse this situation by advocating methodologies that involve students in their own learning, since, as the National Science Education Standards states, "everyone needs to be able to take part in public discourse and debate on the important issues related to science and technology; and everyone deserves to participate in the excitement and personal satisfaction that learning and understanding the natural world can produce" (p. 32) [28].

One of the biggest difficulties in science teaching lies in the predominance of theory over practice [27], causing an apathy of students towards science and an increase in the lack of interest in science. In this sense, the SL methodology could be an important tool to change this paradigm. SL combines experiential learning processes directed towards community service, which has shown to increase learning retention, as well as motivation towards school activities. This is, consequently, reflected in higher classroom attendance and an improvement in average grades [8–10]. Besides, positive results are also seen with respect to a higher ability to analyse and synthesise complex information, showing that service learning has an impact on students' academic development [29,30].

The necessity to create a common awareness that can stop the progressive destruction of nature was highlighted over time [31]. Therefore, education should not be placed on the sidelines of this universal problem [32]. In fact, the university, as an open space for thought, reflection, and action, is called to be at the forefront in performing strategies and methods that enable solutions to the multiple challenges at a global and local level [33]. To achieve this goal, different international organisations and programmes, such as The United Nations Educational, Scientific and Cultural Organization (UNESCO) and the United Nations Environment Programme (UNEP), indicate the importance of intervening at the educational level. Thus, the Education for Sustainable Development (ESD) in Science, Technology, Engineering and Mathematics (STEM) focuses its objectives not only on relating with theoretical aspects, but also on involving citizens through participation. Therefore, ESD combines the promotion of cognitive, socio-emotional, and behavioural learning objectives [6], which is consistent with the new educational paradigm, which understands education as a participatory learning process and a process of construction oriented towards a sustainable ecology [32]. The implementation of the SDGs in education implies the promotion of the development of sustainability competencies, such as critical thinking, ethics and values, and interpersonal skills in students [34–36].

In this context, there are many perspectives and approaches that share the basic aims and principles of ESD, but all agree in the strengthening of civil society [37], i.e., providing participatory structures by conceiving social and educational action as a process of cultural democracy. Nevertheless, as Jickling and Wals [7] state, the real challenge that educators face to achieve a more sustainable future lies in our ability to educate learners differently. In this regard, methodologies such as SL are highlighted. SL, far from being just a tool for educating about and towards sustainability, is also an intrinsically sustainable experience through which students, teachers, and members of the community participate in real, active, and in-depth actions to promote sustainable development.

As a result, the SL methodology is an important tool for the achievement of both curricular competences and environmental awareness, since theoretical knowledge is applied to tangible work to perform a real community service. SL promotes the participants' establishment of links between thought and action, enabling the development of capacities to construct, and applies and transfers knowledge in a meaningful way by placing them in front of problematic situations in their immediate environment, with the intention to promote their active participation in improving them [21].

Only citizen involvement and participation at the local level can broaden horizons, educating towards sustainability and generating sustainable development that goes from the local to the global [38]. Any SL project, regardless of how small it may be, has as its starting point the participation of different entities that coexist in the community. Therefore, SL projects are the backbone of a whole community network that provides a solution to a real problem, acquiring an added value as it conceives communities as a pedagogical resource, promoting development and sustainability as assets in the daily practice of its inhabitants, as socio-educational practices that move from the local to the global. Besides, SL projects offer an opportunity to contribute to the community itself and must also involve the people who are part of it, recognising them as subjects of the action and not as mere objects of attention [31].

1.3. SL Methodology at the University Level

Tackling the lack of motivation among university students is the main challenge facing teachers at this level of education. A way to achieve this goal is implementing a teaching strategy that combines contents and methodology in a practical manner, providing students with the skills and training needed for their future professional career [39].

One way to improve this motivation is to connect theoretical contents with practical situations in which students participate in their own learning. In that way, the interest and the assignment of value to the task make students more cognitively involved [40]. Therefore, the SL methodology contributes to putting the theoretical contents into practice, establishing a connection between the curriculum and the professional reality. The use of SL as an instructional methodology has shown that the students were satisfied with the experience, as they valued it as an advantage in their training, due to being in direct contact with solidarity projects which, in addition to broadening their professional competences, broadened their personal horizons [41]. This being said, and in view of its characteristics and impact on students, SL could be described as a methodology that has attracted the attention of the educational community [42]. This educational community is increasingly focused on bringing theoretical content closer to the realities of the students by creating a new scenario between the university and society, in which the meaning and value of educational practices such as SL, need to be highlighted.

Thus, this research is a descriptive and quantitative evaluation of the students' knowledge of using Service Learning as a teaching methodology for ESD and sustainable education, and their knowledge about SDGs after participating in the SL project as a teaching strategy. In particular, the teaching strategy consisted in the implementation of an SL project through the development of different activities. Accordingly, the hypotheses of this study are:

Hypothesis 1 (H1): *Participating in an SL project allows students to know this methodology and its plausible uses for ESD.*

Hypothesis 2 (H2): *The participation in the proposed SL project increases knowledge about SDGs.*

2. Methodology

2.1. Sample

This research describes the design, implementation, and evaluation of an SL project as an ESD implemented during 2021 in a general science course taught in the Teaching Training School of the University of Extremadura (Spain). Precisely, the research was conducted with pre-service teachers enrolled in the Primary Education programme. A total of 46 students participated in the research on a voluntary basis, with an age range from 19 to 41 years old (mean value 20.2 years old), with a higher proportion of women than men. With regard to the participants' educational background, there were slightly more students with a science background than a social sciences background. Table 1 provides detailed information on the main demographic characteristics of the participating sample.

Table 1. Demographic information of pre-service teachers participating in this research.

N	Gender (%)		Age (%)			Educational Background (%)		
	Male	**Female**	**19–21**	**22–25**	**>25**	**Social Sciences**	**Science**	**Technology**
46	30.43	69.56	89.13	8.69	2.17	41.32	54.34	4.34

2.2. SL Project as Intervention

The SL project was designed and implemented in a general science subject, compulsory in the Primary Education bachelor's degree at the University of Extremadura (Spain). Among other objectives, this subject seeks to provide pre-service teachers with sustainable concepts related to resource preservation and environment preservation. In addition, this course deals with training and teaching strategies to apply these concepts with elementary school students. In this case, the teaching strategy implemented was an SL project that students developed during part of the course. Thus, the project lasted two months, in which different activities were implemented depending on the time and knowledge of the students. As it was mentioned before, an SL project is characterized by creating a link between students and their environment. In this case, the interaction of the SL participants (students) and the city of Cáceres was fostered, to work on the preservation of its historical values and environmental protection. The city of Cáceres is among the cities catalogued as World Heritage Sites by UNESCO. Although the city is particularly well-conserved, many places in the old town must be repaired periodically due to vandalism. Therefore, knowing the heritage value of the city and putting the students' work into its dissemination helps young people to protect and enjoy the city of Cáceres in a more sustainable way.

Thus, the SL project was implemented through different activities that took place following the sequence that it is shown in Figure 2.

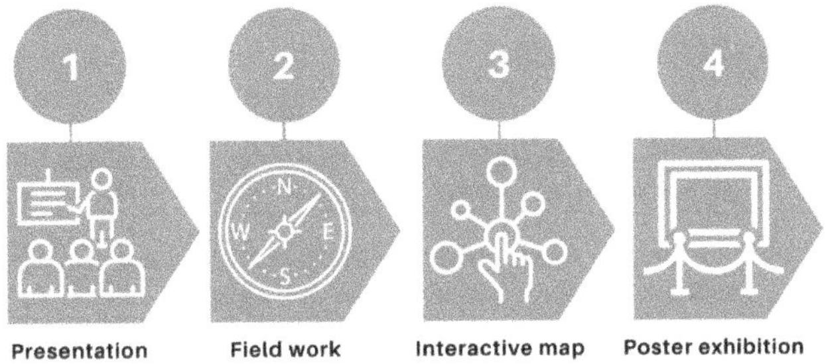

Figure 2. Phases of the project carried out following the SL methodology. Phase 1: Presentation of the project and the SL project; Phase 2: Students' field work about Sustainable Development Goals (SDGs) and the historical town; Phase 3: Students put together all the information and build the service learning goal; and Phase 4: Final discussion and results presentation.

First phase: At the beginning, the basic principles of the SL methodology and its fundamental differences with other experiential practices were explained, as well as a general introduction to the SDGs. For this first activity, the whole group received a lecture with all the activities' details. In addition, students were requested to participate. The whole group took part in this activity and were introduced to the activities that would be developed thereafter, as we must remember that SL projects must be open to the participation and reflection of the students who were involved in them.

Second phase: The students, in a process of enquiry and reflection, had to develop a task that involved studying and visiting the old neighbourhood of the city of Cáceres. Thus, students were able to learn its history, and identify places and situations to work in the context of the SDGs. For the achievement of this activity, students worked in heterogeneous groups. They had to look for information about the cost to the institutions caring for our heritage environment and the numerous acts of vandalism that occur in our city. Once each group had identified different sustainability issues in the city, they were asked to carry out a small project to address them through SL in primary education. In this way, we would ensure that the learners would have an additional tool as future teachers.

Third phase: Following this reflection, in working groups, an interactive map of the old city of Cáceres was created with the help of the Cáceres Historical City Consortium and the City Hall (see Figure 3), which was subsequently disseminated with the aim of raising awareness and valuing our heritage, in order to prevent possible actions that could threaten its conservation. With this activity, students continued working with the same groups, and each of them made a series of videos that were later uploaded to a platform so that the rest of the classmates could watch them. With the information that each group presented in

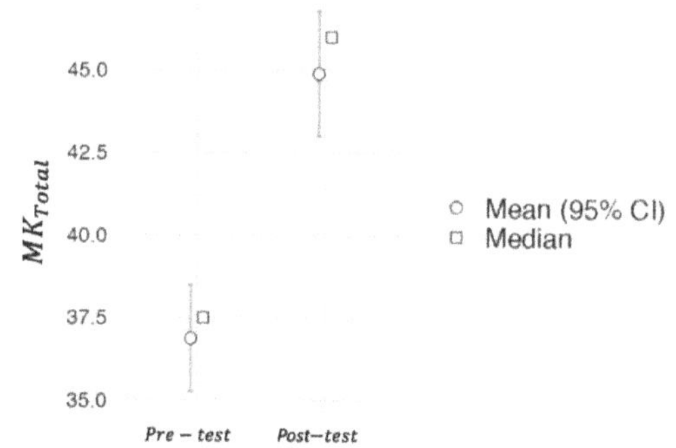

Figure 4. Total score for the MK_{Total} of the participants before (pre-test) and after (post-test) the implementation of the SL project.

3.2. Impact on Knowledge of the SDGs

Another aspect analysed was the extent to which participants were aware of the SDGs. Specifically, Figure 5 summarises the results obtained when participants were asked about the degree of knowledge in relation to the SDGs. As it can be observed in the figure, in all cases, the level of the participants' knowledge about the SDGs increased after the implementation of the designed SL project. A global score for the SDGs scale was calculated as a summatory of the scores of all individual items of this section of the instrument ($SDG_{Total} = \sum_{i=1}^{i=10} SDG_i$). According to the results, the mean value of SDG_{Total} after completing the Project was 28.8 (std dev = 6.95), whereas the mean value of SDG_{Total} before completing the Project was 37.1 (std dev = 5.71). A Student's t-test showed that the difference among both values was significant ($t(82) = 5.93$, $p < 0.001$, $d = 1.30$), suggesting that the implementation of the SL Project made a genuine difference in the students' knowledge about the SDGs. In addition, the Effect Size ($d = 1.30$) denotes a very high effect.

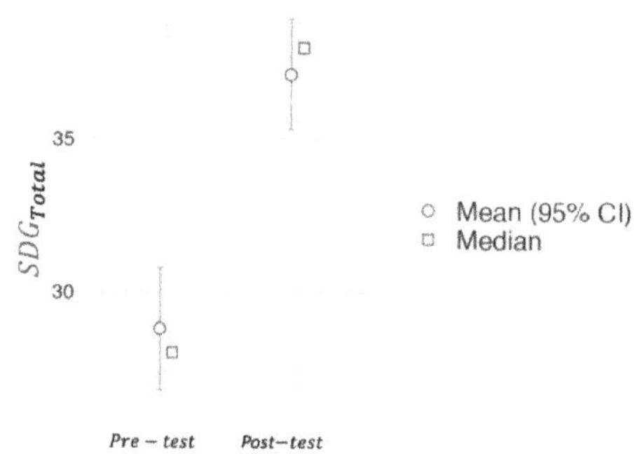

Figure 5. Total score for the participants' SDG knowledge before (pre-test) and after (post-test) the implementation of the SL project.

4. Discussion

The continuous deterioration of our environment is causing a large negative impact on our society that needs to be addressed from different perspectives. Education should not be placed on the sidelines of this universal problem [43]. In fact, different international organisations and programmes, such as UNESCO and the UNEP, indicate the importance of intervening at the educational level [6,32]. Thus, the ESD was set up to involve citizens through participation in educational programmes, in order to foster values and attitudes that enable a more sustainable and fair society for all. Thus, teaching methodologies as SL, which engage students with society by carrying out a service for the community, and at the same time allow them to achieve the academic and transversal competences, must be highlighted [44]. Thus, SL, far from being just a tool for educating about and towards sustainability, is also intrinsically a sustainable experience through which students, teachers, and members of the community participate in real, active, and in-depth actions to promote sustainable development. SL practices are becoming increasingly important in education systems. Different examples of the success of implementing SL projects for ESD are already available in the literature [45–48].

According to the results obtained in this study, the impact of the implementation of the project has been significant in most of the variables of the study in terms of knowledge of SL and active methodologies. Not only was the effect significant, but also the size of the effect was very large ($d = 1.30$). Thus, the implementation and participation in an SL project could be seen as an appropriate strategy in teacher training [39,49], as the students engage effectively in the learning process. Similar results are also reported for active learning methodologies, such as performing project-based Learning (PBL), and activities that can later be applied in future teaching work [50], among other active methodologies [51–53]. The connection of the students to their community has also proved to engage and motivate them more efficiently in the learning process [54]. In this sense, Castro et al. [55] present different SL projects aimed at providing the students skills for the support of a sustainable society. Their study was carried out with pre-service teacher students, concluding with a high degree of satisfaction of all participants. Since students' demotivation is one of the main causes of current school failure [49], SL is a relevant teaching tool, since participation in SL projects favours the motivation of students, where the impact of the SL project was highly significant. In fact, as Hernández-Blanco et al. [39] indicate, SL could be seen as an educational methodology that not only seeks learning, but also a social transformation. SL allows students to learn while they learn and act on the needs of their environment. Similar results were also reported previously with university students participating in SL projects [39,55]. This is reflected in the data obtained after implementing the SL project, since the activities aimed at increasing the knowledge and care of the environment of the community, and at improving the cultural level of the participating students by fostering a bond with their community [54,55]. On the other hand, according to the results collected, the participants showed a low level of knowledge of the history and culture of the environment in which the project took place. Therefore, SL could be seen as an educational practice that helps young people to acquire service-related knowledge, developing academic and professional skills. Moreover, it acts in two directions—on the environment in which the service has an impact and on the institution that promotes the project [17,54]—as it improves the living conditions of the community by fostering active citizenship, while at the same time generating young people with better teaching tools and a higher cultural level.

Despite the efforts made by administrations to promote the achievement of the SDGs, set out in the 2030 Agenda, the reality shows that without the active participation of citizens, which allows them to be aware of the needs of their environment, these proposals are insufficient [33,43]. In this sense, ESD has come to be essential in contributing to the achievement of global sustainable development, not only by providing knowledge, but also by promoting the acquisition of sustainability competences to implement the SDGs [56]. To help to achieve this goal, students must be involved in learning activities aimed at building a connection between students and community [55], as the active participation in an SL project. Thus, the results obtained in this study, which show a significant increase in the interest shown by participants in learning and knowing the SDGs after participating in the SL project, are statistically significant, with a very high effect ($d = 1.30$). This is a fundamental factor for the achievement of the SDGs [57], as the role of young people is crucial in the new global development agenda [56], both for them and for future generations [58,59]. García-Rico et al. [60] report that the use of SL experiences with students boosted and strengthened the knowledge, comprehension, sensitivity, and compromise in relation to the SDGs. In fact, having a solid knowledge of the SDGs allows students to examine their environment in a more critical way. In this way, students are able to understand the path set out by the UN for the sustainable development of our cities [43,61]. In this sense, Howell [62] concludes that to achieve an effective education for sustainable development, appropriate pedagogies are required, highlighting the need to engage learners in transformative learning through collaborative and learner-centred activities. Thus, the activities implemented in the SL project aimed at increasing the students' knowledge about the environment. Working directly with the community contributes to generating an awareness with social projection, as was also reported by Castro et al. [55]. It is important that education starts from the experience of its protagonists, with activities that do not focus on them, but that help and nourish them from their environment, generating a benefit for the community [17,61]. The project has had a significant impact on the knowledge and responsibility of the students with regard to their immediate environment.

Intervening in projects with the characteristics of SL gives students the tools to participate and initiate other activities that have an impact on their environment, since SL practices, unlike other training activities, introduce action in the community with the aim of improving it [16,39,63].

5. Conclusions

In this study, an SL project was designed to implement sustainable education strategies in a university science course. After its implementation, this research aimed to assess the participants' knowledge about the use of innovative teaching methodologies applied to ESD, such as Service Learning (H1), and how designing and implementing an SL projected helped to connect students with their environment, promoting care and responsibility for it, and their awareness of the SDGs (H2). The SL project was implemented in the Teaching Training School of the University of Extremadura (Spain). Precisely, the study was conducted with pre-service teachers enrolled in the Primary Education program.

The SL project designed in this research aimed to increase the teacher trainer students' awareness about the SDGs and their implementation in a local environment, in addition to knowing a new learning strategy to be used for ESD. Thus, the project was implemented in four phases. In the first one, the project was presented to the participants. The second one consisted in students' field work about SDGs and the historical town. In Phase 3, students were asked to put together all the information and build the SL goal. Finally, Phase 4 consisted of a final discussion and results presentation.

A pre- and post-test methodology, submitted before and after the SL implementation, was applied. According to the results, students' involvement in the design and implementation of the SL project applied to ESD had a significant positive effect. Firstly, students' knowledge about innovative teaching strategies to be applied to ESD has significantly increased. Students' engagement in the design and implementation of the SL project was proved to increase their knowledge about the proposed methodology, but also their motivation towards learning. Thus, the SL methodology applied to ESD is a very suitable tool in the university context, and more specifically for trainee teachers, since the participants not only obtain the benefits of the methodology as participating students, but it also allows them to improve their teaching practice. On the other hand, the development of the SL project had a significant impact on the students' knowledge about sustainable education and the SDGs. Precisely, after completing the project, the students w awareness of environment preservation, and the role of education in sustainable development. Moreover, the students' participation in the SL project made them aware of the community implications in maintaining the environment and generating benefit for the whole community.

Thus, this research shows how the SL teaching methodology is an important tool for the achievement of both curricular competences and environmental awareness, since theoretical knowledge is applied to tangible work to perform a real community service, and therefore it is a very suitable teaching strategy to be applied to EDS.

The main limitation of this study is the impossibility of comparing the results to other methodologies besides SL (no control group was used), due to the characteristics of the sample. Besides this, sample size and more information of subsequent years constitute limitations of our findings, which would allow us to prove the potential of SL as a teaching strategy in EDS and sustainable education. However, the results presented constitute the preliminary results of projects in which service learning projects are used in EDS and sustainable education.

Author Contributions: Conceptualization, A.M.-S., D.G.-G. and J.S.J.; methodology, A.M.-S., D.G.-G. and J.S.J.; formal analysis, A.M.-S., D.G.-G. and J.S.J.; investigation, A.M.-S., D.G.-G. and J.S.J.; writing—original draft preparation, A.M.-S.; writing—review and editing, D.G.-G. and J.S.J.; supervision, J.S.J. and D.G.-G.; funding acquisition, J.S.J. and D.G.-G. All authors have read and agreed to the published version of the manuscript.

Funding: This research was funded by Consejería de Economía y Agenda Digital de la Junta de Extremadura (Spain) and the FEDER funds grant numbers IB18004 and GR21047 and Ministry of Science and Innovation grant number PID2020-115214RB-I00/AEI/10.13039/501100011033.

Institutional Review Board Statement: The study was conducted in accordance with the Declaration of Helsinki and approved by the Ethics Committee of University of Extremadura (protocol code 94/2018 and date of approval 06/07/2018).

Informed Consent Statement: Informed consent was obtained from all subjects involved in the study.

Data Availability Statement: The data will be available upon request.

Acknowledgments: Authors are grateful to the Cáceres Historical City Consortium and the Cáceres City Hall for the help provided to accomplish this study.

Conflicts of Interest: The authors declare no conflict of interest.

References

1. Fleming, L.; Leonardi, G.; White, M.; Medlock, J.; Alcock, I.; Macintyre, H.; Maguire, K.; Nichols, G.; Wheeler, B.; Morris, G.; et al. Beyond Climate Change and Health: Integrating Broader Environmental Change and Natural Environments for Public Health Protection and Promotion in the UK. *Atmosphere* **2018**, *9*, 245. [CrossRef]
2. Yusuf, R.; Fajri, I. Differences in Behavior, Engagement and Environmental Knowledge on Waste Management for Science and Social Students through the Campus Program. *Heliyon* **2022**, *8*, e08912. [CrossRef]
3. Sulich, A.; Sołoducho-Pelc, L.; Ferasso, M. Management Styles and Decision-Making: Pro-Ecological Strategy Approach. *Sustainability* **2021**, *13*, 1604. [CrossRef]
4. Keeble, B.R. The Brundtland Report: 'Our Common Future'. *Med. War* **1988**, *4*, 17–25. [CrossRef]
5. Uvalić-Trumbić, S. New Dynamics of Higher Education and Research for Societal Change and Development. *Organ. Znanja* **2009**, *12*, 128–132. [CrossRef]
6. Water, U.N. (Ed.) Water and Climate Change. In *The United Nations World Water Development Report*; UNESCO: Paris, France, 2020; ISBN 978-92-3-100371-4.
7. Jickling, B.; Wals, A.E.J. Globalization and Environmental Education: Looking beyond Sustainable Development. *J. Curric. Stud.* **2008**, *40*, 1–21. [CrossRef]
8. Gallini, S.M.; Moely, B.E. Service-Learning and Engagement, Academic Challende, and Retention. *Mich. J. Community Serv. Learn.* **2003**, *10*, 5–14.
9. Dienhart, C.; Maruyama, G.; Snyder, M.; Furco, A.; McKay, M.S.; Hirt, L., Huesman, R. The Impacts of Mandatory Service on Students in Service-Learning Classes. *J. Soc. Psychol.* **2016**, *156*, 305–309. [CrossRef]
10. Bringle, R.G.; Steinberg, K. Educating for Informed Community Involvement. *Am. J. Community Psychol.* **2010**, *46*, 428–441. [CrossRef]
11. Felten, P.; Clayton, P.H. Service-Learning. *New Dir. Teach. Learn.* **2011**, *2011*, 75–84. [CrossRef]
12. Rodríguez Gallego, M.R. El Aprendizaje-Servicio Como Estrategia Metodológica En La Universidad. *Rev. Complut. Educ.* **2014**, *25*, 95–113. [CrossRef]
13. Sigmon, R.L. *Journey to Service-Learning Experiences from Independent Liberal Arts Colleges and Universities*; Distributed by ERIC Clearinghouse: Place of publication not identified; Council of Independent Colleges: Washington, DC, USA, 1996; ISBN 978-0-937012-12-3.
14. Sotelino Losada, A. *Evaluación y Propuesta de Desarrollo*; Aprendizaje-Servicio En Las Universidades Gallegas: Santiago de Compostela, Spain, 2014; Available online: https://plataformavoluntariado.org/wp-content/uploads/2019/04/tesese-sotelino.pdf (accessed on 6 June 2022).
15. Billig, S.; Furco, A. *Service-Learning through a Multidisciplinary Lens*; Advances in Service-Learning Research; Information Age: Greenwich, CT, USA, 2002; ISBN 978-1-931576-80-2.

16. Mayor Paredes, D. El Aprendizaje-Servicio Como Práctica Educativa Que Promueve Relaciones Colaborativas Entre La Escuela y La Comunidad. Estudio de Caso. Ph.D. Thesis, Universidad de Almeria, Almeria, Spain, 2016.
17. Puig Rovira, J.M.; Palos Rodríguez, J. Rasgos Pedagógicos Del Aprendizaje-Servicio. *Cuad. Pedagog.* **2006**, *357*, 60–63.
18. Martínez-Campillo, A.; Sierra-Fernández, M.d.P.; Fernández-Santos, Y. Service-Learning for Sustainability Entrepreneurship in Rural Areas: What Is Its Global Impact on Business University Students? *Sustainability* **2019**, *11*, 5296. [CrossRef]
19. Carver, R.L. Theoretical Underpinnings of Service Learning. *Theory Pract.* **1997**, *36*, 143–149. [CrossRef]
20. Luna, E. Del Centro Educativo a La Comunidad: Un Programa de Aprendizaje-Servicio Para El Desarrollo de Ciudadanía Activa, Barcelona, Spain. 2010. Available online: https://dialnet.unirioja.es/servlet/tesis?codigo=85430&orden=3&info=link (accessed on 6 June 2022).
21. Mayor Paredes, D.; Rodríguez Martínez, D. Aprendizaje-Servicio y Práctica Docente: Una Relación Para El Cambio Educativo. *Rev. Investig. Educ.* **2016**, *34*, 535. [CrossRef]
22. Rahima, C. *Community Service-Learning: A Guide to Including Service in the Public School Curriculum*; SUNY Series; Democracy and Education; Wade, R.C., Ed.; State University of New York Press: Albany, Australia, 1997; ISBN 978-0-7914-3184-9.
23. Tapia, M.N. Civic Service in South America. *Nonprofit Volunt. Sect. Q.* **2004**, *33*, 148S–166S. [CrossRef]
24. Celio, C.I.; Durlak, J.; Dymnicki, A. A Meta-analysis of the impact of Service-Learning on Students. *J. Exp. Educ.* **2011**, *34*, 164–181. [CrossRef]
25. Welch, M.; Billig, S. *New Perspectives in Service-Learning: Research to Advance the Field*; Advances in Service-Learning Research; Welch, M., Billig, S., Eds.; Information Age Publishing: Greenwich, CT, USA, 2004; ISBN 978-1-59311-158-8.
26. Jeong, J.S.; González-Gómez, D. A web-based tool framing a collective method for optimizing the location of a renewable energy facility and its possible application to sustainable STEM education. *J. Clean. Prod.* **2020**, *251*, 119747. [CrossRef]
27. Jeong, J.S.; González-Gómez, D. A STEM Course Analysis During COVID-19: A Comparison Study in Performance and Affective Domain of PSTs Between F2F and F2S Flipped Classroom. *Front. Psychol.* **2021**, *12*, 669855. [CrossRef]
28. National Committee on Science Education Standards and Assessment; Board on Science Education; Division of Behavioral and Social Sciences and Education; National Research Council. *National Science Education Standards*; National Academies Press: Washington, DC, USA, 1996; p. 4962. ISBN 978-0-309-05326-6.
29. Billig, M. Critical Discourse Analysis and the Rhetoric of Critique. In *Critical Discourse Analysis*; Weiss, G., Wodak, R., Eds.; Palgrave Macmillan UK: London, UK, 2003; pp. 35–46. ISBN 978-0-230-55514-3.
30. González-Gómez, D.; Jeong, J.S. EdusciFIT: A computer-based blended and scaffolding toolbox to support numerical concepts for flipped science education. *Educ. Sci.* **2019**, *9*, 116. [CrossRef]
31. Caride, J.A.; Meira, P.Á. Del Ecologismo Como Movimiento Social a La Educación Ambiental Como Construcción Histórica. *Hist. Educ.* **2019**, *37*, 165. [CrossRef]
32. Molano Niño, A.C.; Herrera Romero, J.F. La Formación Ambiental En La Educación Superior: Una Revisión Necesaria. *Luna Azul* **2014**, *39*, 186–206. [CrossRef]
33. Jeong, J.S.; González-Gómez, D. Assessment of sustainability science education criteria in online-learning through fuzzy-operational and multi-decision analysis and professional survey. *Heliyon* **2020**, *6*, e04706. [CrossRef] [PubMed]
34. Cebrián Bernat, G.; Junyent Pubill, M. Competencias Profesionales En Educación Para La Sostenibilidad: Un Estudio Exploratorio de La Visión de Futuros Maestros. *Enseñ. Las Cienc.* **2014**, *32*, 29–49. [CrossRef]
35. Rieckmann, M. Future-Oriented Higher Education: Which Key Competencies Should Be Fostered through University Teaching and Learning? *Futures* **2012**, *44*, 127–135. [CrossRef]
36. Wiek, A.; Withycombe, L.; Redman, C.L. Key Competencies in Sustainability: A Reference Framework for Academic Program Development. *Sustain. Sci.* **2011**, *6*, 203–218. [CrossRef]
37. Jeong, J.S.; González-Gómez, D. Multi-Criteria Decision Analysis Methods for Sustainability Assessment of Renewable Energy Systems and Its Potential Application to Sustainable STEM Education. In *Energy Systems Evaluation*; Green Energy and Technology; Ren, J., Ed.; Springer: Cham, Switzerland, 2021; Volume 2. [CrossRef]
38. Hewitt, N. ; *International Council for Local Environmental Initiatives. Guía Europea Para la Planificación de las Agendas 21 Locales: Como Implicarse en un Plan de Acción Ambiental a Largo Plazo Hacia la Sostenibilidad*; Bakeaz: Bilbao, Spain, 1998; ISBN 978-84-88949-25-7.
39. Hernández-Barco, M.; Sánchez-Martín, J.; Blanco-Salas, J.; Ruiz-Téllez, T. Teaching Down to Earth—Service-Learning Methodology for Science Education and Sustainability at the University Level: A Practical Approach. *Sustainability* **2020**, *12*, 542. [CrossRef]
40. López Aguado, M. Estrategias de Aprendizaje En Estudiantes Universitarios. Diferencias Por Género, Curso y Tipo de Titulación. *Educ. Knowl. Soc. EKS* **2011**, *12*, 203–233. [CrossRef]
41. Mella-Núñez, Í.; Quiroga-Carrillo, A.; Crespo Comesaña, J. Aprendizaje-Servicio y Desarrollo Cívico-Social En Titulaciones Universitarias Del Ámbito Educativo: Preparando al Alumnado Para La Práctica de Una Educación Inclusiva. *Educar* **2021**, *57*, 363–377. [CrossRef]
42. Álvarez Castillo, J.L.; Martínez Usarralde, M.J.; González González, H.; Buenestado Fernández, M. El Aprendizaje-Servicio En La Formación Del Profesorado de Las Universidades Españolas. *Rev. Esp. Pedagog.* **2017**, *75*, 199–217. [CrossRef]
43. Zamora-Polo, F.; Sánchez-Martín, J. Teaching for a better world. Sustainability and sustainable development goals in the construction of a change-maker university. *Sustainability* **2019**, *11*, 4224. [CrossRef]
44. Seba, D. The impact of the type of project on preservice teacher's conceptualization for service-learning. *Teach. Teach. Educ.* **2013**, *32*, 87–97. [CrossRef]

45. Hebert, A.; Hauf, P. Student learning through service learning: Effects on academic development, civic responsibility, interpersonal skills and practical skills. *Act. Learn. High. Educ.* **2015**, *16*, 37–49. [CrossRef]
46. Kahne, J.; Westheimer, J.; Rogers, B. Service learning and citizenship in higher education. *Mich. J. Community Serv. Learn.* **2000**, *7*, 42–51.
47. Melaugh, C.T.; Kindschuh, T. Engaged in waste: Two case studies from Protland state linking operational sustainability and studente-community engagement. In *University-Community Partnerships*; Wortham-Galvin, B.D., Allen, J.H., Sherman, J.D.B., Eds.; Routledge: London, UK, 2017; Volume 52, pp. 196–206. .
48. Benning, J.L.; Surovek, A.E.; Shearer, C.R. Engagement in practice: A case study on improving community sustainability through service-learning. In Proceedings of the 2018 ASEE Annual Conference, Salt Lake City, UT, USA, 24–27 June 2018; Volume 11, pp. 331–349.
49. Martínez-Valdivia, E.; Burgos-Garcia, A. Academic Causes of School Failure in Secondary Education in Spain: The Voice of the Protagonists. *Soc. Sci.* **2020**, *9*, 11. [CrossRef]
50. Matzembacher, E.D.; Gonzales, R.L.; do Nascimento, L.F. From informing to practicing: Students' engagement through practice-based learning methodology and community services. *Int. J. Mang. Educ.* **2019**, *17*, 191–200. [CrossRef]
51. González-Gómez, D.; Jeong, J.S.; Cañada-Cañada, F. Examining the Effect of an Online Formative Assessment Tool (OFAT) of Students' Motivation and Achievement for a University Science Education. *J. Balt. Sci. Eudc.* **2020**, *19*, 401–414. [CrossRef]
52. Jeong, J.S.; González-Gómez, D. Flipped-OCN method in mathematics learning to analyze the attitudes of pre-service teachers. *Mathemtatics* **2021**, *9*, 607. [CrossRef]
53. González-Gómez, D.; Jeong, J.S. The Flipped Learning Model in General Science: Effects on Students' Learning Outcomes and Affective Dimensions. In *Active Learning in College Science*; Mintzes, J., Walter, E., Eds.; Springer: Cham, Switzerland, 2020. [CrossRef]
54. Foshee Holmens, A.; Webb, K.J.; Albritton, B.R. Connecting students to community: Engaging students through course embedded service-learning activities. *Int. J. Manag. Educ.* **2022**, *20*, 100610. [CrossRef]
55. Castro, P.; Are-Pernas, A.; Depena, A. Service-Learning project in university degrees based on sustainable development goals: Proposals and results. *Susstainability* **2020**, *12*, 7940. [CrossRef]
56. Günther, J.; Overbeck, A.K.; Muster, S.; Tempel, B.J.; Shaa, S.; Schaal, S.; Kühner, E.; Otto, S. Outcome indicator development: Defining education for sustainable development outcomes for the individual level and connecting them to the SDGs. *Global Environ. Change* **2022**, *74*, 102526. [CrossRef]
57. Killian, S.; Lannon, J.; Murray, L.; Avran, G.; Giralt, M.; O'Riordan, S. Social Media for social good: Student engagment for the SDGs. *Int. J. Manag. Educ.* **2019**, *17*, 100307. [CrossRef]
58. Restrepo, E. Cuestiones de Método: «eventualización» y Problematización En Foucault. *Tabula Rasa* **2008**, *8*, 111–132. [CrossRef]
59. Armstrong, A.G.; Mattson, C.A.; Lewis, R.S. Factors leading to sustainable social impact on the affected communities of engineering service-learning projects. *Dev. Eng.* **2021**, *6*, 100066. [CrossRef]
60. García-Rico, L.; Martínez-Muñoz, L.F.; Santos-Pastor, M.L.; Chiva-Bartol, O. Service-learning in physical education teacher education: A pedagogical model towards sustainable development goals. *Int. J. Sustain. High. Educ.* **2021**, *22*, 747–765. [CrossRef]
61. Molderez, I.; Fonseca, E. The efficacy of real-world experiences and service learning for fostering competences for sustainable development in higher education. *J. Clean. Prod.* **2018**, *172*, 4397–4410. [CrossRef]
62. Howell, R.A. Engaging students in education for sustainable development: The benefits of active learning, reflective practices and flipped classroom pedagogies. *J. Clean. Prod.* **2021**, *325*, 129318. [CrossRef]
63. Barth, M.; Adomßent, M.; Fischer, D.; Richter, S.; Rieckmann, M. Learning to change universities from within: A service-learning perspective on promoting sustainable consumption in higher education. *J. Clean. Prod.* **2014**, *62*, 72–81. [CrossRef]

sustainability

MDPI

Review

Examining the Optimal Choice of SEM Statistical Software Packages for Sustainable Mathematics Education: A Systematic Review

Darmaraj Sakaria, Siti Mistima Maat * and Mohd Effendi Ewan Mohd Matore

Faculty of Education, Universiti Kebangsaan Malaysia, Bangi 43600, Malaysia
* Correspondence: sitimistima@ukm.edu.my

Abstract: Intending to analyze structural relationships between measured variables and latent constructs, researchers tend to adopt structural equation modeling (SEM) through either "covariance-based SEM" (CB-SEM) or "variance-based SEM" (VB-SEM)/"partial least squares SEM" (PLS-SEM) by using numerous statistical applications. Nevertheless, the reviews on understanding the optimal choice of proprietary statistical software packages in SEM approaches are scarce despite its immense importance in sustaining education. Therefore, a systematic review would be obligated to scrutinize the empirical studies to fill this gap. By employing the Preferred Reporting Items for Systematic Review and Meta-Analyses (PRISMA) guidelines, a total of 47 publications that met the inclusion criteria were obtained. To extract articles from August 2018 to 2022, Scopus, Web of Science (WoS), and The Education Resources Information Center (ERIC) databases were adopted. The findings imply that six types of proprietary statistical software packages emerged as an optimal choice: Lisrel, Amos, Mplus, SmartPLS, R package (plspm), and WarpPLS. Despite the widespread usage of a variety of statistical applications, SmartPLS and AMOS were rigorously utilized in VB-SEM/PLS-SEM and CB-SEM, respectively. This review is important for practitioners to discover which statistical tools are relevant to use and to identify gaps in order to sustain mathematics education for the future.

Keywords: mathematics education; optimal choice; proprietary statistical software packages; structural equation modeling; systematic review

Citation: Sakaria, D.; Maat, S.M.; Mohd Matore, M.E.E. Examining the Optimal Choice of SEM Statistical Software Packages for Sustainable Mathematics Education: A Systematic Review. *Sustainability* **2023**, *15*, 3209. https://doi.org/10.3390/su15043209

Academic Editor: Rosabel Roig-Vila

Received: 19 December 2022
Revised: 23 January 2023
Accepted: 6 February 2023
Published: 9 February 2023

1. Introduction

Sustainable Development Goals (SDGs) are relevant in multiple industries; however, there is no disputing that the education sector is as significant as other industries [1]. The fourth United Nations Sustainable Development Goal (SDG4) emphasized the importance of high-quality education to support the sector's sustainability as it prepares for the year 2030 [2,3]. To accomplish that aim, research regarding structural equation modeling, or SEM for short, has gained greater traction in mathematics education. As the latest method for multivariate data analysis [4,5], SEM was pioneered by Sewall Wright in 1934. The analysis method was developed as an alternative to regression modeling through the Ordinary Least Square (OLS) method [6,7]. Since this method was found to be able to overcome the limitations found in the regression modeling analysis, it is often referred to as the Second-Generation multivariate analysis method [6–8] in the mathematical landscape.

Despite analyzing structural relationships among constructs and indicators, SEM permits researchers to simultaneously model and estimate complex relationships among multiple dependent and independent variables. In the mathematical area, prior researchers used SEM as their preference over the traditional multivariate method. This is because SEM provides explicit evaluation of measurement error, estimates the latent variables via observed/manifest variable, and is capable of model testing, in which a structure may be imposed and the data's fit can be verified. In addition, most multivariate approaches, by not explicitly modeling measurement error, accidently disregard it, whereas the SEM model estimates these error variance parameters for both exogenous and endogenous variables. Another reason why previous mathematics scholars preferred SEM is that it offers the estimation of latent variables from observed variables such that the creation of composites considers measurement error. Eventually, completely developed models can be tested against data using SEM as a conceptual or theoretical framework or mathematical model, and their fit to the sample data can be evaluated. Researchers generally draw on two appropriate statistical methods to estimate SEM—CB-SEM or VB-SEM/PLS-SEM [4,8–12]. Looking across the literature, PLS-SEM is also referred to as PLS path modeling (PLS-PM) [13,14]. The study in [15] noted that researchers must be aware of the variations in order to implement the appropriate methodology, since each estimation is ideal for a different research circumstance. Neither method is generally preferable over the other, and neither is acceptable in every context [16] of mathematics.

This is because CB-SEM in mathematics research, in fact, is generally employed to confirm (or reject) theories and the underlying hypotheses. It tends to confirm or reject hypotheses by evaluating how closely a suggested theoretical model can replicate the covariance matrix for an observed sample dataset [8]. On the contrary, PLS has been established as a "causal-predictive" approach to SEM, which focuses on explaining the variance in the models' dependent variables [17]. In short, CB-SEM is the prominent method for testing more established theories, while VB-SEM/PLS-SEM is a more applicable strategy that permits exploratory research in causal research, especially building and predicting new theory [11,18]. Consequently, the choice between both SEM approaches is highly considered around the globe due to the need for this methodology in sustainable mathematics education. As a result, SEM statistical software packages are expanding quickly to serve in this matter. This is because the SEM tool and its method are closely associated and have an impact on the outcomes and the analysis carried out [19,20] by mathematics scholars.

Current literature has underlined the statistical applications, such as EQS [6,7,9,21,22], Amos [4,6,7,9,21–25], Lisrel [4,6,7,9,21,22,25], Stata (Builder, Sem, Gsem) [9,22], STAT PROC CALIS (SAS) [6,7,9,22], Statistica (Sepath) [9,22], Simplis [6,7], Systat (Ramona) [22], Ωinyx [9,22], Prelis, and a number of R-based packages (Lavaan, Sem, OpenMx, Lava, and Systemfit) [9,22]. These are most widely used for mathematical research applying CB-SEM studies. While users who prefer open-source environments may discover the R packages appealing, there are also other commercial software packages in CB-SEM that are rapidly evolving, particularly Mplus [4,6,7,9,22,25,26]. VB-SEM/PLS-SEM, the second approach, concentrates on the analysis of variance and can be performed by applying PLS-Graph [25], SmartPLS [4,9,25,27], WarpPLS [9,25,28], Adonco [25,29], Xlstat [25], Stata [4], SAS, LVPLS, and VisualPLS. Research has brought forward several packages for the R environment such as SEMinR [25,30,31], cSEM [9,25,30,32], semPLS [33], plspm [34], and Matrixpls, the use of which has recently become more ubiquitous. It can also be carried out using the PLS module in the "R" statistical software package.

Since statistical applications have been established as a significant predictor of SEM outcomes, extensive mathematics research has been executed to elevate the SEM statistical software package features and to identify the nature and elements of it. Aligned with increasing mathematics research on SEM statistical software packages, limited literature reviews were conducted to synthesize prior studies from different circumstances. The studies included a review and comparative study on software packages for SEM [21], a review of eight software packages for SEM [35], and a review of PLS-SEM statistical programs [25]. The process of synthesis is lacking in the articles; for instance, most systematic reviews did not mention the databases that were used to retrieve the articles, which led to unreliable results. To counter these limitations, it has been recommended to use Scopus and WoS databases, as they are well known for being the complete data sources for many applications and bibliometric analysis [36]. In addition, ERIC is another choice as it offers mathematics academic scholars, educators, academicians, or the general public with a comprehensive, rich, searchable, user-friendly, internet-based bibliographic and full-text database of mathematics education research and information. Generally, those sources used high-quality systematic reviews, meta-analyses, meta-synthesis, and bibliometric studies in the context of mathematics.

Furthermore, it has been noticed that the PRISMA guidelines are not promoted in those studies, resulting in misinterpretation and inadvertent bias [37]. While these studies were carried out prior to this review and informed the direction for future studies through basic guidelines, they could not provide overall current research trends on the optimal choice of CB-SEM and VB-SEM/PLS-SEM statistical applications through comparing perspectives via a single review article in the field of mathematics, thereby leaving insufficient systematic reviews as a source of reference for the mathematics researchers. Apart from systematic reviews, previous empirical studies looked at numerous SEM statistical applications from various angles [30,38–42]. Those studies highlighted various SEM statistical applications used for analysis of SEM. However, given that there are not many mathematics researchers that systematically reviewed the current research patterns on the optimal choice of statistical applications in mathematics education, those studies are still inadequate.

In a case where the existing review lacks a solid synthesis in the mathematical context, a systematic review might be a useful technique [43]. This is because it gathers pertinent data about a specific subject that met the pre-established eligibility requirements and provided an accurate solution to the research questions that had been posed [44]. Therefore, a systematic review of previous mathematics research is essential to identify the current research trends by examining the optimal choices of proprietary statistical software packages in SEM approaches, respectively, for sustainable mathematics education. By detecting these gaps, this study attempts to conduct a systematic review to evaluate the latest research trends in mathematics education with the following research question:

RQ: What are the optimal choices of proprietary statistical software packages in SEM approaches for sustainable mathematics education?

With this purpose in mind, we started to compile the scholarly articles that used SEM statistical applications within the mathematical area. To do so, the method of systematic review was employed, targeting main articles published in Scopus, WoS, and ERIC. Since systematic literature reviews (SLRs) are deemed to be highly transparent and are seen as a rigorous search procedure, identifying the protocol or the guidelines before carrying out the review process is crucial. As such, the authors adopted the PRISMA guidelines involving four robust steps: identification, screening, eligibility, and inclusion. Every step was performed correctly, such as the keywords and the selection of articles, which are described in detail. Beginning with the identification steps, the authors identified 396 publications in total, of which the sample was reduced to 47 papers over the next three phases. This review took up the challenges to contribute to the existing body of knowledge by conducting an SLR on the optimal choice of SEM tools in mathematics research. Based on these ideas, this study aims to accommodate the gap by careful review of the past studies to acquire a better knowledge of the optimal choice of SEM tools in the context of mathematics learning.

In this regard, a few contributions were configured. First, this review serves as a guide for mathematics users to comprehend the distinctive features of each SEM statistical application and make informed decisions on the most appropriate application for future mathematics research. In addition, it may aid the upcoming mathematics researchers and educators in planning future studies by assisting them in understanding research patterns in this field. This is in line with [45], who emphasized that publishing trends are a crucial signal for identifying a field's development. Furthermore, the stakeholders, non-government, and government parties, such as mathematics educators and experts or curriculum developers, can now understand the necessities of arranging the forum or seminar with hands-on activity dealing with SEM statistical tools, instead of conducting SEM methodological class virtually for upcoming mathematics scholars. On top of that, this study provides specific areas that should be the target of future mathematics researchers and beginners. The basic structure of this paper follows these phases: introduction, methods, results, discussion, and conclusion.

2. Methodology

An overview of the research methods used in the current study will be provided in this section. Five main sub-sections—PRISMA, resources, inclusion and exclusion criteria, systematic review process, and data abstraction and analysis—will be discussed comprehensively.

2.1. The Review Protocol (PRISMA)

PRISMA is a peer-reviewed standard approach that employs a checklist of recommendations to ensure the consistency and quality of the revision process [46]. PRISMA suits SLR in the social sciences with three main tangible benefits [47]. First, it is able to define clear research questions that permit systematic research. Second, it can identify inclusion and exclusion criteria. Lastly, PRISMA can examine a large database of scientific literature in a defined time. Since this approach is deemed suitable to identify the relevant statistical applications used in SEM approaches in mathematics coherently, it was adopted in this review. In order to produce a well-organized and transparent systematic review, a few phases of identification, screening, eligibility, and exclusion were carried out [48,49].

2.2. Resources

Acknowledged as premier journal databases amongst bibliographic databases [50] and for being the most comprehensive data sources for numerous applications and bibliometric analysis [36], Scopus and WoS were applied as the primary databases. As for journal coverage, both complemented each other in terms of impact, prestige, and influence [51]. WoS was the first comprehensive international bibliographic database. As a result, it evolved into the most significant source of bibliographic data for tasks like journal selection, research evaluation, bibliometric analysis, and others over time [52]. The WoS Core Collection comprises about 74.8 million academic data and datasets, 1.5 billion referenced references (going back to 1900), and 254 topic disciplines, according to the most recent figures from 2020 [36].

Nevertheless, Scopus has proven to be reliable and, in some ways, even better than WoS, earning its place as a comprehensive bibliographic data source [53]. The authors of [54] added it is an international database of peer-reviewed publications from all over the world. Scopus has updated its content coverage guide, which now lists 206,000 books from more than 5000 international publishers, about 120,000 conferences, and 23,452 current journal titles [36]. Because it is primarily a resource for education-related texts (80%) with the largest index of articles, Eric was included as a supporting database to determine more empirical research regarding SEM in the mathematics field. Over 250 journals are currently available in ERIC, and they were chosen for inclusion in the database based on certain standards. The database search cannot be limited to a single database [43]. Thus, employing these three databases is deemed sufficient.

2.3. Systematic Review Process

As it is less stringent [55] with a reliable, reproducible, and methodical approach, extensive applications, and explicit execution [56], SLR is known as the most reputable research method. It acts as a replicable methodology for finding, evaluating, and synthesizing information with a high level of objectivity [43]. The study in [57] emphasized that when conducting systematic reviews, adherence to a pre-established and explicitly stated protocol is crucial. Due to this, this SLR consolidates four phases as per the PRISMA 2020 guidelines as recommended by [58], reflected in Figure 1.

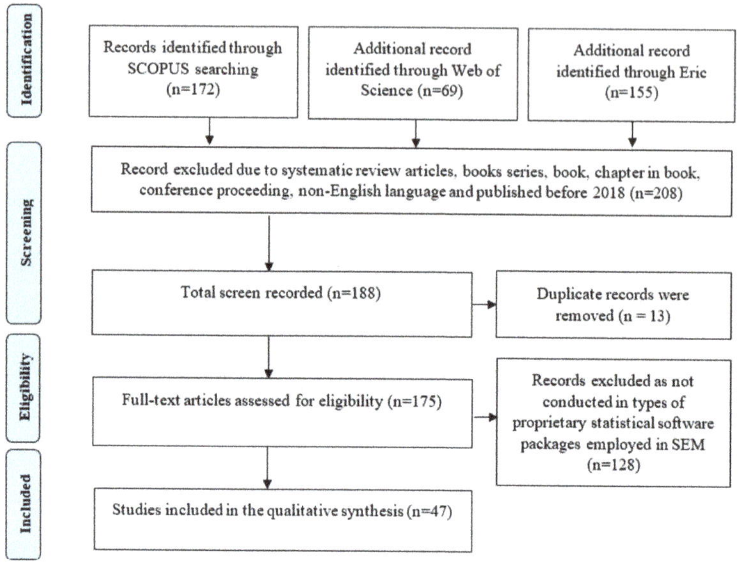

Figure 1. Study flow diagram (adapted from [58]).

2.3.1. Identification

Identification is the first phase of applying several procedures to enrich major keywords so that articles from the database may be retrieved as widely as feasible. By the time of the protocol, reviewers should have determined the keywords that dictate which documents are obtained [59]. Consultations with experts might help find more keywords [43]. Since one author in this systematic review is an expert in the field, a discussion was held. A cross-reading of a few articles served to identify missing keywords for the search strings [43]. On top of that, to improve the quality of the main keywords, a list of synonyms and similar terms was garnered using the electronic dictionary Thesaurus and Oxford Lexico, dictionaries, and encyclopedia, resulting in 23 keywords.

An advanced search was employed in this review to enable extensive search queries employing field codes, Boolean, and proximity operators to limit the search. Thus, Boolean AND OR were used to broaden the scope of the search as recommended by [60]. Employing informatics phrase searching, wild cards, truncation, and mixed Boolean operators, Table 1 displays the search terms for Scopus. The identical search term was then entered utilizing the title search (TS) feature into the WoS database. However, the search term varied for ERIC due to its different features. A systematic search of "Structural equation modeling" OR "SEM" AND "Mathematics" OR "Mathematics education" was used for ERIC. The specified databases produced a total of 396 possible articles (Scopus, N = 172, WoS, N = 69 & ERIC, N = 155).

Table 1. The search string used for the systematic review process.

Database	Keywords Used
Scopus	TITLE-ABS-KEY (["structural equation modeling" OR "SEM"] AND ["covariance-based SEM" OR "CB-SEM" OR "variance-based SEM" OR "VB-SEM" OR "partial least square" OR "partial least square-SEM" OR "partial least square structural equation modeling" OR "PLS-SEM" OR "proprietary statistical software package*" OR "statistical application*" OR "statistical program*" OR "statistical software*" OR "SEM software*" OR "software package*" OR "software program"*] AND ["mathematic*" OR "mathematic* education" OR "mathematic* teach* and learning" OR "mathematic* literacy" OR "mathematic* subject" OR "mathematic* discipline"])

2.3.2. Screening

The process of including or excluding articles generated by databases is known as screening in the second phase. The inclusion and exclusion of studies determines the scope and validity of systematic review results [61] in order to find suitable documents [62]. Thus, multiple sets of inclusion and exclusion criteria were applied. Practically, the review must always be constrained to studies relevant to the research question [63]. Since this review attempted to gain more data on empirical documents, the journal articles that used empirical evidence were chosen. This action was the first criterion. On the contrary, systematic reviews, non-empirical articles, book series, chapters in books, and conference proceedings were excluded because they were deemed as unnecessary articles [48]. The study in [64] mentioned that conference proceedings and book chapters are less comprehensive. Second, concentrating on English-language journal articles made it possible to avoid the prospect of difficult or ambiguous translations [65] and reduce misunderstandings.

Before reviewing the articles, writers should ascertain the time period covered by the articles [66]. Thus, for the third inclusion, authors set the time range. Although a short time span could significantly reduce the number of research articles that are eligible [61], authors considered articles published within the last five years. The criterion was set because years from 2017 to 1990 produced less than ten articles each year for Scopus and Wos, respectively. This proved that many articles pertaining to SEM statistical tools in mathematics were published in numbers reaching over ten each year after 2017. The review's scope may be restricted to a few high-quality journals or only include publications in a particular field of study [67]. Therefore, articles that were distributed globally in mathematics education/subject/discipline/literacy/Teaching and Learning (TnL) were concentrated in the final part of the inclusion and exclusion procedure. At the completion of the review procedure, 175 articles were discovered after the exclusion of 208 articles that were unrelated to the criteria/topic and 13 duplicate articles, as shown in Figure 1. Table 2 featured the summary of the included and excluded criteria.

Table 2. The exclusion and inclusion criteria.

Criterion (C)	Inclusion (I)	Exclusion (E)
Type of article/literature	Journal (research articles/empirical articles)	Journals (systematic review/non-empirical articles), book series, chapter in book, and conference proceeding
Language	English	Non-English
Timeline	Between 2018 and 2022	<2018
Country/region	All	-
Field	Mathematics education/subject/discipline/literacy/TnL	Non-mathematics education/subject/discipline/literacy/TnL

2.3.3. Eligibility

The third step is eligibility. A total of 175 articles were verified to see if they fulfilled the requirements for inclusion and were compatible with the objectives of the current research. This was achieved by reading through the titles, abstracts, methods, results, and discussions. Only 47 potential publications were ready for further analysis after the elimination of 128 papers. Those 128 articles were not in the mathematics context, although they elaborated upon SEM statistical applications. This was also applied for Science, Technology, Engineering, and Mathematics (STEM), since it was referred to as multidisciplinary [68,69] and was not focused on just the field of mathematics, although the use of publications from other disciplines [43] leads to a broad view and provides the foundation to synthesize the research field from different perspectives [70]. Surprisingly, six studies were eligible in the targeted field; however, methodological flaws existed, as they never mentioned the SEM statistical applications [71–76]. Since those studies were deemed as useful for this systematic review, authors contacted the authors via email. By prompt reply from them, those studies were included.

2.3.4. Inclusion Criteria

The articles were retained and met the criteria for analysis during the final phase, known as the inclusion criteria. The 47 articles for this systematic review revolved around SEM statistical applications in mathematics education. Previous literature underscored that the number of articles to be included in SLRs is never more than 50 [77]. Thus, 47 articles are deemed sufficient to carry out a systematic review in line with the statement. The availability of adequate literature on the issue to support a synthesis serves as the foundation for writing an SLR [78]. Thus, it was believed that those 47 articles are sufficient to produce a holistic finding.

2.4. Data Abstraction and Analysis

Those 47 articles were evaluated, reviewed, and analyzed; the results are explained in depth in this report.

3. Results

3.1. General Findings

3.1.1. Distribution of Publications Based on Countries

Figure 2 displays the distribution of articles by country. Only the nation of the primary author can receive the maximum score of 1 for each publication. By producing 27.7% or 13 articles, Malaysia was identified as a prominent country that embraced the trend of SEM applications in mathematics education, followed by Indonesia (N = 8 and 17.0%), West Africa (N = 5 and 10.6%), and East Africa (N = 3 and 6.4%). There are three countries with two papers on SEM statistical applications: (Cyprus) Southeast Europe, Spain, and USA. United Arab Emirates, Turkey, Philippines, Taiwan, South Korea, Australia, India, Southeast and Central Finland, Israel, South Africa, Switzerland, and China, the other 12 remaining countries, published one paper each. It was discovered that the top seven nations contributed about 66.0% of all relevant publications related to the target field over the previous five years.

3.1.2. Distribution of Publications Based on Years

Based on the results from Figure 2, the authors have demonstrated the growth of SEM tool publications in the period from June 2018 to June 2022 via Figure 3. Although there have been fluctuations in the number of papers issued, the findings revealed that the interest in the area has been stable from 2018 to 2020, with eight papers written each year. At the peak of targeted research publications in 2021, Indonesia (N = 5) had more papers published than Malaysia (N = 3), Spain (N= 2), West Africa (N = 1), (Cyprus) Southeast Europe (N = 1), USA (N = 1), and Australia (N = 1). This year contributed 31.9% to the field. The number of reviewed articles increased and declined by 53.3% in the last three consecutive years (2020, 2021, 2022).

3.1.3. Distribution of Publications Based on Research Design

The relationship/correlations shown by SEM correspond to the researchers' hypotheses. Thus, the SEM approach entails correlational or causal assumptions together with quantitative data in the model [6,7]. Parallel to the above statements, the findings of the study underscored that almost all researchers (N = 45 and 95.7%) employed quantitative design in their study, followed by qualitative (N = 0 and 0%) and mixed method (N = 2 and 4.3%). The trends in quantitative methods appear to be applied consistently throughout the last five years, even though the number of papers in this field has fluctuated, as shown in Figure 4. Years 2019 and 2021 are the only two years focused on mixed method. This shows more studies should utilize this technique in the upcoming year. However, there are no studies undertaken that use a qualitative research method.

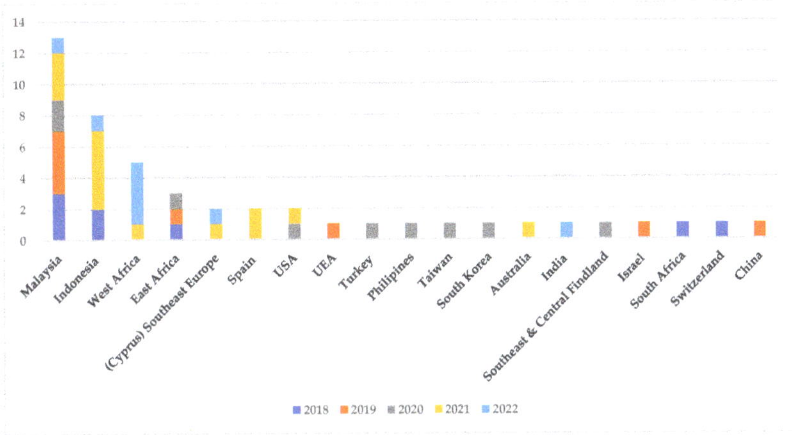

Figure 2. Distribution of articles based on countries.

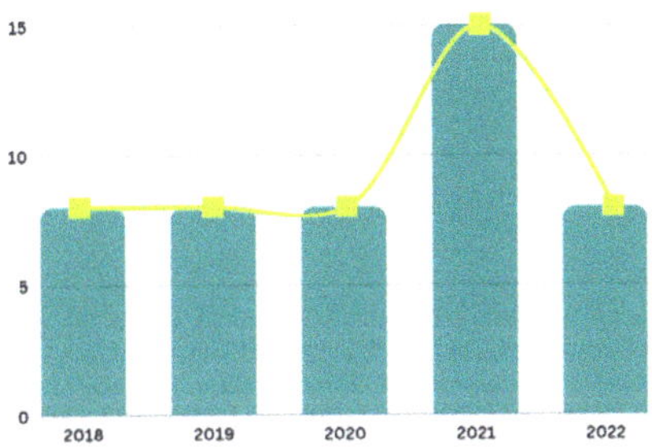

Figure 3. Distribution of articles based on years.

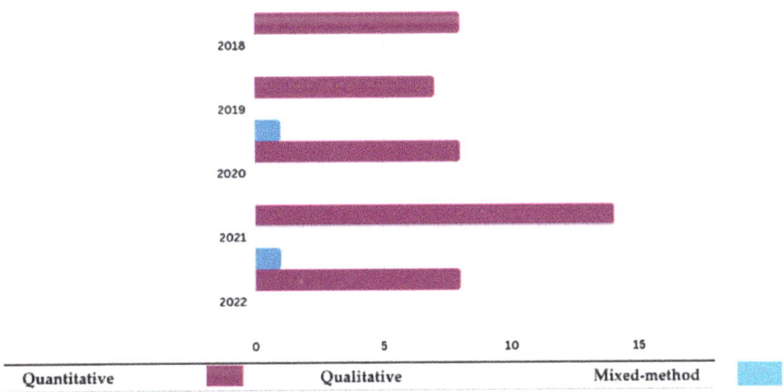

Figure 4. Distribution of articles based on research design.

3.1.4. Distribution of Publications Based on Samples

The distribution of papers by sample is shown in Figure 5. Articles that prioritize the students from secondary/middle/high school (N = 16) are commonly found, followed by students/undergraduate students from university (N = 15), mathematics teachers from secondary/middle/high school (N = 9), students from primary/elementary school (N = 7), prospective/pre-service/graduate mathematics teachers from university (N = 3), mathematics teachers from primary/elementary school (N = 2), principal (N = 2), and parents (N = 1). Samples from students were high (N = 28) compared to others. All eight categories of samples were fully discussed in 2019, but not in other years. Other samples might appear, such as learners, educators, pre-school students, pre-school mathematics teachers, novice/in-service mathematics teachers, lectures, professors, doctors, and so on, but this review solely covered empirical articles from 2018 to 2022.

3.2. Main Findings

Proper SEM statistical applications are considered vital for good SEM results. This notwithstanding, SEM statistical applications in mathematics are a cause for concern. The question "Which statistical application should I use for my data analysis in SEM?" is commonly asked by numerous scholars [25]. There are no trends that attempt to steer clear of direct answers to such inquiries; instead, recommendations are made that readers read papers with proper direction. By implementing the proper SEM statistical applications in SEM approaches in mathematics education, these issues could be resolved. Thus, we feel a little pressured and motivated by the requirement to offer a thorough evaluation of these applications by a systematic review in order to comply with requests, as well as to close the gap.

The study in [79] highlighted that systematic reviews are inspiring, valuable, and essential for identifying the precedence of future research and the range of human knowledge in order to reach an appropriate and authoritative conclusion. Thus, it is believed that the findings of this systematic review can aid researchers in better comprehending and choosing appropriate statistical analysis tools in the field of mathematics. Therefore, for the thematic analysis, 47 articles were reviewed over two themes, namely CB-SEM and VB-SEM/PLS-SEM statistical applications, to overcome the issues undertaken in mathematics education. The studies included are displayed in Table 3. A total of six sub-themes were found based on the two themes reviewed in this systematic review. The following sub-topics provide an overview for each of the themes and their sub-themes.

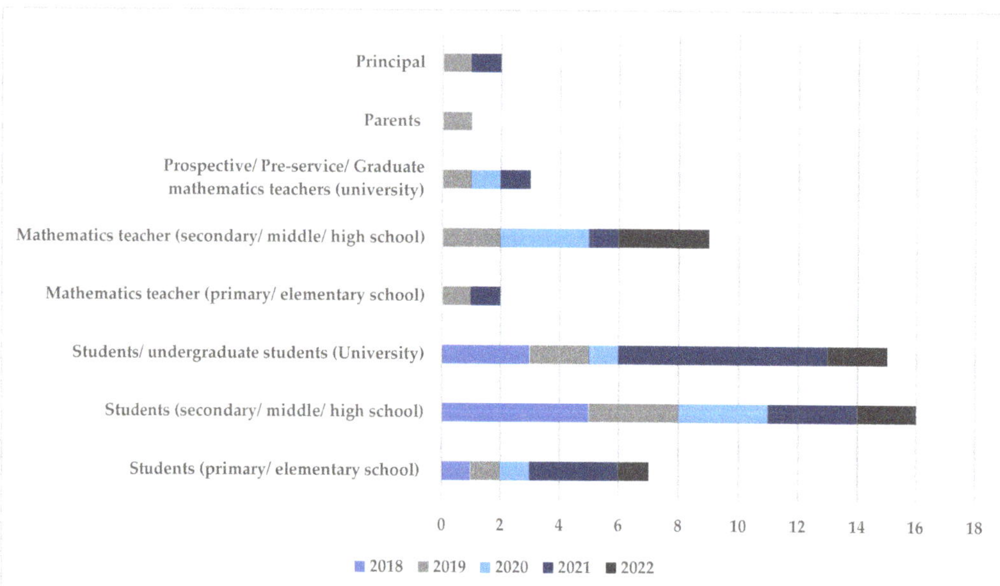

Figure 5. Distribution of articles based on samples.

Table 3. The findings regarding the types of proprietary statistical software packages according to SEM approaches.

No	Study	Research Design	Countries	Sample and Level	Types of Proprietary Statistical Software Packages According to SEM Approaches					
					CB-SEM			VB-SEM/PLS-SEM		
					Lisrel	Amos	Mplus	SmartPLS	R Package	WarpPLS
1	[71]	QN	Malaysia	Students (International secondary school)	X	X	X	√	X	X
2	[72]	QN	Malaysia	Students (International secondary school)	X	X	X	√	X	X
3	[73]	QN	Malaysia	Students (International secondary school)	X	X	X	√	X	X
4	[80]	QN	West Africa	Core and elective mathematics teachers (Secondary school)	X	√	X	X	X	X
5	[81]	QN	West Africa	Undergraduate students (University)	X	√	X	X	X	X
6	[82]	QN	West Africa	Undergraduate students (University)	X	√	X	X	X	X
7	[83]	QN	West Africa	Senior students (High school)	X	√	X	X	X	X
8	[84]	QN	UEA	Parents, Mathematics teachers, and students (Elementary school)	√	X	X	X	X	X
9	[85]	QN	Turkey	Prospective mathematics teachers (University)	X	X	X	√	X	X
10	[86]	QN	Philippines	Mathematics teachers (High school)	X	X	X	X	X	√
11	[87]	QN	Taiwan	Students (Vocational high school)	X	X	X	√	X	X
12	[88]	QN	Malaysia	Mathematics teachers (Secondary school)	X	X	X	√	X	X
13	[89]	QN	Malaysia	Students (Secondary school)	X	X	X	√	X	X
14	[90]	QN	Indonesia	Students (University)	√	X	X	X	X	X
15	[91]	QN	Indonesia	Students (University)	X	√	X	X	X	X
16	[92]	QN	Indonesia	Students (University)	X	√	X	X	X	X
17	[93]	QN	Indonesia	Students (University)	X	√	X	X	X	X

Table 3. *Cont.*

No	Study	Research Design	Countries	Sample and Level	Types of Proprietary Statistical Software Packages According to SEM Approaches					
					CB-SEM			VB-SEM/PLS-SEM		
					Lisrel	Amos	Mplus	SmartPLS	R Package	WarpPLS
18	[94]	QN	South Korea	Students (Elementary school)	X	X	√	X	X	X
19	[95]	QN	(Cyprus) Southeast Europe	Students (Primary school)	X	X	X	√	X	X
20	[74]	QN	Spain	Pre-service mathematics teachers (University)	X	X	X	√	X	X
21	[96]	QN	Australia	Students (University)	X	X	X	√	X	X
22	[97]	QN	(Cyprus) Southeast Europe	Principal, Mathematics teachers, and students (Primary school)	X	√	X	X	X	X
23	[75]	QN	India	Undergraduate students (University)	X	X	X	√	X	X
24	[98]	MM	East Africa	Students (Secondary school)	X	X	√	X	X	X
25	[99]	QN	Indonesia	Students (University)	√	X	X	X	X	X
26	[100]	MM	USA	Students (Elementary school, charter school, and home-school groups)	X	X	√	X	X	X
27	[101]	QN	Indonesia	Mathematics teachers (Secondary school)	X	X	X	√	X	X
28	[102]	QN	Malaysia	Undergraduate students (University)	X	X	X	√	X	X
29	[103]	QN	Malaysia	Graduate mathematics teachers (University)	X	X	X	√	X	X
30	[104]	QN	Malaysia	Undergraduate students (University)	X	X	X	√	X	X
31	[105]	QN	Malaysia	Undergraduate students (University)	X	X	X	√	X	X
32	[106]	QN	East Africa	Students (Lower secondary school)	X	X	√	X	X	X
33	[107]	QN	Southern and central Finland	Students (Lower and upper secondary school)	X	X	√	X	√	X

Table 3. *Cont.*

No	Study	Research Design	Countries	Sample and Level	Types of Proprietary Statistical Software Packages According to SEM Approaches					
					CB-SEM			VB-SEM/PLS-SEM		
					Lisrel	Amos	Mplus	SmartPLS	R Package	WarpPLS
34	[76]	QN	West Africa	Mathematics teachers (Secondary school)	X	X	X	√	X	X
35	[108]	QN	Israel	Principals, mathematics teachers, and students (Middle school)	X	X	√	X	X	X
36	[109]	QN	South Africa	Students (Public university)	X	√	X	X	X	X
37	[110]	QN	Malaysia	Students (Primary school)	X	X	X	√	X	X
38	[111]	QN	Indonesia	Students (Secondary school)	√	X	X	X	X	X
39	[112]	QN	Switzerland	Students (Primary and secondary school)	X	X	√	X	X	X
40	[113]	QN	Malaysia	Students (Private high school)	X	X	X	√	X	X
41	[114]	QN	Malaysia	Students (Private high school)	X	X	X	√	X	X
42	[115]	QN	Malaysia	Students (Private lower-level high school)	X	X	X	√	X	X
43	[116]	QN	Spain	Undergraduate students (University)	X	X	X	√	X	X
44	[117]	QN	East Africa	Mathematics teachers (Secondary school)	X	√	X	X	X	X
45	[118]	QN	Indonesia	Mathematics teachers (Secondary school)	X	X	X	√	X	X
46	[119]	QN	China	Students (University)	X	√	X	X	X	X
47	[120]	QN	USA	Mathematics teachers and students (Middle school)	X	X	X	X	X	√

QN—Quantitative; QL—Qualitative; MM—Mixed method.

According to Table 3, the authors offer a systematic review to explore the optimal choice of SEM statistical software packages for sustainable mathematics education. The authors used the PRISMA protocol for the systematic review, which has four phases. Starting with the identification phase in three databases (Scopus, WoS, and ERIC), the authors identified 396 studies in total, from which the sample was reduced over the next three phases to a total of 47 articles. Based on the author's findings, there are six software packages that emerged: Lisrel (N = 4), Amos (N = 11), Mplus (N = 7), SmartPLS (N = 23), the R package (plspm) (N = 1), and WarpPLS (N = 2). Nevertheless, the findings led to the identification that two packages, namely SmartPLS (VB-SEM/PLS-SEM) and Amos (CB-SEM), have been highly considered by researchers in mathematics education.

3.2.1. CB-SEM Statistical Applications

The CB-SEM statistical software applications in this systematic review were classified into three categories—Lisrel, Amos, and Mplus. For an appropriate categorization of statistical software package implementation in the context of mathematics fields based on the literature review, these subgroups were designed. Table 4 depicts the different types of categorizations together with the associated articles (N = 22) that were employed in this study. As displayed in Table 4, four articles were connected to the implementation of Lisrel in the field of mathematics. Results from Lisrel v.9.20 highlighted the significance of the attitudes and behavior of teachers, parents, and students in TIMSS scores [84]. Using Lisrel v.8.80, the factors that affect the difficulty of students learning mathematics (campus environment, family environment, community environment, and seating) significantly affected the students' self [90]. The data analyses in [99] used descriptive statistics, confirmatory factor analysis (CFA), and SEM. Furthermore, findings illustrated that the levels of all aspects of students' mathematics anxieties (N = 109) in online mathematics learning were at the medium level. The authors included further outcomes that showed that other than physiological and behavioral, two anxiety aspects—affective and cognitive—affected mathematics learning achievement. As such, ref. [111] noted a positive correlation between blended professional training on mathematics teachers' creativity and their teaching effectiveness through the Lisrel program.

Table 4. Findings regarding CB-SEM statistical applications.

Study	CB-SEM Statistical Applications
[84,90,99,111]	Lisrel (N = 4 studies)
[80–83,91–93,97,109,117,119]	Amos (N = 11 studies)
[94,98,100,106–108,112]	Mplus (N = 7 studies)

On top of that, also included in this review were 11 studies that demonstrated Amos as another type of proprietary statistical software package used in mathematics (Table 4). The SEM was run in Amos (v.23) to test the various hypotheses in the first four studies [80–83]. Yarhands Dissou Arthur, being a main author for these studies, emphasized that the results of the studies were positive. For instance, school-related factors and ICT training significantly enhanced the perceived ease of use and perceived usefulness of ICT by mathematics teachers [80]. Among senior high school students (N= 321) in Ghana, peer tutoring, teaching quality, and motivation had significant positive effects on mathematics achievement [83]. The path analysis by using Amos in another study exhibits that the mediating role of student learning interest partially mediated the relationships between learning motivation and mathematics performance, as well as between teaching quality and mathematics performance [82]. In [81], the additional variables, among which was airline, however, show that there is no significant effect on mathematics achievement. On the contrary, the effect of peer-assisted mathematics learning on mathematics performance was fully mediated by students' learning interest. The Amos version 23, continued by [117] with findings, shows that learning intentions, success criteria, and peer assessments are significant predictors of teachers' evaluating skills.

In [91], the CFA method by AMOS (v.18) software verified that the questionnaire of achievement goal was appropriate for the context of Indonesian students. The authors also mentioned that 538 Indonesian students adopted other avoidance and self-approach goals in order to determine the nature of achievement goals in mathematics education programs. Likewise, refs. [92,93] handled AMOS (v.18). A study from [92] demonstrates that while task and self-approach had a significant and positive impact on mathematical modeling competency, task avoidance goals had a significant and detrimental impact. Furthermore, the findings depicted that metacognition and mathematical modeling had a direct correlation that was statistically significant [93]. Additionally, the authors argued that the interrelationships between mathematical modeling competency and metacognitive strategies are considerably moderated by academic year level, which acts as a partial moderator. By utilizing the other version of Amos, which is v.22, the results from [109] reflect how TAM constructs significantly influence the acceptance of e-books among mathematics and statistics students at universities. A discussion in [119] was related to the interrelationships between these factors and the learning effects on advanced mathematics. The result of the study revealed that the affective support of teachers is positively correlated with students' learning cognition and learning self-efficacy. These data were analyzed using software AMOS (v.19). Although the author used an old version of Amos, version 7.0, the outcome was stable by illustrating that effective teaching strategies and student achievement are positively and significantly impacted by principal evaluation [97].

Mplus is the other type of statistical software that is considered in this review apart from Lisrel and Amos. Most of the studies revealed a positive finding regarding the integration of the Mplus tool in SEM analysis. For instance, there is a significant connection between students' cognitive appraisals with enjoyment and mathematics anxiety [106], and punishment sensitivity with psychological strain (motivational appraisals and task achievement) [107]. Moreover, in fourth grade, students who participate in more early numeracy activities at home are more likely to perform well in mathematics [94]. Next, authors ran an SEM mediation path analysis using MPlus software [100]. Based on this review, the output from Mplus indicated significant direct and indirect effects for all pathways (math attitude, math pre-test, math digital game use, helping affordance perception, hindering affordance perception, post-test performance) for all 187 children. In [112], the cross-lagged models were computed using Mplus v.7.3 to test the reciprocal effects between self-determined motivation (intrinsic and identified motivation) and negative emotions (anxiety, anger, and boredom) in mathematics. The authors claimed that regarding the influence of students' emotions on their motivation, the relationship had a consistent direction for all emotions. Furthermore, the results from Mplus v.7.02 emphasized the value of parental interactions and the necessity of putting into action effective strategies for fostering parental interactions [108]. According to survey data analyzed using SEM by Mplus v.7.31, students' perceptions of the effectiveness of teachers' feedback delivery and perceived scaffolding positively predicted students' use of feedback, whereas perceived monitoring adversely predicted the use of feedback [98].

3.2.2. VB-SEM/PLS-SEM Statistical Applications

The authors of 26 articles were quite enthusiastic regarding VB-SEM/PLS-SEM statistical software applications, as seen by the fact that most of them embraced it in their articles. The articles that addressed VB-SEM/PLS-SEM statistical applications in the context of mathematics are presented in Table 5. Findings indicate that most authors discussed the smartPLS as an analysis tool for SEM analysis. Based on the data gathered from the sample, each study (N = 23) using SmartPLS showed a significant and positive relationship between endogenous and exogenous variables, as illustrated in Table 6. Another type of VB-SEM/PLS-SEM statistical software application is plspm from R packages. The authors in [107] proved from their studies that the integration of the R package (plspm) in the mathematics classroom showed a significant relationship between eighth graders' temperamental reward and punishment sensitivities and their motivational appraisals (interest, strain, effort). As illustrated in Table 5, two articles mentioned the choice of WarpPLS, especially versions 5.0 and 6.0, to analyze SEM in their studies. These two articles [86,120] have similar findings whereby the implementation of WarpPLS in mathematics education provides benefits to the research findings of SEM. For example, the individual institutional leadership model had a significant and favorable impact on scientific output [86]. The findings from [120] revealed that because teachers offer greater possibilities for content engagement, in-class computer use was indirectly related to mathematical achievement.

Table 5. Findings regarding VB-SEM/PLS-SEM statistical applications.

Study	VB-SEM/PLS-SEM Statistical Applications
[71–76,85,87–89,95,96,101–105,110,113–116,118]	SmartPLS (N = 23 studies)
[107]	R package (plspm) (N = 1 study)
[86,120]	WarpPLS (N = 2 studies)

Table 6. Findings regarding SmartPLS.

Study	Findings
[71]	F1: a significant relationship between performance expectancy, effort expectancy, and student attitude toward the use of an online mathematics homework tool. F2: a significant relationship between student attitudes and their actual use of online homework.
[72]	F1: a significant relationship between perceived usefulness, perceived ease of use, and attitude toward the use of a web-based mathematics homework tool. F2: a significant relationship between attitude and mathematics self-efficacy factor.
[73]	F1: perceived usefulness and perceived ease of use are predictors of attitude toward the use of OHW.
[85]	F1: direct effects of technological content knowledge (TCK), technological pedagogical knowledge (TPK21), and pedagogical content knowledge (PCK21) on TPACK-21. F2: teachers' content knowledge (CK), technological knowledge (TK), and pedagogical knowledge (PK21) directly affect technological content knowledge (TCK).
[87]	F1: perceived usefulness significantly affected attitude toward use and behavioral intention to use. F2: attitude toward use significantly affected behavioral intention to use. F3: attitude toward use exhibited significant mediating effects between perceived usefulness and behavioral intention to use.
[88]	F1: infrastructure support and system quality affect teachers' intention to use geometer's sketchpad.
[89]	F1: teacher affective support and classroom instruction predict attitude towards mathematics more than parental influences.
[95]	F1: the mathematical mindset of students could directly and moderately describe their mathematical knowledge. F2: mathematical knowledge and mathematical mindset can both directly and to a significant extent be used to describe mathematical imagination.
[74]	F1: component relation effects of OB, ATP, and ATN of pre-service teachers toward mathematics learning and the influence of their educational background. F2: science and technology background were positively correlated after the flipped-OCN method compared with the rest of pre-service teachers.
[96]	F1: a significant relationship between students' self-efficacy, self-regulated learning strategies, and epistemological beliefs about mathematics as well as their perceptions of the learning environment.
[75]	F1: learning through constructivist Digital Learning Heutagogy supported academic achievement, learning engagement, and positive emotions F2: peer relationship not supported by the intervention.
[101]	F1: attitude toward E-learning use and E-learning experience were the two most significant constructs in predicting E-learning use.
[102]	F1: a significant relationship between teaching quality and students' academic performance.
[103]	F1: a significant relationship between Program Education Objectives (PEOs) and Program Learning Outcomes (PLOs).

Table 6. *Cont.*

Study	Findings
[104]	F1: a significant relationship between statistical reasoning and students' academic performance.
[105]	F1: students' attitude and belief toward statistics, statistical reasoning, self-efficacy, motivation, and the relationship with academic performance are statistically important.
[76]	F1: a significant relationship between the will, skill, tool, and pedagogy parameters and the stages of adoption of teachers' use of ICT. F2: Tool strongly predicts ICT integration.
[110]	F1: a significant relationship between cognitive factors (symbol sense, pattern sense, number sense, and operation sense) and algebraic thinking.
[113]	F1: task value and critical thinking skills predicts students' performance in mathematical reasoning. F2: critical thinking skills fully mediated with the relationship of mastery goal orientation on the students' abilities to solve the reasoning tasks.
[114]	F1: students' formative performance predicts their summative performance. F2: formative performance significantly mediates the relationship between self-confidence and summative performance.
[115]	F1: behavioral regulations (self-observation, self-judgment, and self-reaction) significantly influence student academic achievement and mathematical reasoning ability. F2: cognition regulation significantly mediates the relationship between motivational regulation and reasoning ability. F3: behavioral, cognition regulation, and students' reasoning ability significantly mediates the relationship between motivational regulation and academic achievement.
[116]	F1: Format and depth of the video tutorials predict performance learning and promoting autonomy.
[118]	F1: a significant relationship between perceived ease of use and subjective norm influence (PEU and SN) with teachers' microgame usage behaviors and intentions.

F1—Finding 1; F2—Finding 2; F3—Finding 3.

4. Discussion

The review's main findings betokened the various types of proprietary statistical software packages used in analysis of CB-SEM and VB-SEM/PLS-SEM in the field of mathematics. Ultimately, the results led to the identification of numerous statistical applications: Lisrel, Amos, Mplus, SmartPLS, R package (plspm), and WarpPLS. This notwithstanding, the results of this review revealed that, when compared to the other five statistical applications, smartPLS was the most extensively used in the domain of mathematics. This demonstrates that smartPLS is increasingly commonly employed to solve the SEM analysis that mathematics researchers encountered, especially in VB-SEM/PLS-SEM. Findings indicated that 23 studies showed a significant positive correlation among the variables employed. Through a sophisticated reporting feature, a user-friendly interface, advanced reporting capabilities, and availability at no cost to academics and researchers, the software has gained popularity in mathematics. Being a freely available and graphical user interface software, it was designed in a contemporary Java-based programming environment [25]. Following the 2003 launch of the initial online version, SmartPLS 2 was released in 2005. The program was updated and extended in 2015 [27]. The software has been developed to be very applicable and user-friendly to assist experts and beginners in creating scientifically sound and state-of-the-art VB-SEM/PLS-SEM analyses [121].

To enhance modeling and analysis capabilities, regular upgrades and additions are offered. Current versions of Apple and Microsoft operating systems are also compatible with the application. In addition, many analytical functions were automated in the subsequent iteration of SmartPLS, and PLS-SEM applications in journals expanded substantially [4]. Since it is known as a scientifically grounded software, it strives to provide complete transparency on how results are computed mathematically in order to ensure the repeatability of outcomes. Results from SmartPLS are presented in neatly organized tables and, in certain cases, in illuminating results graphics [122,123]. Additionally, users can save the outcomes or reports in Excel, HTML, and R formats for subsequent use or collaboration with others [25]. If there are subgroups spanning the entire theoretical model simultaneously, the SmartPLS software has numerous methods for finding them. These methods belong to the broad category of models known as latent class methods [4]. Although smartPLS was identified as the most prevalent statistical application in the mathematical perspective, other types of proprietary statistical software packages play an important part in SEM analysis, and their contributions cannot be discounted. For instance, becoming a growing trend in mathematics next to SmartPLS, Analysis of Moment Structures, or Amos for short, revealed a significant and favorable impact on the variables that were examined in 11 studies. Amos is known by another name: IBM SPSS AMOS. This software (Amos Development Corporation, 1983–2013) is for Windows computers. The authors in [22] emphasized that Amos is composed of two primary components: Amos Graphics and a separate Program Editor for working with Amos syntax. Having a fantastic graphical user interface, Amos can be quickly accessible and has an organized output format. It incorporates special features, such as a search for specifications in the absence of theory, diverse bootstrapping options, and a restricted application of Bayesian estimation. A highly intriguing feature of AMOS has been created within the Microsoft Windows interface, enabling mathematics researchers to either directly write the equation statements via AMOS graphics, or to specify the model by drawing a path diagram illustrating the relationships between variables.

However, since AMOS graphics offer all the tools that will ever be required for developing and dealing with SEM path diagrams, researchers will always choose to leverage it to easily detect relationships between the variables [21]. This software is most suitable and practical to use for post-graduates because they can convert the research framework into IBM-SPSS-AMOS graphics for analysis [6,7]. In addition to being a module in SPSS, AMOS is one of the first SEM tools that largely relies on a graphical interface for all functions, so researchers never have to use syntactic commands or computer code [4]. Through AMOS software, mathematics researchers can test the validity and reliability of a construct measurement model built by using the CFA procedure. After completing the CFA report, the mathematics researcher can model all these constructs into a structural model for analysis. Therefore, this resource is the best and most user-friendly method for analyzing and testing a theory [6,7]. The literature has underlined that although Amos is under CB-SEM, somehow it can handle analysis of PLS-SEM [124]. Although SmartPLS and Amos dominated among other software packages in mathematics education, Lisrel, Mplus, R package (Plpsm), and WarpPLS still have different strengths, special features, areas of improvement, estimation methods, and limitations that could influence a researcher's choice. For instance, like Amos, the other two packages, Lisrel and Mplus, allow users to estimate parameters for models with well-defined structures [35] and present a specific version of the maximum likelihood for incomplete data files that operate in the way just explained [22]. The main difference between these packages is the presence of a graphical interface for model specification and results presentation. Historically, the first commercial CB-SEM program to become widely used was LISREL [9,125,126], although it was not the first software to perform path analysis or SEM [126]. The name was derived from LInear Structural RELations [4]. LISREL is a versatile program that may be applied in a wide range of contexts (including cross-sectional, experimental, quasi-experimental, and longitudinal investigations) and at one point nearly entirely replaced SEM.

Next, a modeling program with numerous approaches, called Mplus, also includes a graphical user interface [127]. Despite the fact that there are several software packages available today, most require data that are continuous. In several types of analysis, Mplus permits the use of binary, ordinal, and censored variables. As if that were not enough, Mplus integrates some types of analysis that are difficult to access in other statistical packages (such as latent class analysis) and enables the researcher to use novel approaches that are not available elsewhere, like exploratory structural equation modeling (ESEM) [26]. Furthermore, Mplus has a very active community where mathematics researchers can get troubleshooting assistance if needed. The other VB-SEM/PLS-SEM software program that is still under active development is WarpPLS. Like SmartPLS, WarpPLS is very user-friendly with its modeling interface, and analyses may be carried out without any programming knowledge. The program has a graphical user interface for variance-based and factor-based SEM, employing classic composite-based PLS and more modern factor-based techniques [25]. Unlike SmartPLS, WarpPLS is categorized under commercial software. Hence, it was developed by Ned Kock in 2009. WarpPLS's capacity to recognize and model non-linearity among variables in path models, whether these variables are measured as latent variables or not, is one of its key advantages. This capability results in parameters that take relevant underlying heterogeneity into consideration. However, plspm is an R package including a collection of functions for doing PLS-PM analysis on both metric and non-metric data, as well as REBUS analysis. The project began in the fall of 2005 [14]. The first R package for PLS-PM was released in 2009, almost four years later. The absence of a graphical interface for creating path diagrams is one of the primary distinctions between plspm and other PLS-PM applications [14].

In a nutshell, although the first commercial software that appeared on the market was LISREL, other statistical software programs and approaches have emerged over time [9]. SEM is a versatile approach to examining how things are related to each other in the context of mathematics. Therefore, SEM statistical applications can appear quite different [10]. Findings of a review revealed that among 47 articles, a total of 26 studies were analyzed through VB-SEM/PLS-SEM statistical applications, whereas 22 publications were examined through the CB-SEM software packages. The trends reflects that these six SEM tools were the optimal choices among researchers in mathematics education at present. To ensure inclusive and equitable quality education and create opportunities for lifelong learning in the mathematical landscape, each mathematics researcher used relevant statistical software to undertake their analysis, although their features are varied. Some statistical applications are more established and provide a wide range of mathematical analysis, whilst the rest are newer and focus on a specific sort of analysis. Thus, trying to identify which is best would be an incorrect approach, since there are several advocates and users of each application. Each package differs in terms of strengths, limitations, areas of improvement, and special features that may dictate the choice of selection in the field of mathematics, as indicated by the findings of this systematic review.

5. Conclusions

To explore the latest trends of multiple CB-SEM and VB-SEM/PLS-SEM statistical applications in mathematics education was the aim of this systematic review. This research thereby fills the gap left by the paucity of systematic reviews on SEM statistical applications in mathematics. Additionally, this research covers a knowledge gap in lifelong learning through research trends on optimal choices of proprietary statistical software packages in SEM approaches, which may be essential for achieving the SDG4. Three databases have been included to retrieve 47 potential articles in accordance with the guidelines issued by PRISMA. Dealing with those studies showed the same results according to each statistical application. No study shows significantly different results, and suspicion was not raised [128]. The main findings have sought to explain that there are six different commercial/free software packages with "stand alone"/"packages" CB-SEM and VB-SEM/PLS-SEM applications, respectively, with graphical user interfaces that were identified, namely Lisrel, Amos, Mplus, SmartPLS, WarpPLS, and R package (plspm). This indicated that those statistical applications have been the optimal choices for researchers in the field of mathematics.

Although six SEM statistical application trends emerged from this review, the results add to our understanding that SmartPLS (VB-SEM/PLS-SEM) and Amos (CB-SEM) have been highly considered by researchers in mathematics education. This result was in tandem with earlier research indicating that SmartPLS and Amos were the best software packages for SEM approaches [9]. Recently, in a bibliometric review, ref. [129] analyzed 164 documents from Brazilian journals from the EBSCO and PROQUEST database during 1996–2015. They pointed out that the nearly 50% of publications (N = 82) used Amos, followed by 30% of articles using the SmartPLS (N = 31) and Lisrel (N = 19). AMOS (part of SPSS) was the most frequently used software, with 13 papers (40.6%) found in other systematic reviews that were conducted in 2009 [130]. On the contrary, Mplus, with 23 studies (37.0%), dominated Amos and Lisrel, with 14 studies (24.0%) each in another review [131]. Furthermore, findings from other reviews portrayed SmartPLS [121] and Amos [132] as comprehensive software programs compared to others. This implies that in order to facilitate diverse analyses, SmartPLS and Amos will, in essence, be gradually developed in the coming years. By gaining notoriety, both can be predicted to have a successful future and will keep solidifying their position as premier standard statistical software solutions for research in mathematics education.

This study has several limitations. The first limitation of the study is that only six statistical applications were the subject of this paper. The other SEM statistical applications were not mentioned as a trend in this review. This limitation undoubtedly creates new possibilities in the future, particularly in terms of choosing mathematics research for other different SEM statistical applications. Second, this review was conducted only using three databases: Scopus, WoS, and Eric. Consequently, incorporating articles from additional databases could produce various outcomes; therefore, further studies can investigate the same subjects by including more articles from different databases such as Mendeley, Google Scholar, Semantic scholar, Dimension.ai, PsycInfo, Science Direct, Ebsco, Proquest, Social Science Citation Index (SSCI), and so on. The study's exclusion of dissertations, theses, and manuscripts produced in languages other than English is its third limitation. Fourth, this review used literature from the last five years. Therefore, a longer time span should be used in literature review studies in the future. It was observed that contrasting CB-SEM and VB-SEM/PLS-SEM applications in a single review may produce a comparison between the statistical applications and their respective analytical techniques. Consequently, the future review should focus on either CB-SEM or VB-SEM/PLS-SEM applications separately in mathematics education.

Next, other statistical applications, such as R packages (sem, Lavaan and OpenMx) [21,35], EQS, PROC CALIS (SAS) [35], ADANCO [25], EQS [130], Stata, DEA Warwick, EBTs, and EQS [129], have been discussed in the respective prior reviews. Some of the articles examined in this review did not describe those statistical applications. Future studies could examine the relationships between learners, educators, academics, and context in order to further assess those statistical applications in mathematics education. Despite its limitations, this study makes a few vital contributions and implications to the body of knowledge relating to mathematics education and to practical concerns. The outcomes of this review function as a manual to help mathematics users and practitioners understand the unique characteristics of each SEM software program and decide which is best for their research, resulting in closing the gap found in the field of mathematics. Even though there are numerous statistical analysis programs in SEM approaches available on the market, when they are applied properly with the suitable research design and procedures, there is little difference between these programs in terms of their findings. Thus, it was recommended that mathematics researchers and interested parties should comprehend the most recent SEM statistical application adaptations and strategically select an appropriate application based on their research questions and design, as well as their pertinence to the system of education in the respective country for sustainable mathematics education. We hope that this systematic review will act as a springboard for understanding the various applications in SEM approaches before making the best decision in mathematics education, as well as contribute to the knowledge gap in promoting lifelong learning via SEM statistical applications, which could be valuable for achieving the fourth SDG, especially in the area of mathematics.

Author Contributions: Conceptualization, D.S., S.M.M. and M.E.E.M.M.; methodology, S.M.M. and M.E.E.M.M.; software, D.S.; validation, D.S., S.M.M. and M.E.E.M.M.; writing—original draft preparation, D.S.; writing—review and editing, D.S., S.M.M. and M.E.E.M.M.; visualization, D.S.; supervision, S.M.M. and M.E.E.M.M. All authors have read and agreed to the published version of the manuscript.

Funding: This research was funded by the Universiti Kebangsaan Malaysia, grant number GG-2022-021 and GG-2020-026.

Institutional Review Board Statement: Not applicable.

Informed Consent Statement: Not applicable.

Data Availability Statement: Not applicable.

Acknowledgments: The authors would like to thank the anonymous reviewers for the valuable comments and suggestions regarding improving the quality, content, and presentation of this paper.

Conflicts of Interest: The authors declare no conflict of interest.

References

1. Ramalingam, S.; Yunus, M.M.; Hashim, H. Blended Learning Strategies for Sustainable English as a Second Language Education: A Systematic Review. *Sustainability* **2022**, *14*, 8051. [CrossRef]
2. Boeren, E. Understanding Sustainable Development Goal (SDG) 4 on "quality education" from micro, meso and macro perspectives. *Int. Rev. Educ.* **2019**, *65*, 277–294. [CrossRef]
3. Hogan, D.; O'Flaherty, J. Exploring the nature and culture of science as an academic discipline: Implications for the integration of education for sustainable development. *Int. J. Sustain. High. Educ.* **2022**, *23*, 120–147. [CrossRef]
4. Hair, J.J.F.; Black, W.C.; Babin, B.J.; Anderson, R.E. *Multivariate Data Analysis*, 8th ed.; Cengage Learning EMEA: Andover, UK, 2019.
5. Mondiana, Y.Q.; Pramoedyo, H.; Sumarminingsih, E. Structural Equation Modeling on Likert Scale Data with Transformation by Successive Interval Method and with No Transformation. *Int. J. Sci. Res. Publ.* **2018**, *8*, 398–405. [CrossRef]
6. Awang, Z. *SEM Made Simple, a Gentle Approach to Learning Structural Equation Modelling*; MPWS Rich Publication: Selangor, Malaysia, 2015.
7. Awang, Z.; Lim, S.H.; Awang, N.F. *Pendekatan Mudah SEM: Structural Equation Modelling*; MPWS Rich Publication: Selangor, Malaysia, 2018.
8. Hair, J.F.J.; Hult, G.T.M.; Ringle, C.M.; Sarstedt, M.; Danks, N.P.; Soumya, R. *Review of Partial Least Squares Structural Equation Modeling (PLS-SEM) Using R: A Workbook*; Springer Nature: Cham, Switzerland, 2021.
9. Alcaraz, J.L.G.; Vargas, A.R. *Algorithms and Computational Techniques Applied to Industry*; Springer Nature: Cham, Switzerland, 2022.
10. Hair, J.F.; Hult, G.T.M.; Ringle, C.M.; Sarstedt, M. *A Primer on Partial Least Squares Structural Equation Modeling (PLS-SEM)*, 2nd ed.; Sage: Thousand Oaks, CA, USA, 2014.
11. Jaya, I.G.M.; Hermina, N.; Sunengsih, N. CB-SEM and VB-SEM: Evaluating Measurement model of Business Strategy of Internet Industry in Indonesia. *Int. J. Sci. Eng. Res.* **2019**, *10*, 878–883.
12. Mohamad, M.; Afthanorhan, A.; Awang, Z.; Mohammad, M. Comparison between CB-SEM and PLS-SEM: Testing and confirming the Maqasid Syariah quality of life measurement model. *J. Soc. Sci. Res.* **2019**, *5*, 608–614. [CrossRef]
13. Henseler, J.; Hubona, G.; Ray, P.A. Using PLS path modeling in new technology research: Updated guidelines. *Ind. Manag. Data Syst.* **2016**, *116*, 2–20. [CrossRef]
14. Sanchez, G. PLS Path Modeling with R. R Package Notes 235. 2013. Available online: http://gastonsanchez.com/PLS_Path_Modeling_with_R.pdf (accessed on 10 October 2022).
15. Rigdon, E.E.; Sarstedt, M.; Ringle, C.M. On comparing results from CB-SEM and PLS-SEM: Five perspectives and five recommendations. *Mark. ZFP* **2017**, *39*, 4–16. [CrossRef]
16. Petter, S. "Haters gonna hate": PLS and information systems research. *ACM SIGMIS Database DATABASE Adv. Inf. Syst.* **2018**, *49*, 10–13. [CrossRef]
17. Chin, W.W.; Cheah, J.-H.; Liu, Y.; Ting, H.; Lim, X.-J.; Cham, T.H. Demystifying the role of causal-predictive modelling using partial least squares structural equation modelling in information systems research. *Ind. Manag. Data Syst.* **2020**, *120*, 2161–2209. [CrossRef]
18. Nejati, M.; Rabiei, S.; Chiappetta Jabbour, C.J. Envisioning the invisible: Understanding the synergy between green human resource management and green supply chain management in manufacturing firms in Iran in light of the moderating effect of employees' resistance to change. *J. Clean. Prod.* **2017**, *168*, 163–172. [CrossRef]
19. Albright, J.J.; Park, H.M. *Confirmatory Factor Analysis using Amos, LISREL, Mplus, SAS/ STAT CALIS*; Working Paper; The University Information Technology Services (UITS) Center for Statistical and Mathematical Computing, Indiana University, Indiana University Publishing: Bloomington, IN, USA, 2009; Available online: www.indiana.edu/~statmath (accessed on 15 October 2022).

20. Marôco, J. *Análise de Equações Estruturais: Fundamentos Teóricos, Software & Aplicações*, 2nd ed.; Rolo & Filhos II, SA: Mafra, Portugal, 2014.
21. El-Sheikh, A.A.; Abonazel, M.R.; Gamil, N. A Review of Software Packages for Structural Equation Modeling: A Comparative Study. *Appl. Math. Phys.* **2017**, *5*, 85–94.
22. Kline, R.B. *Principles and Practice of Structural Equation Modelling*, 4th ed.; The Guilford Press: New York, NY, USA, 2016.
23. Arbuckle, J.L. *Amos 7.0 User's Guide*; SPSS Inc.: Chicago, IL, USA, 2014.
24. Arbuckle, J.L. *Amos (Version 25.0)*; Computer; IBM SPSS: Chicago, IL, USA, 2017.
25. Memon, M.A.; Ramayah, T.; Cheah, J.-H.; Ting, H.; Chuah, F.; Cham, T.H. PLS-SEM Statistical Programs: A Review. *J. Appl. Struct. Equ. Model.* **2021**, *5*, i–xiv. [CrossRef] [PubMed]
26. Kelloway, E.K. *Using Mplus for Structural Equation Modeling*, 2nd ed.; SAGE: Thousand Oaks, CA, USA, 2015.
27. Ringle, C.M.; Wende, S.; Becker, J.M. *SmartPLS 3*; SmartPLS GmbH.: Boenningstedt, Germany, 2015.
28. Kock, N. WarpPLS 6.0 User Manual. 2017. Available online: http://warppls.com/ (accessed on 21 October 2022).
29. Henseler, J.; Dijkstra, T.K. *ADANCO*; GmbH & Co.: Kleve, Germany, 2015. Available online: http://www.compositemodeling.com (accessed on 27 October 2022).
30. Chuah, F.; Memon, M.A.; Ramayah, T.; Cheah, J.H.; Ting, H.; Cham, T.H. PlS-SEM using R: An introduction to cSEM and SEMinR. *J. Appl. Struct. Equ. Model.* **2021**, *5*, 1–35. [CrossRef] [PubMed]
31. Ray, S.; Danks, N.P.; Calero Valdez, A. R Package Seminr: Domain-Specifc Language for Building and Estimating Structural Equation Models Version 2.1.0 [Computer Software]. 2021. Available online: https://cran.r-project.org/web/packages/seminr/ (accessed on 25 October 2022).
32. Rademaker, M.E.; Schuberth, F.; Schamberger, T.; Klesel, M.; Dijkstra, T.K.; Henseler, J. R Package cSEM: Composite-Based Structural Equation Modeling Version 0.3.0 [Computer Software]. 2020. Available online: https://cran.r-project.org/web/packages/cSEM/ (accessed on 24 October 2022).
33. Monecke, A.; Leisch, F. R package semPLS: Structural Equation Modeling using Partial Least Squares Version 1.0–10 [Computer Software]. 2013. Available online: https://cran.r-project.org/web/packages/semPLS/ (accessed on 2 November 2022).
34. Sanchez, G.; Trinchera, L.; Russolillo, G. *Plspm: Tools for Partial Least Squares Path Modeling (PLS-PM)*; R package: Boston, MA, USA, 2015.
35. Narayanan, A. A review of eight software packages for structural equation modeling. *Am. Stat.* **2012**, *66*, 129–138. [CrossRef]
36. Singh, V.K.; Singh, P.; Karmakar, M.; Leta, J.; Mayr, P. The journal coverage of Web of Science, Scopus, and Dimensions: A comparative analysis. *Scientometrics* **2021**, *126*, 5113–5142. [CrossRef]
37. Selcuk, A.A. A Guide for Systematic Reviews: PRISMA. *Turk. Arch. Otorhinolaryngol.* **2019**, *57*, 57–58. [CrossRef]
38. Mirazchiyski, P.V. RALSA: The R analyzer for large-scale assessments. *Large-Scale Assess. Educ.* **2021**, *9*, 21. [CrossRef]
39. Ringienė, L.; Žilinskas, J.; Jakaitienė, A. ILSA Data Analysis with R Packages. In *Modelling, Computation and Optimization in Information Systems and Management Sciences*; Le Thi, H.A., Pham Dinh, T., Le, H.M., Eds.; MCO 2021. Lecture Notes in Networks and Systems; Springer: Cham, Switzerland, 2022; Volume 363.
40. Serang, S.; Jacobucci, R.; Stegmann, G.; Brandmaier, A.M.; Culianos, D.; Grimm, K.J. Mplus Trees: Structural Equation Model Trees Using Mplus. *MPLUS TREES* **2021**, *28*, 127–137. [CrossRef]
41. Wang, W.; Tien, L.; Du, Y. The Construction of the Influence Model of Artistic Creativity Based on AMOS Data Analysis. In *2021 International Conference on Big Data Analytics for Cyber-Physical System in Smart City. BDCPS 2021*; Atiquzzaman, M., Yen, N., Xu, Z., Eds.; Lecture Notes on Data Engineering and Communications Technologies; Springer: Singapore, 2022; Volume 102.
42. Wong, K.K.-K. Partial Least Squares Structural Equation Modeling (PLS-SEM) Techniques Using SmartPLS. *Mark. Bull.* **2013**, *24*, 1–32.
43. Kraus, S.; Breier, M.; Dasi-Rodriguez, S. The art of crafting a systematic literature review in entrepreneurship research. *Int. Entrep. Manag. J.* **2020**, *16*, 1023–1042. [CrossRef]
44. Mengist, W.; Soromessa, T.; Legese, G. Method for conducting systematic literature review and meta-analysis for environmental science research. *MethodsX* **2020**, *7*, 100777. [CrossRef] [PubMed]
45. Subramaniam, S.; Maat, S.M.; Mahmud, M.S. Computational thinking in mathematics education: A systematic review. *Cypriot J. Educ.* **2022**, *17*, 2029–2044. [CrossRef]
46. Abelha, M.; Fernandes, S.; Mesquita, D.; Seabra, F.; Ferreira-Oliveira, A.T. Graduate employability and competence development in higher education: A systematic literature review using PRISMA. *Sustainability* **2020**, *12*, 5900. [CrossRef]
47. Sierra-correa, P.C.; Ricardo, J.; Kintz, C. Ecosystem-based adaptation for improving coastal planning for sea-level rise: A systematic review for mangrove coasts. *Mar. Policy* **2015**, *51*, 385–393. [CrossRef]
48. Mohamed, R.; Ghazali, M.; Samsudin, M.A. A Systematic Review on Mathematical Language Learning Using PRISMA in Scopus Database. *Eurasia J. Math. Sci. Technol. Educ.* **2020**, *16*, em1868. [CrossRef]
49. Moher, D.; Shamseer, L.; Clarke, M.; Ghersi, D.; Liberati, A.; Petticrew, M.; Shekelle, P.; Stewart, L.A. Preferred reporting items for systematic review and meta-analysis protocols (PRISMA-P) 2015 statement. *Syst. Rev.* **2015**, *4*, 1. [CrossRef]
50. Pranckute, R. Web of Science (WoS) and Scopus: The Titans of Bibliographic Information in Today's Academic World. *Publications* **2021**, *9*, 12. [CrossRef]
51. Joshi, A. Comparison between Scopus & ISI Web of Science. *J. Glob. Values* **2016**, *7*, 976–9447.

52. Li, K.; Rollins, J.; Yan, E. Web of Science use in published research and review papers 1997–2017: A selective, dynamic, cross-domain, content-based analysis. *Scientometrics* **2018**, *115*, 1–20. [CrossRef]
53. Harzing, A.W.; Alakangas, S. Google Scholar, Scopus, and the Web of Science: A longitudinal and cross-disciplinary comparison. *Scientometrics* **2016**, *106*, 787–804. [CrossRef]
54. Perevochtchikova, M.; De la Mora-De la Mora, G.; Hernández Flores, J.Á.; Marín, W.; Langle Flores, A.; Ramos Bueno, A.; Rojo Negrete, I.A. Systematic review of integrated studies on functional and thematic ecosystem services in Latin America, 1992–2017. *Ecosyst. Serv.* **2019**, *36*, 100900. [CrossRef]
55. Snyder, H. Literature Review as a Research Methodology: An Overview and Guidelines. *J. Bus. Res.* **2019**, *104*, 333–339. [CrossRef]
56. Xiao, Y.; Watson, M. Guidance on Conducting a Systematic Literature Review. *J. Plan. Educ. Res.* **2019**, *39*, 93–112. [CrossRef]
57. Mariano, D.C.B.; Leite, C.; Santos, L.H.; Rocha, R.E.; de Melo-Minardi, R.C. A guide to performing systematic literature reviews in bioinformatics. *arXiv* **2017**, arXiv:1707.05813.
58. Page, M.J.; McKenzie, J.E.; Bossuyt, P.M.; Boutron, I.; Hoffmann, T.C.; Mulrow, C.D.; Shamseer, L.; Tetzlaff, J.M.; Akl, E.A.; Brennan, S.E.; et al. The PRISMA 2020 statement: An updated guideline for reporting systematic reviews. *BMJ* **2021**, *372*, 105906.
59. Thornley, C.; Gibb, F. Meaning in philosophy and meaning in information retrieval (IR). *J. Doc.* **2009**, *65*, 133–150. [CrossRef]
60. Grewal, A.; Kataria, H.; Dhawan, I. Literature search for research planning and identification of research problem. *Indian J. Anaesth.* **2016**, *60*, 635–639. [CrossRef]
61. Meline, T. Selecting Studies for Systemic Review: Inclusion and Exclusion Criteria. *Contemp. Issues Commun. Sci. Disord.* **2006**, *33*, 21–27. [CrossRef]
62. Nang, A.F.M.; Maat, S.M.; Mahmud, M.S. Teacher Technostress and Coping Mechanisms during COVID-19 Pandemic: A Systematic Review. *Pegem Egit. Ve Ogretim Derg.* **2022**, *12*, 200–212.
63. Dawson, K.; Ferdig, R.E. Commentary: Expanding notions of acceptable research evidence in educational technology: A response to Schrum et al. *Contemp. Issues Technol. Teach. Educ.* **2006**, *6*, 133–142.
64. González-Albo, B.; Bordons, M. Articles vs. proceedings papers: Do they differ in research relevance and impact? A case study in the Library and Information Science field. *J. Informetr.* **2011**, *5*, 369–381. [CrossRef]
65. Rahman, N.A.; Rosli, R.; Rambely, A.S.; Halim, L. Mathematics teachers' practices of stem education: A systematic literature review. *Eur. J. Educ. Res.* **2021**, *10*, 1541–1559.
66. Okoli, C. A guide to conducting a standalone systematic literature review. *Commun. Assoc. Inf. Syst.* **2015**, *37*, 879–910. [CrossRef]
67. Singh, G.; Haddad, K.M.; Chow, C.W. Are articles in "top" management journals necessarily of higher quality? *J. Manag. Inq.* **2007**, *16*, 319–331. [CrossRef]
68. Bybee, R.W. *The Case for STEM Education: Challenges and Opportunities*; National Science Teachers Association: Arlington, VI, USA, 2013.
69. Vasquez, J.; Sneider, C.; Comer, M. *STEM Lesson Essentials, Grades 3–8: Integrating Science, Technology, Engineering, and Mathematics*; Heinemann: Portsmouth, NH, USA, 2013.
70. Pittaway, L.; Holt, R.; Broad, J. (Eds.) *Synthesising Knowledge in Entrepreneurship Research: The Role of Systematic Literature Reviews (Handbook of Research on Small Business and Entrepreneurship)*; Edward Elgar: London, UK, 2014.
71. Albelbisi, N.A.; Yusop, F.D. Secondary School Students' Use of and Attitudes toward Online Mathematics Homework. *TOJET Turk. Online J. Educ. Technol.* **2018**, *17*, 144–153.
72. Albelbisi, N.A. The Effect of Web-based Homework on Student's Mathematics Self-Efficacy. *Int. J. Technol. Math. Educ.* **2021**, *27*, 181–190.
73. Albelbisi, N.A.; Al-Adwan, A.S.; Habibi, A.; Rasool, S. The relationship between students' attitudes toward online homework and mathematics anxiety. *Int. J. Math. Educ. Sci. Technol.* **2022**. [CrossRef]
74. Jeong, J.S.; González-Gómez, D. Flipped-ocn method in mathematics learning to analyze the attitudes of pre-service teachers. *Mathematics* **2021**, *9*, 607. [CrossRef]
75. Khan, T.; Thomas, S. Promoting Positive Education through Constructivist Digital Learning Heutagogy: An Intervention Outcome. *J. Learn. Dev.* **2022**, *9*, 305–316. [CrossRef]
76. Sawyerr, A.; Agyei, D.D. Mathematics teachers' use of ICT in classroom instruction: Exploring the will-skill-tool-pedagogy model in the Ghanaian context. *Educ. Inf. Technol.* **2022**. [CrossRef]
77. Robinson, P.; Lowe, J. Literature reviews vs systematic reviews. *Aust. N. Z. J. Public Health* **2015**, *39*, 103. [CrossRef] [PubMed]
78. Hodgkinson, G.P.; Ford, J.K. What makes excellent literature reviews excellent? A clarification of some common mistakes and misconceptions. *J. Organ. Behav.* **2015**, *36*, S1–S5. [CrossRef]
79. Higgins, J.P.T.; Thomas, J.; Chandler, J.; Cumpston, M.; Li, T.; Page, M.J.; Welch, V.A. *Cochrane Handbook for Systematic Reviews of Interventions*, 2nd ed.; John Wiley & Sons: Hoboken, NJ, USA, 2019.
80. Arthur, Y.D. Mathematics Teachers' Acceptance of ICT in Teaching and Learning: An Extended Technology Acceptance Model. *Probl. Educ. 21st Century* **2022**, *80*, 408–425. [CrossRef]
81. Arthur, Y.D.; Dogbe, C.S.K.; Asiedu-Addo, S.K. Modeling students' mathematics achievement and performance through teaching quality: SERVQUAL perspective. *J. Appl. Res. High. Educ.* **2021**, *14*, 1509–1523. [CrossRef]
82. Arthur, Y.D.; Dogbe, C.S.K.; Asiedu-Addo, S.K. Enhancing Performance in Mathematics Through Motivation, Peer Assisted Learning, And Teaching Quality: The Mediating Role of Student Interest. *Eurasia J. Math. Sci. Technol. Educ.* **2022**, *18*, em2072. [CrossRef]

83. Arthur, Y.D.; Boadu, S.K.; Asare, B. Effects of Peer Tutoring, Teaching Quality and Motivation on Mathematics Achievement in Senior High Schools. *Int. J. Educ. Sci.* **2022**, *37*, 35–43.
84. Badri, M. School emphasis on academic success and TIMSS science/math achievements. *Int. J. Res. Educ. Sci.* **2019**, *5*, 176–189.
85. Başaran, B. Investigating science and mathematics teacher candidate's perceptions of TPACK-21 based on 21st century skills. *Elem. Educ. Online* **2020**, *19*, 2212–2226.
86. Cardona, R.S. The Enablers and Outcomes of Research Productivity among Junior High School Mathematics Teachers: A Structural Model. *Eurasia J. Math. Sci. Technol. Educ.* **2020**, *16*, em1901. [CrossRef]
87. Chen, C.L.; Wu, C.C. Students' behavioral intention to use and achievements in ICT-Integrated mathematics remedial instruction: Case study of a calculus course. *Comput. Educ.* **2020**, *145*, 103740. [CrossRef]
88. Cob, Z.C.; Ali, N.; Ramli, R. An empirical evidence of geometers' sketchpad acceptance. *Int. J. Eng. Adv. Technol.* **2019**, *9*, 3436–3439. [CrossRef]
89. Davadas, S.D.; Lay, Y.F. Factors affecting students' attitude toward mathematics: A structural equation modeling approach. *Eurasia J. Math. Sci. Technol. Educ.* **2018**, *14*, 517–529.
90. Elastika, R.W.; Sukono; Dewanto, S.P. Analysis of factors affecting students' mathematics learning difficulties using sem as information for teaching improvement. *Int. J. Instr.* **2021**, *14*, 281–300.
91. Hidayat, R.; Syed Zamri, S.N.A.; Zulnaidi, H. Exploratory and confirmatory factor analysis of achievement goals for indonesian students in mathematics education programmes. *Eurasia J. Math. Sci. Technol. Educ.* **2018**, *14*, em1648. [CrossRef]
92. Hidayat, R.; Zamri, S.N.A.S.; Zulnaidi, H. Does mastery of goal components mediate the relationship between metacognition and mathematical modelling competency? *Educ. Sci. Theory Pract.* **2018**, *18*, 579–604.
93. Hidayat, R.; Syed Zamri, S.N.A.; Zulnaidi, H.; Abdullah, M.F.N.L.; Adnan, M. The Interrelationships between Metacognition and Modeling Competency: The Moderating Role of the Academic Year. *Eur. J. Educ. Res.* **2021**, *10*, 1853–1866. [CrossRef]
94. Hwang, S. Examining the effect of students' early numeracy activities at home on later mathematics achievement via early numeracy competencies and self-efficacy beliefs. *Int. Electron. J. Elem. Educ.* **2020**, *13*, 47–56. [CrossRef]
95. Irakleous, P.; Christou, C.; Pitta-Pantazi, D. Mathematical imagination, knowledge, and mindset. *ZDM Math. Educ.* **2022**, *54*, 97–111. [CrossRef]
96. Ji, R.; Yue, X.; Zheng, X. Using PLS-SEM to Examine the Structure of First-year University Students' Mathematics-related Beliefs. *High. Educ. Stud.* **2021**, *11*, 7. [CrossRef]
97. Kazi, M. Instructional leadership: Teaching evaluation as a key element for 6th grade student's achievement in mathematics. *Int. J. Educ. Manag.* **2021**, *35*, 1191–1204. [CrossRef]
98. Kyaruzi, F.; Strijbos, J.W.; Ufer, S.; Brown, G.T.L. Students' formative assessment perceptions, feedback use and mathematics performance in secondary schools in Tanzania. *Assess. Educ. Princ. Policy Pract.* **2019**, *26*, 278–302. [CrossRef]
99. Lailiyah, S.; Hayat, S.; Urifah, S.; Setyawati, M. Levels of students' mathematics anxieties and the impacts on online mathematics learning. *Cakrawala Pendidik.* **2021**, *40*, 107–119. [CrossRef]
100. Litster, K.; Lommatsch, C.W.; Novak, J.R.; Moyer-Packenham, P.S.; Ashby, M.J.; Roxburgh, A.L.; Bullock, E.P. The Role of Gender on the Associations Among Children's Attitudes, Mathematics Knowledge, Digital Game Use, Perceptions of Affordances, and Achievement. *Int. J. Sci. Math. Educ.* **2021**, *19*, 1463–1483. [CrossRef]
101. Mailizar, M.; Almanthari, A.; Maulina, S. Examining teachers' behavioral intention to use e-learning in teaching of mathematics: An extended tam model. *Contemp. Educ. Technol.* **2021**, *13*, ep298. [CrossRef]
102. Mohamed, Z.; Ubaidullah, N.H. Evaluation of the teaching quality model and its relationship with students' academic performance through partial least squares-structural equation model. *Int. J. Recent Technol. Eng.* **2019**, *8*, 4035–4039. [CrossRef]
103. Mohamed, Z.; Zain, H.H.M.; Yahya, R.A.S.; Damanhuri, M.I.M. Assessing the relationship between program education objectives and program learning outcomes in outcome-based education using partial least squares-structural equation model. *Int. J. Innov. Technol. Explor. Eng.* **2019**, *8*, 1709–1713. [CrossRef]
104. Mustafa, N.; Mohamed, Z.; Ubaidullah, N.H. Modeling of statistical reasoning and students' academic performance relationship through partial least squares-structural equation model (PLS-SEM). *Univers. J. Educ. Res.* **2020**, *8*, 3519–3526. [CrossRef]
105. Mustafa, N.; Ubaidullah, N.H.; Ahmad, M.Z.; In'am, A. PLS-SEM Model of Students' Attitudes toward Statistics, Reasoning of Statistics, Self-Efficacy, Motivation and Academics Performance. *Rev. Int. Geogr. Educ. Online* **2021**, *11*, 992–1003.
106. Muwonge, C.M.; Ssenyonga, J.; Kwarikunda, D. Cognitive Appraisals, Achievement Emotions, and Motivation towards Learning Mathematics among Lower Secondary Students. *Afr. J. Res. Math. Sci. Technol. Educ.* **2018**, *22*, 243–253. [CrossRef]
107. Rawlings, A.M.; Tapola, A.; Niemivirta, M. Temperamental Sensitivities Differentially Linked with Interest, Strain, and Effort Appraisals. *Front. Psychol.* **2021**, *11*, 551806. [CrossRef] [PubMed]
108. Shapira-Lishchinsky, O.; Zavelevsky, E. Multiple Appearances of Parental Interactions and Math Achievement on TIMSS International Assessment. *Int. J. Sci. Math. Educ.* **2020**, *18*, 145–161. [CrossRef]
109. Smeda, A.; Shiratuddin, M.F.; Wong, K.W. A structural equation modelling approach for adoption of e-book amongst mathematics and statistics (MAS) students at higher education institutions in Libya. *Int. J. Inf. Learn. Technol.* **2018**, *35*, 240–254. [CrossRef]
110. Somasundram, P. The Role of Cognitive Factors in Year Five Pupils' Algebraic Thinking: A Structural Equation Modelling Analysis. *Eurasia J. Math. Sci. Technol. Educ.* **2021**, *17*, em1935. [CrossRef] [PubMed]
111. Suryanti, S.; Arifani, Y. The Relationship between Blended Mathematics Professional Training and Teachers' Creativity and Effectiveness. *Int. J. Instr.* **2021**, *14*, 139–154. [CrossRef]

112. Sutter-Brandenberger, C.C.; Hagenauer, G.; Hascher, T. Students' self-determined motivation and negative emotions in mathematics in lower secondary education-Investigating reciprocal relations. *Contemp. Educ. Psychol.* **2018**, *55*, 166–175. [CrossRef]
113. Tee, K.N.; Leong, K.E.; Abdul Rahim, S.S. The Mediating Effects of Critical Thinking Skills on Motivation Factors for Mathematical Reasoning Ability. *Asia-Pac. Educ. Res.* **2018**, *27*, 373–382. [CrossRef]
114. Tee, K.N.; Leong, K.E.; Rahim, S.S.A. Modeling relationships of affective and metacognitive factors on grade eleven students' mathematics achievement. *Int. J. Res. Educ. Sci.* **2019**, *5*, 295–308.
115. Tee, K.N.; Leong, K.E.; Abdul Rahim, S.S. A Self-Regulation Model of Mathematics Achievement for Eleventh-Grade Students. *Int. J. Sci. Math. Educ.* **2020**, *19*, 619–637. [CrossRef]
116. Valls Martínez, M.D.C.; Martín-Cervantes, P.A.; Sánchez Pérez, A.M.; Martínez Victoria, M.D.C. Learning mathematics of financial operations during the COVID-19 era: An assessment with partial least squares structural equation modeling. *Mathematics* **2021**, *9*, 2120. [CrossRef]
117. Wafubwa, R.N.; Csíkos, C. Formative Assessment as a Predictor of Mathematics Teachers' Levels of Metacognitive Regulation. *Int. J. Instr.* **2020**, *14*, 983–998. [CrossRef]
118. Wijaya, T.T.; Cao, Y.; Bernard, M.; Rahmadi, I.F.; Lavicza, Z.; Surjono, H.D. Factors influencing microgame adoption among secondary school mathematics teachers supported by structural equation modelling-based research. *Front. Psychol.* **2022**, *13*, 952549. [CrossRef] [PubMed]
119. Xiang, J.; Wan, Y.; Zhou, J. Factors affecting the learning effect of advanced mathematics among Chinese college students in social science majors. *Eurasia J. Math. Sci. Technol. Educ.* **2019**, *15*, em1770. [CrossRef]
120. Zhang, Y.; Wang, Q. Content learning opportunities, computer-based instruction, and students' mathematics and science achievement. *Int. J. Math. Educ. Sci. Technol.* **2020**, *51*, 1164–1180. [CrossRef]
121. Sarstedt, M.; Cheah, J.H. Partial Least Squares Structural Equation Modeling Using SmartPLS: A Software Review. *J. Mark. Anal.* **2019**, *7*, 196–202. [CrossRef]
122. Hair, J.F.; Hult, G.T.M.; Ringle, C.M.; Sarstedt, M. *A Primer on Partial Least Squares Structural Equation Modeling (PLS-SEM)*, 3rd ed.; SAGE: Thousand Oaks, CA, USA, 2022; forthecoming.
123. Ramayah, T.; Jun-Hwa, C.; Chuah, F.; Ting, H.; Memon, M.A. *Partial Least Squares Structural Equation Modeling (PLS-SEM) using SmartPLS 3.0: An Updated Guide and Practical Guide to Statistical Analysis*, 2nd ed.; Pearson: Kuala Lumpur, Malaysia, 2018.
124. Zamir, S.; Yang, Z.; Wenwu, H.; Sarwar, U. Assessing the attitude and problem-based learning in mathematics through PLS-SEM modeling. *PLoS ONE* **2022**, *17*, e0266363. [CrossRef] [PubMed]
125. Jöreskog, K.G.; Sörbom, D. *LISREL 9.20 for Windows*; Scientific Software International, Inc.: Skokie, IL, USA, 2015.
126. Hair, J.F.; Black, W.C.; Babin, B.J.; Anderson, R.E. *Structural Equations Modeling Overview: Multivariate Data Analysis*; Pearson Education Ltd.: London, UK, 2014.
127. Finch, H.; Bolin, J. *Multilevel Modeling using Mplus*; CRC Press: Boca Raton, FL, USA, 2017.
128. Rousseau, D.M.; Manning, J.; Denyer, D. Chapter 11: Evidence in management and organizational science: Assembling the Field's full weight of. *Acad. Manag. Ann.* **2008**, *2*, 475–515. [CrossRef]
129. Rivera, J.R.D.; Jacomossi, R.R.; Barrichello, A.; Morano, R.S. Using structural equation modeling: Patterns and trends of publications in Brazilian journals. *Rev. De Gest.* **2018**, *25*, 291–302. [CrossRef]
130. Guo, B.; Brian, E.P.; David, F.G. A systematic review of structural equation modelling in Social Work Research. *Br. J. Soc. Work* **2009**, *39*, 1556–1574. [CrossRef]
131. Okech, D.; Kim, J.; Little, T.D. Recent developments in structural equation modelling research in social work journals. *Br. J. Soc. Work* **2013**, *45*, 685–704. [CrossRef]
132. Owolabi, H.O.; Ayandele, J.K.; Olaoye, D.D. A Systematic Review of Structural Equation Model (SEM). *Open J. Educ. Dev.* **2020**, *1*, 27–39. [CrossRef]

MDPI
St. Alban-Anlage 66
4052 Basel
Switzerland
www.mdpi.com

Sustainability Editorial Office
E-mail: sustainability@mdpi.com
www.mdpi.com/journal/sustainability

www.ingramcontent.com/pod-product-compliance
Lightning Source LLC
Chambersburg PA
CBHW061502210926
51850CB00008B/1791
9783725800247